Qualitative Methods in Mathematical Analysis

TRANSLATIONS OF
MATHEMATICAL
MONOGRAPHS

VOLUME 12

L. È. Èl'Sgol'c

Qualitative Methods in Mathematical Analysis

American Mathematical Society

Л. Э. ЭЛЬСГОЛЬЦ

КАЧЕСТВЕННЫЕ МЕТОДЫ
В МАТЕМАТИЧЕСКОМ АНАЛИЗЕ

Государственное Издательство
Технико-Теоретической Литературы
Москва 1955

Translated from the Russian by A. A. Brown and J. M. Danskin

Publication aided by grant NSF-GN 57
from the
NATIONAL SCIENCE FOUNDATION

Text composed on Photon, partly subsidized by NSF Grand G21913

2010 *Mathematics Subject Classification.* Primary 34-XX.

Library of Congress Cataloging-in-Publication Data

Èl′sgol′t͡s, L. È. (Lev Èrnestovich)

Kachestvennye metody v matematicheskom analize. English.

Qualitative methods in mathematical analysis, by L. È. Èl′sgol′c. [Translated from the Russian by A. A. Brown and J. M. Danskin].

p. cm. — (Translations of mathematical monographs ; v. 12)

Includes Bibliography.

ISBN 978-0-8218-1562-5

1. Mathematical analysis. 2. Differential equations. I. Title.

QA303.E413
517 64016170

---

10 9 8 7 21 20 19 18 17 16

Table of Contents

Foreword

The purpose of the present book "Qualitative methods in mathematical analysis" is to introduce the reader to the ideas which are characteristic for qualitative methods and to throw some light on the details of certain new questions not yet fully developed.

The author, desiring to present to the reader material of a very wide and diverse character, was obliged to report many results without proof or with a mere outline of the proof. In these cases he has given references to the literature.

Chapters I, II and in part Chapter III require from the reader a knowledge of the basic elements of topology. Chapters IV, V, and VI, with the exclusion of one section in each chapter, do not suppose any acquaintance with topology and do not require that Chapters I, II, and III be studied first.

Several questions are presented in this book for the first time. Many other questions have appeared up till now only in scientific papers. Thus the first more or less systematic exposition of this material will necessarily contain a number of lacunae. The author nevertheless hopes that in spite of its deficiencies this book will enable a wide circle of mathematicians to acquaint themselves with the fundamental ideas and some of the problems in qualitative methods in analysis.

Introduction

Qualitative methods in mathematics are methods which make it possible in the absence of a quantitative solution of a mathematical problem to indicate a number of qualitative properties of the desired solution.

Sometimes the qualitative analysis of a mathematical problem is only the first step of an investigation, in which one proves the existence of a solution, estimates the number of solutions, and establishes some peculiarities of the solutions, thus facilitating in the future their exact or approximate solution.

However, one not infrequently has to deal with problems in which the question is from the beginning purely qualitative, and the finding of an exact or approximate solution of the equations of the problem does not make it possible to answer the question as posed, and frequently does not even help in finding the solution of that question.

We present a number of examples of problems which are solved by qualitative methods:

(1) Given a differential equation, to determine whether its solutions are stable, whether it has periodic solutions, whether its solutions oscillate, whether it has singular points, and if so, what are their types.

(2) To estimate the number of solutions of a given problem of finding extreme values of a function or of a functional, in terms of the properties of the space on which this function or functional is defined.

(3) Without calculating the roots of the equation $f(z) = 0$, to indicate various peculiarities of their distribution, for example to determine whether the real parts of all the roots of this equation are negative or not.

(4) To estimate the number of singular points of a function $f(z)$ which is analytic at every point of a region D with the exception of a finite set of singular points, in terms of the properties of the region D.

(5) To prove the existence of solutions of a given equation and to estimate their number.

Particular questions of qualitative analysis have been encountered for a long time. However, the methods of qualitative analysis have become widespread only in the twentieth century, and are nowadays applied in all branches of mathematics.

Qualitative methods have been developed considerably in the theory of differential equations, in extremal problems, and in the proof of existence theorems.

CHAPTER I

Qualitative Methods in Extremal Problems

1. **Fundamental method of estimation of the number of critical points.** In the investigation of extreme values of functions and particularly of functionals, it is generally necessary to solve complex systems of equations or to integrate differential equations which only in exceptional cases can be integrated in finite form.

The first and basic problem in investigations by qualitative methods is the estimation of the number of solutions of an extremal problem. If the question concerns the extremum of a function, then in the simplest case the question may be posed as follows: on an n-dimensional three-times differentiable closed manifold M^n there is given a twice continuously differentiable function f. It is required to find a lower bound of the number of critical points of this function, i.e., of the points where $df = 0$.

It turns out that the greatest lower bound of the number of critical points of a continuous function depends on the topological properties of the set on which the function is given. A partial, but far from complete, answer to this question is given by the Weierstrass theorem on the existence of a maximum and a minimum for a continuous function given on a closed set. This theorem guarantees the existence of two critical points, the maximum and the minimum, for each continuous function given on a closed manifold. However this estimate of the number of critical points for manifolds with complex topological structure is very rough. Although on the sphere one can define a function which has only two critical points, for functions given on the torus or on the projective plane the number of critical points is not less than three, and for n-dimensional projective space or the n-dimensional torus it is not less than $n+1$. The basic results in the estimation of the number of critical points for functions and functionals were obtained by L. A. Ljusternik [**136-151**], L. G. Šnirel'man [**151; 322**], and M. Morse [**188-191**].

We represent the increment Δf of the function f in the neighborhood of a point p in the following form:

$$\Delta f = \sum_{i=1}^{n} \frac{\partial f(p)}{\partial x_i} \Delta x_i + \frac{1}{2!} \sum_{i,j=1}^{n} \frac{\partial^2 f(p)}{\partial x_i \partial x_j} \Delta x_i \Delta x_j + R.$$

where x_i $(i = 1, 2, \cdots, n)$ are local coordinates.

If p is a critical point, then the first term on the right side disappears, and up to quantities of higher than second order with respect to $\rho = \sqrt{\sum_{i=1}^{n} \Delta x_i^2}$ the increment of the function may be represented in the form

$$\Delta f \approx \frac{1}{2!} \sum_{i,j=1}^{n} \frac{\partial^2 f(p)}{\partial x_i \partial x_j} \Delta x_i \Delta x_j. \tag{1}$$

Suppose that at the critical point in question the nth order determinant $|\partial^2 f / \partial x_i \partial x_j| \neq 0$. Such a point is called a *nondegenerate critical point.* By means of an affine transformation of the coordinates we may bring the quadratic form on the right side of (1) into the canonical form

$$\Delta f \approx -\sum_{i=1}^{k} \Delta y_i^2 + \sum_{j=k+1}^{n} \Delta y_j^2.$$

The number of negative coefficients on the squares of the increments of the second differential in canonical form is called *the type number of a nondegenerate critical point.* If the type number is equal to k, then the critical point is said to be *a point of type k.* In this terminology a maximum point will be a point of type n, and a minimum a point of type 0.

The set of points of the manifold M^n at which the function f is not larger than a given number x will be denoted by $(f \leqq x)$ and called *a region of smaller values of the function f.* The set of points at which $f = x$ will be called a *level surface* and denoted by $(f = x)$. In the neighborhood of a critical point of type k the level surfaces are approximately represented by hypersurfaces of second order $-\sum_{i=1}^{k} \Delta y_i^2 + \sum_{j=k+1}^{n} \Delta y_j^2 = c$, and the critical level surface $\Delta f = 0$ is approximately a conical surface.

$$-\sum_{i=1}^{k} \Delta y_i^2 + \sum_{j=k+1}^{n} \Delta y_j^2 = 0.$$

In Figure 1 are shown level surfaces in the neighborhood of critical points of the simplest types: (a) a point p of minimum or maximum type, (b) a point q of type 1, and (c) a point q of type 1 or 2.

If $k = 0$ or $k = n$, then in the neighborhood of a critical point the level surfaces are closed and are approximately ellipsoids. For other values of k the level surfaces in the neighborhood of a critical point are approximately hyperboloids. The region of smaller values $(f \leqq x)$, and also the level surfaces, can change their topological and even their isotopy[1] properties only when x passes through a critical value, i.e., the value of the function at a critical point.

[1] Two sets A and B are said to be *isotopic* in the space M^n containing them if there exists a continuous deformation $\varphi_t(A)$ of the set A into the set B in the space M^n such that for each t, $0 \leqq t \leqq 1$, $\varphi_t(A)$ is homeomorphic to A, $\varphi_0(A) = A$, $\varphi_1(A) = B$.

In fact, at the noncritical points the direction of the gradient of the function f is defined and its coordinates are given by $\partial f/\partial x_1, \partial f/\partial x_2, \cdots, \partial f/\partial x_n$. More over, the equations $dx_1/(\partial f/\partial x_1) = dx_2/(\partial f/\partial x_2) = \cdots = dx_n/(\partial f/\partial x_n)$, which satisfy the conditions of the existence and uniqueness theorem, define the orthogonal trajectories to the level surfaces. We introduce on the orthogonal trajectories a deformation parameter for the x,[2] equal to the value of the function f at the corresponding point. Then,

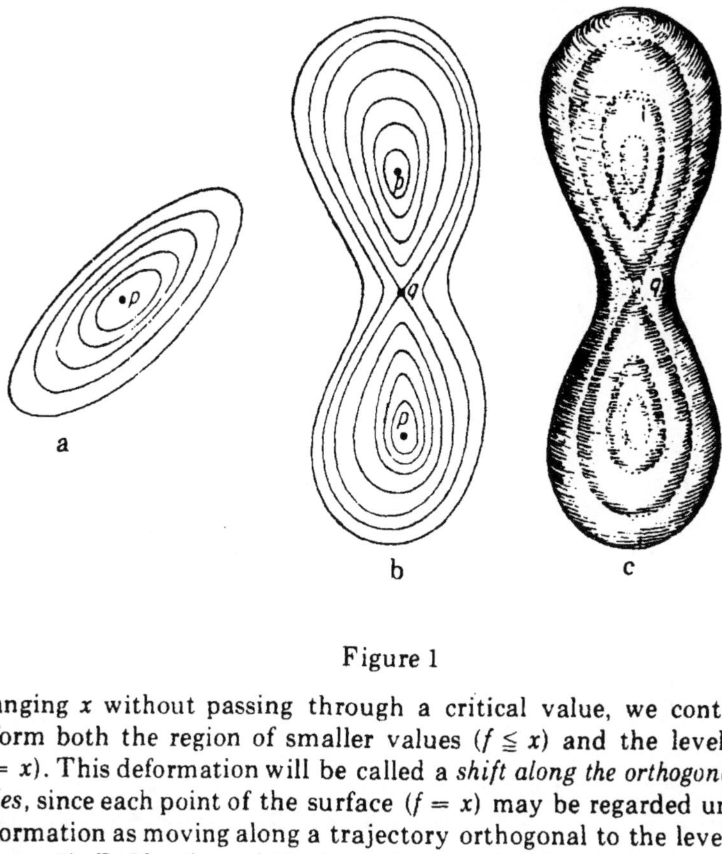

Figure 1

changing x without passing through a critical value, we continuously deform both the region of smaller values ($f \leqq x$) and the level surface ($f = x$). This deformation will be called a *shift along the orthogonal trajectories*, since each point of the surface ($f = x$) may be regarded under this deformation as moving along a trajectory orthogonal to the level surface (Figure 2). Evidently, a shift along the orthogonal trajectories represents a topological and indeed isotopic mapping of the set $f \leqq \bar{x}$ into the set $f \leqq \bar{\bar{x}}$, provided there are no critical values on the segment $[\bar{x}, \bar{\bar{x}}]$.

But if on a given level surface there is an isolated critical point p, the direction of the gradient is not defined there. Then by means of a shift

[2] In the case at hand the deformation parameter does not change from 0 to 1 as usual but rather from $\bar{x}$ to $\bar{\bar{x}}$.

along the orthogonal trajectories everywhere outside an arbitrarily small neighborhood of the critical point we may isotopically deform the set $(f \leqq f(p) + \epsilon)$ into the set $(f \leqq f(p) - \epsilon) + U(p)$,

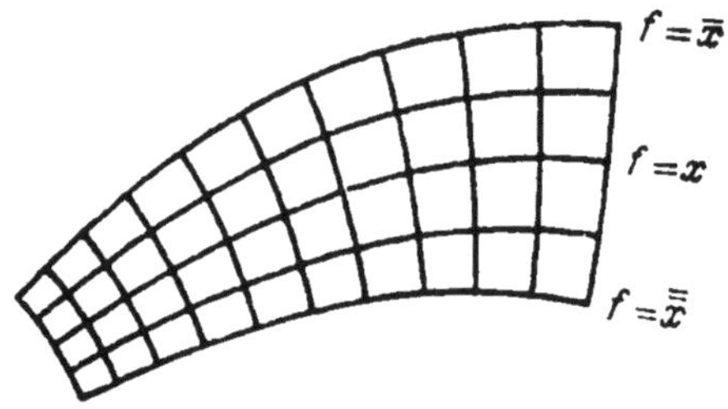

Figure 2

where $\epsilon > 0$ is sufficiently small and $U(p)$ is a neighborhood of the critical point which is homeomorphic to a sphere. The shift along the orthogonal trajectories must be carried out with attenuation on approaching the neighborhood of the critical point in order to preserve the continuity of the deformation (Figure 3).

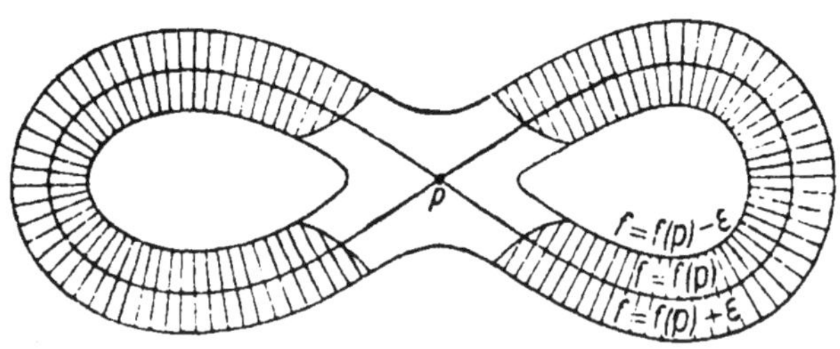

Figure 3

Thus, *on the inclusion into the region of smaller values* $(f \leqq x)$ *of one isolated critical point, which may be either degenerate or nondegenerate, a cell is attached to the region of smaller values* $(f \leqq f(p) - \epsilon)$. *If the critical point included in* $(f \leqq x)$ *is nondegenerate and has type* k, *then the cell attached to* $(f \leqq f(p) - \epsilon)$ *intersects the subcritical region of smaller values in a neighborhood of a* $(k-1)$*-dimensional sphere.*

Indeed, in the neighborhood of a critical point of type k the critical level surface approximately coincides with a cone:

$$-\sum_{i=1}^{k} \Delta y_i^2 + \sum_{j=k+1}^{n} \Delta y_j^2 = 0,$$

and this cone intersects the sphere $\sum_{i=1}^{n} \Delta y_i^2 = \epsilon^2$, which is the boundary of an ϵ-neighborhood of the critical point, in such a way that in the region of smaller values ($\Delta f < 0$) the surface of the sphere $\sum_{i=1}^{n} \Delta y_i^2 = \epsilon^2$ contains a sphere of dimension $k - 1$, obtained as the intersection of the sphere $\sum_{i=1}^{n} \Delta y_i^2 = \epsilon^2$ with the k-dimensional plane $\Delta y_{k+1} = \Delta y_{k+2} = \cdots = \Delta y_n = 0$, and also contains the neighborhood of this $(k-1)$-dimensional sphere, which is the topological product of the $(k-1)$-dimensional sphere with a cell of dimension $n - k$. In order to estimate the number of critical points of a function defined on the manifold M^n, one may follow the changes of some topological, or preferably isotopy invariant of the set $(f \leqq x)$ as x changes from the absolute minimum of the function f to its absolute maximum. Each change of an isotopy invariant will indicate the inclusion into $(f \leqq x)$ of at least one critical point. Thus the minimal number of changes of a given isotopy invariant which necessarily occur in the passage, as x varies from a value less than the absolute minimum of f to its absolute maximum, from the empty set $(f \leqq x)$ to the entire manifold $(f \leqq x)$ will be a lower bound for the number of critical points of f on the manifold $M^{(n)}$.

2. Estimate of the number of analytically distinct critical points.

1. *Estimate of the number of critical points by means of the Betti numbers.* We first consider the changes in the Betti numbers mod 2 of the region of smaller values.

To this end we use the following addition theorem for Betti numbers mod 2:

$$p^r(K_1 + K_2) = p^r(K_1) + p^r(K_2) - p^r(K_1 \cdot K_2) + n^r + n^{r-1},$$

where K_1 and K_2 are complexes, $K_1 \cdot K_2$ is their intersection, $p^r(K_i)$ is the Betti number mod 2 of the complex K_i, and n^r and n^{r-1} are the ranks of the groups of linking cycles, i.e., cycles lying in the intersection $K_1 \cdot K_2$ and homologous to zero both in K_1 and in K_2. This theorem asserts that in $K_1 + K_2$ there exist homology classes of cycles of the following types:

(a) homology classes of cycles generated by cycles lying in K_1 or in K_2 separately;

(b) classes generated by covering cycles, i.e., cycles obtained by stretching onto the linking cycle films in K_1 and in K_2. More precisely, if z_{r-1} is a linking cycle, i.e., $z^{r-1} = \Delta C_1^r$,[3] where $C_1^r \subset K_1$, and $z^{r-1} = \Delta C_2^r$, where $C_2^r \subset K_2$, then $C_1^r + C_2^r$ is a covering cycle.

The inclusion into $(f \leqq x)$ of one nondegenerate point of type k leads, as a result of this addition theorem, to the following change of the Betti number mod 2:

[3] Here the sign ΔC_1^r denotes the boundary of C_1^r

$$p^r(f \leqslant f(p)+\varepsilon) - p^r(f \leqslant f(p)-\varepsilon) = p^r(E^n) - p^r(S^{k-1}) + n^r + n^{r-1},$$

since $(f \leqq f(p)+\epsilon)$ is homeomorphic to $(f \leqq f(p)-\epsilon)+U(p)$, where the neighborhood $U(p)$ of the critical point is a cell, denoted by E^n, intersecting $(f \leqq f(p)-\epsilon)$ in a neighborhood of a $(k-1)$-dimensional sphere S^{k-1}. Accordingly, if $k>1$, then only for $r=k-1$ and $r=k$ can the right side be different from zero. We consider two cases:

(a) the cycle S^{k-1} is a linking cycle, i.e., $S^{k-1} \sim 0$ in $(f \leqq f(p)-\epsilon)$. Then for $r=k-1$ the first and third terms of the right side are equal to zero, and $p^{k-1}(S^{k-1}) = n^{k-1} = 1$, so that for the increment in the Betti number we have

$$\Delta p^{k-1}(f \leqslant x) = p^{k-1}(f \leqslant f(p)+\varepsilon) - p^{k-1}(f \leqslant f(p)-\varepsilon) = 0.$$

For $r=k$ the only term on the right side which is not equal to zero is the last term $n^{k-1}=1$, so that $\Delta p^k(f \leqq x) = 1$ (Figure 4 on the left).

(b) the cycle S^{k-1} is not a linking cycle, i.e., $S^{k-1} \nsim 0$ in $(f \leqq f(p)-\epsilon)$. Then only for $r=k-1$ is the right side different from zero: $p^{k-1}(S^{k-1})=1$, so that $\Delta p^{k-1}(f \leqq x) = -1$ (Figure 4, right).

For $k=0$ and $k=1$ the separate terms on the right side change differently: $p^0(E^n)=1$, $p^0(S^0)=2$. But, as one easily verifies, the final result remains unchanged: *on the addition to the region of smaller values of one critical point of type k, only one of the Betti numbers* mod 2

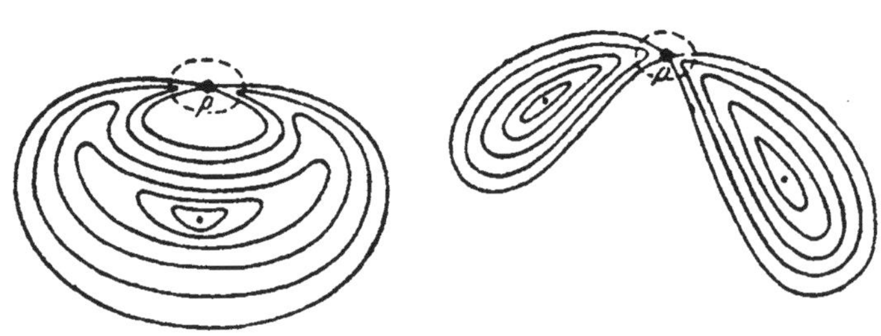

Figure 4

of the region of smaller values changes: either the k-dimensional Betti number goes up by one (such a point will be called a point of increase of type k), or else the (k − 1)-dimensional Betti number goes down by one (such a point is called a point of decrease of type k).[4]

[4] Morse [189].

This result makes it possible to estimate the number of critical points, if all the points are nondegenerate, as follows: *the minimal number of nondegenerate critical points* $\tilde{K}(M^n)$ *of a function given on the manifold* M^n *is not less than the Betti number* mod 2 *of the manifold* M^n:

$$\tilde{K}(M^n) \geqslant \sum_{i=0}^{n} p^i(M^n).$$

If one denotes by m_+^k the number of increasing points of type k and by m_-^k the number of critical points of decreasing type k, then

$$\begin{aligned} p^0 &= m_+^0 - m_-^1, \\ p^1 &= m_+^1 - m_-^2, \\ &\cdots\cdots\cdots \\ p^{n-1} &= m_+^{n-1} - m_-^n, \\ p^n &= m_+^n, \end{aligned}$$

from which one immediately obtains the inequalities

$$\begin{gathered} p^0 \leqslant m^0, \\ p^0 - p^1 \geqslant m^0 - m^1, \\ p^0 - p^1 + p^2 \leqslant m^0 - m^1 + m^2, \\ \cdots\cdots\cdots\cdots\cdots \\ \sum_{i=0}^{n} (-1)^i p^i = \sum_{i=0}^{n} (-1)^i m^i. \end{gathered}$$

For example, every function (with no degenerate critical points) defined on the torus has at least four nondegenerate critical points, one of type 0, two of type 1, and one of type 2. Such a function defined on n-dimensional projective space P^n has at least $n+1$ nondegenerate points, one for each type, since $p^i = 1 \quad (i = 0, 1, \cdots, n)$.

The results indicated above generalize to the case of degenerate critical points and to the case of critical sets of more complex type if we ascribe to the critical set a certain multiplicity $(m^0, m^1, \cdots, m^n)$, indicating to what minimal number of critical points of the various types the given critical set is equivalent, in the sense that the inclusion of the given critical set into a region of smaller values would call for the same change in the Betti number mod 2 as would be obtained by inserting into the region of smaller values m^0 critical points of type 0, m^1 critical points of type $1, \cdots, m^n$ critical points of type n (see [**190**; **191**]). It is impossible to prove that for a small change of the function f a critical set of multiplicity $(m^0, \cdots, m^n)$ decomposes, generally speaking, into m^0 critical points of type 0, m^1 critical points of type 1, $\cdots$, m^n points of type n.

In Figure 5 we exhibit the level surfaces

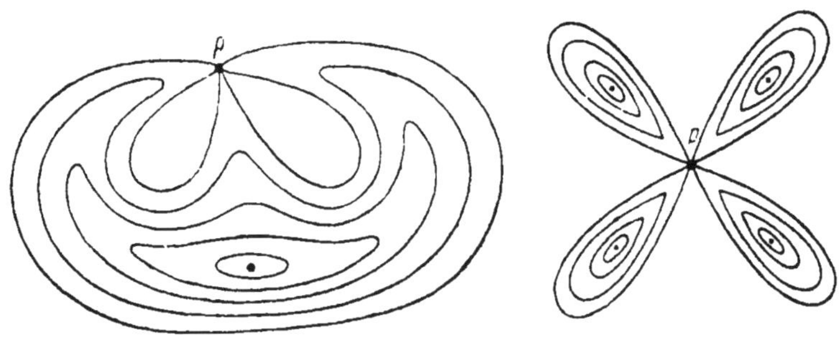

Figure 5

in the neighborhood of degenerate critical points p of multiplicity $(0, 2, 0)$ and $(0, 3, 0)$.

Now suppose that every cycle z^k of a given homology class contains cycles z^r of some other homology class, and consider then the nondegenerate critical point of type k whose inclusion in the region of smaller values $(f \leqq x)$ led to the first appearance of a cycle of the class of z^k. Obviously, such a critical point can appear only after the point of type r whose inclusion in the region of smaller values $(f \leqq x)$ led to the first appearance of a cycle homologous to the cycle z^r. This evident remark establishes a certain order of inclusion of critical points of the various types.

For greater precision in the estimates of the number of nondegenerate critical points we consider the change of other invariants of the region of smaller values.

2. *The change of the fundamental group of the region of smaller values.* We recall two addition theorems for the fundamental group, which we shall have to use in the sequel.

(1) The fundamental group of the sum of two complexes K_1 and K_2, under the assumption of connectivity for their sum $K_1 + K_2$ and their intersection $K_1 \cdot K_2$ is isomorphic to the free product of the fundamental groups of the terms factored over the fundamental group of their intersection. More precisely, if the fundamental groups of the complexes K_1 and K_2 are given by generators and relations, then the fundamental group of the sum $K_1 + K_2$ will be obtained if to the generators and relations of the fundamental group of the complex K_1 one adjoins the generators and relations of the fundamental group of the complex K_2, and also adjoins relations of the type

$$R_i(a_j) = R_s(b_r),$$

expressing the fact that the elements of the fundamental group of the intersection, expressed by the generators a_i of the fundamental group of the complex K_1, and the same elements expressed in terms of the generators b_r of the fundamental group of the complex K_2 will coincide in $K_1 + K_2$.

(2) The fundamental group of the sum of a complex K^n and a cell E^n, intersecting K^n in two components homeomorphic to the cell, has a fundamental group differing from the fundamental group of the complex K^n by the presence of one free generator (see [**105**]).

Suppose that in each component of the set $(f \leqq x)$ the fundamental group is given by generators and relations, and consider the change of the fundamental group $F(f \leqq x)$ of the region of smaller values which comes about when one nondegenerate critical point of type k with the critical value c is included in $(f \leqq x)$.

We consider the following cases:

1. $k > 2$. Then, taking into account that the supercritical region of smaller values $(f \leqq c + \epsilon)$ is homeomorphic to the union of the subcritical region of smaller values $(f \leqq c + \epsilon)$ and a cell E^n whose intersection $(f \leqq c - \epsilon) \cdot E^n$ is homeomorphic to the product of a $(k - 1)$-dimensional sphere S^{k-1} with the cell E^{n-k+1} and using the first of the above addition theorems, we conclude that the fundamental group $F(f \leqq x)$ of the region of smaller values remains unchanged, since the fundamental group of the intersection $(f \leqq c - \epsilon) \cdot E^n$ is trivial.

2. $k = 2$. The fundamental group of the intersection $(f \leqq c - \epsilon) \cdot E^n$ is a free cyclic group. If the generating element S^1 of this group shrinks to a point in $(f \leqq c - \epsilon)$, then, again using the first of the addition theorems, we discover that the fundamental group $F(f \leqq x)$ remains unchanged after the inclusion of the critical point in question. If now S^1 does not contract to a point in $(f \leqq c - \epsilon)$, then from the same addition theorem the group $F(f \leqq x)$ acquires a new relation: $\widetilde{S}^1 = R(a_i) = 1$, where $\widetilde{S}^1$ is an element of the fundamental group homotopic to S^1, and $R(a_i)$ is its expression in terms of the generating elements a_i of the fundamental group $F(f \leqq c - \epsilon)$. The presence of this new relation may lead to a decrease in the number of generators of the fundamental group.

3. $k = 1$. If the neighborhood of the critical point, $U(p) = E^n$, intersects with only one component of the region of smaller values $(f \leqq c - \epsilon)$, as in Figure 6 left, then, from the second of the above addition theorems, we find that the fundamental group $F(f \leqq x)$ acquires a new free generator.

But if $U(p)$ intersects with two components of the region of smaller values $(f \leqq c - \epsilon)$, as in Figure 6 right, a double application of the first

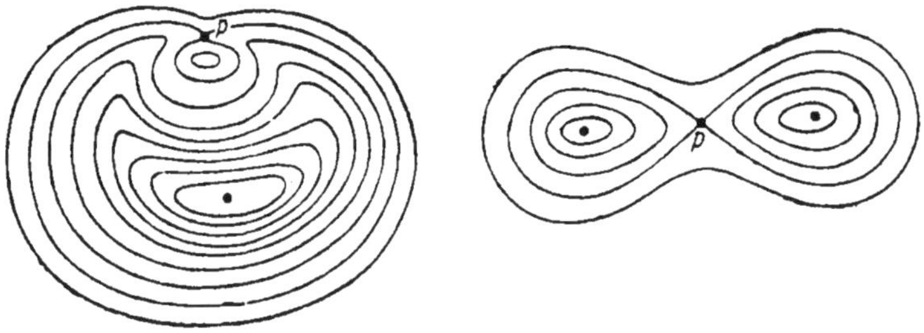

Figure 6

addition theorem shows us that on the attachment to one of those two components of the cell E^n the fundamental group of that component does not change, and on subsequent attachment of the second component the fundamental group becomes isomorphic to the free product of the fundamental groups of the separate components.

4. $k = 0$. In this case a new component with trivial fundamental group is attached to $(f \leqq x)$.

The result of an investigation of the changes in the fundamental group $F(f \leqq x)$ makes it possible in certain cases to improve the estimates obtained by means of the Betti numbers.

For example, if a twice continuously differentiable function having only nondegenerate critical points is defined on the Poincaré sphere, then, if we take account only of the change of the Betti numbers, we can only guarantee the presence of two critical points, while on taking the fundamental group into account, we can be certain of the existence of six critical points: the minimum and maximum points, two points of type 1 (since the fundamental group, being noncommutative, has at least two generators), and two points of type 2 in view of the inequalities on page 7.

3. *Change of the Betti group of the region of smaller values.* The change in the Betti group of the region of smaller values can be investigated by means of the addition carried cut with the addition theorem of Bokšteĭn [**35**]:

$$B_S^r \cong N^{r-1} \times N^r \times B_K^r \times B_L^r / B_D^r,$$

where $B_K^r, B_L^r, B_S^r, B_D^r$ are the Betti groups, with respect to a prime modulus or with respect to the field of rational numbers, of the complexes, $K, L, S = K + L$ respectively, and N^{r-1} and N^r are the $(r-1)p$ and r-dimensional groups of linking cycles, i.e., cycles of D which are homologous to zero both in K and in L; here $\cong$ denotes isomorphism, and $\times$ denotes the direct product. If we are dealing with Betti groups over an integer field of coefficients, then B_S^r is defined by the following isomorphisms: $N^{r-1} \cong B_S^r / B_{K,L}^r$, where $B_{K,L}^r$ is the group of homology classes of cycles of the form $z_1 + z_2$, where z_1 and z_2 are cycles from K and L respectively (if N^{r-1} does not contain elements of finite order, then $\widetilde{B}^r_S \cong N^{r-1} \times B^r_{K.L}$) and $B^r_{K.L} \cong B^r_K \times B^r_L/\widetilde{B}^r_D$, where $\widetilde{B}^r_D \cong B^r_D/N^r$. We shall carry out the investigation over an integer field of coefficients.

Theorem. *If in the region of smaller values ($f \leqq x$) of the twice continuously differentiable function f, defined on a manifold, we include one nondegenerate critical point of type k, then there are the following possibilities for the change in the Betti group over an integer field of coefficients:*

(1) *the k-dimensional Betti group $B^k(f \leqq x)$ of the region of smaller values acquires an infinite-cyclic factor. No other changes in the Betti groups $B^i(f \leqq x)$ occur.*

(2) *The group $B^k(f \leqq x)$ acquires an infinite cyclic factor, and the group $B^{k-1}(f \leqq x)$ loses torsion. No other changes in the groups $B^i(f \leqq x)$ occur.*

(3) *The group $B^k(f \leqq x)$ acquires a free cyclic factor, one of the torsion coefficients in the group $B^k(f \leqq x)$ decreases, being divided by an integer.*

(4) *In the group $B^{k-1}(f \leqq x)$ an infinite cyclic factor is lost. No other changes in the groups $B^i(f \leqq x)$ occur.*

(5) *In the group $B^{k-1}(f \leqq x)$ an infinite cyclic factor is lost, but torsion appears.*

Points of type (2) and (3) can appear in ($f \leqq x$) only after points (5).

The reader can easily construct examples of cases (1), (2), (4) and (5). However, the question as to the existence of an example of case (3) still remains open. In any case, if an example of case (3) can be found, it must be very complicated, since no such example can exist in Euclidean space or on the homology sphere, as easily follows from the duality law of Alexander.

Proof. With the above notations we have:

$$K = (f \leqslant c - \varepsilon), \quad L = U(p) = E^n,$$
$$D = (f \leqslant c - \varepsilon) \cdot E^n = U(S^{k-1}),$$

$S = (f \leqq c - \epsilon) + U(p)$, and S is isotopic in M^n to the region of smaller values ($f \leqq c + \epsilon$) inclusive of the critical region. We suppose that $k \geqq 2$.

(a) If the cycle S^{k-1}, lying in $\widetilde{(f} \leqq c - \epsilon) \cdot E^n$ and not homologous to zero there, is a linking cycle, then $\widetilde{B}^k_D$ and B^k_D are trivial and $B^k_{K.L} \cong B^k_K \times B^k_L \cong B^k_K$, N^{k-1} is an infinite cyclic group. We also have $B^k_S \cong N^{k-1} \times B^k_K$, i.e., $B^k_S \cong B^k(f \leqq c + \epsilon)$ acquires an infinite cyclic factor in comparison with $B^k_K = B^k(f \leqq c - \epsilon)$. Also B^{k-1}_D and N^{k-1} are infinite cyclic groups, and $\widetilde{B}^{k-1}_D$ is trivial or is a cyclic group of finite order:

$$B^{k-1}_{K,L} \cong B^{k-1}_K / \widetilde{B}^{k-1}_D, \quad B^{k-1}_S \cong B^{k-1}_{K,L},$$

so that either $B^{k-1}_S \cong B^{k-1}_K$, or B^{k-1}_S loses torsion in comparison with B^{k-1}_K, or the torsion coefficient is divided by t, where t is the order of the group $\widetilde{B}_D^{k-1}$, $B_S^r = B_K^r$ for $r \neq k$ and $r \neq k-1$.

(b) If the cycle S^{k-1} is not a linking cycle, then the group B_D^{k-1} is trivial, $B_{K,L}^k \cong B_K^k$, N^{k-1} is trivial and accordingly $B_S^k \cong B_K^k$, $\widetilde{B}_D^{k-1} \cong B_D^{k-1}$, $B_{K,L}^{k-1} \cong B_K^{k-1}/\widetilde{B}_D^{k-1}$, where $\widetilde{B}_D^{k-1}$ is an infinite cyclic group, $B_S^{k-1} \cong B_{K,L}^{k-1}$. Thus the group B_S^{k-1} loses an infinite cyclic factor in comparison with the group B_K^{k-1}, while it is possible that torsion appears in its place.

For $k = 1$, in case (a), by the same argument as for $k \geqq 2$ we see that B_S^1 acquires an infinite cyclic factor. $B_S^0 \cong B_{K,L}^0 \cong B_K^0 \times B_L^0/\widetilde{B}_D^0$, B_L^0 is an infinite cyclic group, $\widetilde{B}_D^0 \cong B_D^0/N^0$, where B_D^0 is the direct product of two infinite cyclic groups, and N^0 is an infinite cyclic group. $B_S^0 \cong B_K^0$, since B_S^0 does not contain elements of finite order.

In case (b) for $k = 1$ we obtain $B_S^1 \cong B_K^1$, and in the group B_S^0 an infinite cyclic factor disappears, since $B_S^0 \cong B_K^0 \times B_L^0/B_D^0$, where B_L^0 is an infinite cyclic group, and B_D^0 is the direct product of two infinite cyclic groups, while B_S^0 does not contain elements of finite order. For $k = 0$ only one case is possible: the group B_S^0 acquires an infinite cyclic factor in comparison with the group B_K^0, since $B_S^0 \cong B_K^0 \times B_L^0$, where B_L^0 is an infinite cyclic group.

If the manifold has complicated torsion groups, the changes in the Betti groups, and also in the Betti numbers with respect to various moduli, make it possible to sharpen the estimates obtained by using Betti numbers mod 2.

The question naturally arises, how does one obtain a greatest lower bound for the number of nondegenerate critical points of a function defined on a given manifold?

The answer to this question can be obtained by introducing a new isotopic invariant, namely the number of elements of the manifold with regular adjunction.

4. *The number of cells with regular adjunction.* As we indicated above (page 4), the inclusion, as x increases, of one nondegenerate critical point of type k into the region of smaller values ($f \leqq x$), is isotopically equivalent to the attachment to the subcritical region of a cell intersecting the subcritical region in a neighborhood of some $(k-1)$-dimensional sphere. As x varies from the absolute minimum of the function f on the manifold M^n to its absolute maximum, the inclusion of each critical point attaches a cell to the region of smaller values.

Thus, the entire manifold may be considered as a union of cells into which the manifold can be decomposed, and the minimal number of such cells will be a lower bound for the number of critical points. This lower bound can be improved if one requires that the cells into which the manifold is decomposed be alloted additional properties, belonging to those cells which are attached to the region of smaller values on the inclusion number q of closed cells E_i $(i = 1, \cdots, q)$ with the following properties:

(1) $\sum_{i=1}^{q} E_i = A$.

(2) The intersection $E_{s+1} \cdot \sum_{i=1}^{s} E_i = D_{s+1}$ $(s = 1, 2, \cdots, (q-1))$ is the topological product of the p-dimensional sphere with an $(n-1-p)$-dimensional cell.

If $p = n$, then D_{s+1} is an $(n-1)$-dimensional sphere.

(3) If $p \neq n$, there exists an arbitrarily small neighborhood $\overline{D}_{s+1}$ of the set D_{s+1} taken with respect to the boundary of the set $\sum_{i=1}^{s} E_i$, monotonically isotopic [5] to the set D_{s+1} on the boundary of the sum $\sum_{i=1}^{s} E_i$ $(s = 1, 2, \cdots, (q-1))$.

(4) If $p \neq n$, then there exists a proper subset $\widetilde{D}_{s+1}$ of the set D_{s+1} which is monotonically isotopic to D_{s+1} on the boundary of the sum

$$\sum_{i=1}^{s} E_i \quad (s = 1, 2, \ldots, (q-1)).$$

The number of cells with regular adjunction of the set A is denoted by $\widetilde{\text{El}}\, A$. If A is a manifold, then $\widetilde{\text{El}}\, A$ is automatically defined, since on any manifold one can define a function f having only nondegenerate critical points, and the inclusion of one nondegenerate critical point into the region of smaller values $(f \leqq x)$ of the function f elevates $\widetilde{\text{El}}\,(f \leqq x)$ by not more than unity. Hence it also follows that $\widetilde{\text{El}}\, M \leqq \widetilde{K}(M)$, where $K(M)$ is a lower bound of the number of nondegenerate critical points of functions given on the manifold M.

THEOREM 1. $\widetilde{\text{El}}\, M = \widetilde{K}(M)$, *where M is a closed n-dimensional manifold.*

SKETCH OF THE PROOF. Since $\widetilde{\text{El}}\, M \leqq \widetilde{K}(M)$, it is sufficient to construct

---

[5] The set A is said to be monotonically isotopic to the set B in the set M, if there exists in M a deformation $\varphi_t(A)$ such that $\varphi_t(A)$ is homeomorphic to B for each t, $0 \leqq t \leqq 1$, and $\varphi_0(A) = A$, $\varphi_1(A) = B$, while $\varphi_t(A) \supset \varphi_{\bar{t}}(A)$ if $t > \bar{t}$. The deformation inverse to φ and the identical deformation are also called monotonically isotopic. Inclusion is understood in the strict sense: the boundary of $\varphi_{\bar{t}}(A)$ does not intersect the boundary of $\varphi_t(A)$ if $t \neq \bar{t}$. The passage with increasing x from the region of smaller values $(f \leqq x)$ to the region $(f \leqq \bar{x})$, is an example, provided there are no critical values on the interval $[x, \bar{x}]$, of the monotonically isotopic mapping of the set $(f \leqq x)$ into $(f \leqq \bar{x})$.

on the manifold M a continuous function having $\check{E}l\, M$ nondegenerate critical points.

Suppose that $\check{E}l\, M = q$, so that $M = \sum_{i=1}^{q} E_i$, where E_i $(i = 1, \cdots, q)$ are cells with the properties indicated in the definition of the number of cells with regular adjunction. We topologically map the system of cells E_i onto an abstractly given system of n-dimensional balls $\bar{E}_i$. Suppose that $\varphi(E_i) = \bar{E}_i$. In what follows, by the radius of the cell E_i we shall mean the image of the radius of the ball $\bar{E}_i$ under the mapping φ^{-1}, and by the center of the cell E_i the image of the center of $\bar{E}_i$ under the mapping φ^{-1}.

Suppose that on $M_s = \sum_{i=1}^{s} E_i$ there is given a continuous function f having s nondegenerate interior critical points and taking on a constant value on the boundary M_s, equal to $\max f$ on M_s. For $s = 1$ the possibility of defining such a function is evident: it is sufficient to take $f(\bar{p})$ equal to the distance of the point $\bar{p}$ from the center $\bar{o}_1$ of the ball $\bar{E}_1$, and then define f on E_1 by means of the mapping φ^{-1}, taking $f(p) = f[\varphi(p)]$, where $p \in E_1$.

Consider the set $M_{sl} = (M_s + P_{s+1}) - U(o_{s+1}, \epsilon_l)$, where P_{s+1} is the image under the mapping φ^{-1} of the set of points $\bar{P}_{s+1}$, lying on radii projecting the set $\varphi(E_{s+1} \cdot \sum_{i=1}^{s} E_i)$ from the center $\bar{o}_{s+1}$ of the n-dimensional ball $\bar{E}_{s+1}$; and $U(o_{s+1}, \epsilon_l)$ is the image of the ϵ_l-neighborhood $\bar{U}(\bar{o}_{s+1}, \epsilon_l)$ of the point $\bar{o}_{s+1}$ under the mapping φ^{-1}, ϵ_l sufficiently small.

We shall show that the set M_{sl} is homeomorphic and even isotopic in M to the set M_S.

Suppose that $\tilde{D}_{s+1}$ is the subset of D_{s+1} monotonically isotopic to the set D_{s+1} mentioned in the definition of the number of elements with regular adjunction.

We consider the following isotopic deformation: the points of $\tilde{D}_{s+1}$ move along radii of the cell E_{s+1} to the boundary of M_{sl}, points of the set $D_{s+1} - \tilde{D}_{s+1}$ also move along radii of the cell E_{s+1}, but at a decreasing rate on the approach to the boundary of D_{s+1}.[6]

More precisely, to each point of the set $D_{s+1} - \tilde{D}_{s+1}$ there corresponds a number t equal to the value of the parameter of the monotonically isotopic deformation ψ_t of the set $D_{s+1} = \psi_0(D_{s+1})$ into $D_{s+1} = \psi_1(D_{s+1})$, under which p is the image of some point of the boundary of the set D_{s+1}. The magnitude of shift of the points of $D_{s+1} - \tilde{D}_{s+1}$ along the radii of the cell E_{s+1} must be a monotonic function of t only, equal to zero for $t = 0$ and equal to $R - \epsilon_l$ for $t = 1$, where R is the radius of the ball $\bar{E}_{s+1}$ and the

[6] We have described the isotopic deformation only on the boundary $\sum_{i=1}^{s} E_i$ and assume that this deformation is extended with attenuation to $\sum_{i=1}^{s} E_i$, as is easily done since the boundary of $\sum_{i=1}^{s} E_i$ is a noncritical level surface.

magnitude of the shift is measured in $\bar{E}_{s+1}$.

Now by deforming D_{s+1} isotopically into $\tilde{D}_{s+1}$, we extend this deformation along the radii of the cell E_{s+1}, as is possible, since the intersections of the set obtained from M_s by a shift along radii, as described above, with the preimages of spheres with center o_{s+1} and radius less than R are projections of the sets $\psi_t(D_{s+1})$. As a result the set M_s is isotopically deformed into a set differing from M_{sl} in the fact that the truncated cone $P_{s+1} - U(o_{s+1}, \epsilon_l)$ is constructed on $\tilde{D}_{s+1}$ rather than on D_{s+1}.

Now we perform an isotopic deformation of the set $\tilde{D}_{s+1}$ into D_{s+1} and extend this deformation along radii of the cell E_{s+1}. Under this deformation the truncated cone constructed on $\tilde{D}_{s+1}$ goes into a truncated cone constructed on D_{s+1}, which completes the isotopic deformation of M_s into M_{sl}.

Because of the homeomorphism between M_{sl} and M_s we can define on the set M_{sl} a continuous function f_l, having only s interior critical points and assuming a maximum at all points of the boundary.

Consider a monotonically decreasing sequence $\epsilon_1, \epsilon_2, \cdots, \epsilon_l, \cdots$ such that $\epsilon_l > 0$ and $\lim_{l\to\infty} \epsilon_l = 0$ and the corresponding sequence of sets $M_{s1}, M_{s2}, \cdots, M_{sl}, \cdots$, on which the continuous functions $f_1, f_2, \cdots, f_l, \cdots$, are defined, the f_l satisfying the conditions stated above and so chosen that $\lim_{l\to\infty} f_l = f$ exists and is a continuous function on $M_s + P_{s+1}$. For this it is sufficient to vary f_l in comparison with f_{l-1} only in a neighborhood of the point o_{s+1} which decreases as l increases. The function f has s interior nondegenerate critical points and also one singular point or critical point at o_{s+1}.

It is easy to see that the sum $M_s + P_{s+1} + U(o_{s+1})$, where $U(o_{s+1})$ is a closed sufficiently small neighborhood of the point $\bar{o}_{s+1}$ which is the preimage of a ball of radius ϵ with center at $\bar{o}_{s+1}$, is homeomorphic and even isotopic to M_{s+1}. This assertion follows immediately from the structure of the complement $M - (M_s + P_{s+1} + U(o_{s+1}))$, which is the sum $\sum_{i=s+2}^{q} E_i$ with a truncated cone attached in the same way as we attached it to M_s in obtaining M_{sl}. The proof of this homeomorphism reduces to the proof of the isotopy of the complementary spaces, and therefore repeats word for word the proof of the homeomorphism of the sets M_s and M_{sl}. One needs only to replace the set $\tilde{D}_{s+1}$ by the set $\bar{D}_{s+1}$.

Accordingly, it suffices to define on $M_s + P_{s+1} + U(o_{s+1})$ the function f with $s + 1$ nondegenerate interior critical points and assuming a maximum at all points of the boundary.

Consider the section of the cone P_{s+1} by the preimages S_{η_1} and S_{η_2} under the mapping φ of the sphere with center $\bar{o}_{s+1}$ and sufficiently small radii η_1 and η_2, and with sufficiently small difference $(\eta_1 - \eta_2) > 0$.

The level surface $f = f(o_{s+1}) = C$ and the level surfaces sufficiently close to it excise on the preimages of the spheres S_{η_1} and S_{η_2} two topologically equivalent families of $(n-2)$-dimensional surfaces, corresponding to the values of the function from $f(o_{s+1})$ to $f(o_{s+1})-\delta$, where $\delta>0$ is sufficiently small. With the $(n-2)$-dimensional level surfaces on the sphere S_{η_1} corresponding to the values $f(o_{s+1})-\delta/2-t, 0\leqq t\leqq\delta/2$ of the function, we put in correspondence the $(n-2)$-dimensional level surfaces on the sphere S_{η_2} corresponding to the values of the function $f(o_{s+1})-\delta$, $0\leqq t\leqq\delta/2$. Since the parts of the family of $(n-2)$-dimensional level surfaces on the spheres S_{η_1} and S_{η_2} are topologically equivalent, and each curve of the family is isotopic to another on the sphere S_{η_1}, in the interval between the spheres S_{η_1} and S_{η_2} we may join the corresponding $(n-2)$-dimensional level surfaces on the spheres S_{η_1} and S_{η_2}, i.e., with one and the same value of the parameter t, by $(n-1)$-dimensional level surfaces without singularities.

To the level surfaces thus obtained and to their prolongation inside the sphere S_{η_2} we assign the value $c-\delta/2-t$ of the function.

Thus, the value of the function corresponding to the critical level surface passing through the point o_{s+1} is now not c but $c-\delta/2$. The level surfaces $f=c-t$, defined outside the ball bounded by the sphere S_{η_1}, without singularities to the preimage $S_{R(t)}$ under the mapping φ of the sphere $\bar{S}_{R(t)}$ with center at $\bar{o}_{s+1}$ and sufficiently large radius $R(t)$ so that on this sphere the surfaces $f=c-t$ excise an $(n-2)$-dimensional level surface isotopic to the intersection of $f=c-\delta/2$ with $S_{R(t)}$, and at the same time we require that the continuous function $R(t)\to 0$ as $t\to 0$. For this it suffices to require that the function $R(t)$ should have an order lower than the maximum distance of the level surface $f=c-\delta/2$ from the

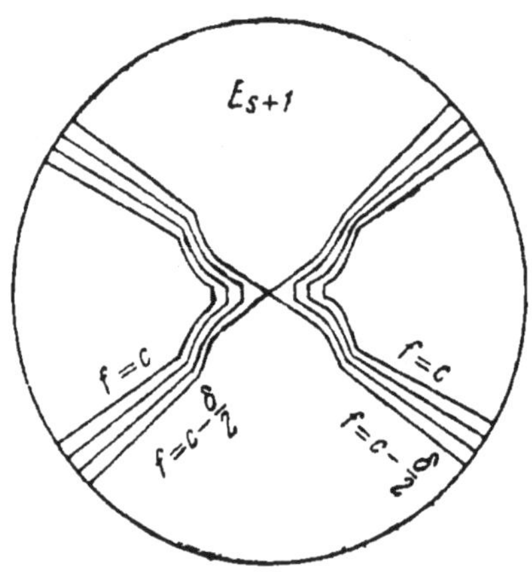

Figure 7

surface $f = c - \delta/2$ in the neighborhood of the point o_{s+1}.

The level surface $f = c - t$ defined outside the sphere $\bar{S}_{R(t)}$ is completed by the part of that sphere lying outside the region in which the function f (Figure 7) was already defined.

The mapping φ^{-1} makes it possible for us to define the function f on the set $M_s + P_{s+1} + U(o_{s+1})$, homeomorphic to $\sum_{i=1}^{s+1} E_i$, having $s+1$ critical points, s of which by hypothesis are situated in the set $\sum_{i=1}^{s} E_i$ and a further critical point coinciding with o_{s+1} and having type $p+1$ (if $E_{s+1} \cdot \sum_{i=1}^{s} E_i$ is the product of the p-dimensional sphere by an $(n-p-1)$-dimensional cell).

When we have attached the element E_q the above construction is inadmissible, but in this case the function f may be extended to E_q without difficulty. The level surfaces will be the preimages of spheres with centers at $\bar{o}_q$. At the center of the element E_q we will have a maximum point.

If it is necessary to obtain a function differentiable once or several times, then of course we must somewhat modify the construction of f.

Remark. Evidently $\widetilde{\text{El}}\, M \geqq \sum_{i=0}^{n} p^i$, where M is a closed manifold, and p^i is its Betti number mod 2. One can prove (see [**334**]) that if $\widetilde{\text{El}}\, M = \sum_{i=0}^{n} p^i$ and the number of critical points of the various types is given: $m^0, m^1, \cdots, m^n$, where m^k is the number of critical points of type k, satisfying the Morse inequalities:

$$m^0 \geqslant p^0,$$
$$m^0 - m^1 \leqslant p^0 - p^1,$$
$$m^0 - m^1 + m^2 \geqslant p^0 - p^1 + p^2,$$
$$\cdots\cdots\cdots\cdots\cdots$$
$$\cdots\cdots\cdots\cdots\cdots$$
$$\sum_{i=0}^{n} (-1)^i m^i = \sum_{i=1}^{n} (-1)^i p^i,$$

then there exists a continuous function defined on the manifold M and having the critical points of the preassigned types.

Example. Let us estimate the minimal number of critical points of a twice-differentiable function defined on the topological product of two spheres $S_1^m \times S_2^n$. The Betti numbers of this product, if $m \neq n$, are as follows: $p^0 = 1$, $p^n = 1$, $p^m = 1$, $p^{m+n} = 1$, and the remaining $p^i = 0$. Accordingly, for each function, having only nondegenerate critical points, defined on the given product of two spheres, there must be not less than four critical points, one of each of the types 0, n, m, $n+m$.

We shall show that this estimate is the best possible. To this end we represent each of the spheres in the form of a sum of two hemispheres:

$$S_1^m = E_{11}^m + E_{12}^m \text{ and } S_2^n = E_{21}^n + E_{22}^n.$$

It is not difficult to verify that the cells

$$E_{11}^m \times E_{21}^n, \quad E_{11}^m \times E_{22}^n, \quad E_{12}^m \times E_{21}^n, \quad E_{12}^m \times E_{22}^n$$

give a decomposition of $S_1^m \times S_2^n$ into cells with regular adjunction and accordingly the minimal number of nondegenerate critical points of a function defined on the product of the two spheres is equal to 4.

Although the number of cells with regular adjunction gives an exact estimate of the number of nondegenerate critical points, the calculation of $\widetilde{\mathrm{El}}\, M$ is highly difficult. Thus less complete estimates of the number of nondegenerate critical points, using the Betti numbers, the Betti groups, and the fundamental group, do not lose their significance.

3. Estimate of the number of geometrically distinct critical points.

1. *Method of estimation of the number of geometrically distinct critical points.* In the preceding section we estimated the number of nondegenerate critical points, and if degenerate critical points were admitted, we assigned to them a certain multiplicity. Thus, we estimated not the number of geometrically distinct critical points, but rather the sum of their multiplicities, or the so-called number of analytically distinct critical points.

In the present section we shall present methods of estimation of the number of geometrically distinct critical points. The founders of this study are L. A. Ljusternik and L. G. Šnirel'man, who have carried out the investigation by a somewhat different method (see page 38, §6).

Consider an integer-valued function $n(A)$ of the closed set A, contained in the manifold M^n and satisfying the following conditions:

(1) $n(E) = 1$, if E is a cell;

(2) $n(A)$ is an isotopic invariant;[7]

(3) $n(a + b) \leq n(A) + n(B)$.

THEOREM 2. *An integer-valued function $n(A)$, satisfying the three conditions enumerated above, bounds from below the number of geometrically distinct critical points of a twice differentiable function given on a manifold M^n:*

$$n(M^n) \leqslant K(M^n),$$

[7] It is sufficient to require that $n(A)$ remain unchanged under monotonically isotopic deformations of the set A.

where $K(M^n)$ is a lower bound of the number of critical points of twice continuously differentiable functions defined on the manifold.

Proof. Let us study the changes of $n\ (f \leqq x)$ on the inclusion into the region of smaller values $(f \leqq x)$ of one critical point which may be degenerate or nondegenerate.

Since outside a neighborhood of the critical point p the set $(f \leqq x)$ may be isotopically deformed by a displacement along orthogonal trajectories, $(f \leqq f(p) + \epsilon)$ is isotopic for sufficiently small ϵ to the set $(f \leqq f(p) - \epsilon) + \overline{U(p)}$, where $\overline{U(p)}$ is a closed neighborhood of the critical point, homeomorphic to the n-dimensional ball (see page 4). Accordingly, by successive use of properties (2), (3), and (1) of the function n, we will have:

$$n(f \leqslant f(p) + \varepsilon) = n[(f \leqslant f(p) - \varepsilon) + \overline{U(p)}]$$
$$\leqslant n(f \leqslant f(p) - \varepsilon) + n(\overline{U(p)}) = n(f \leqslant f(p) - \varepsilon) + 1.$$

Thus, when we include one critical point in the region of smaller values $(f \leqq x)$, the function $n(f \leqq x)$ goes up by not more than unity, so that

$$n(f \leqq x) \leqq K(f \leqq x),$$

where $K(f \leqq x)$ is the number of critical points of the function f situated in $(f \leqq x)$. In particular,

$$n(M^n) \leqq K(M^n).$$

We note that if property (3) of the function n is valid only under certain additional restrictions referring to the intersection of the sets A and B, then the theorem remains valid if the intersection of the sets $(f \leqq f(p) - \epsilon)$ and $\overline{U(p)}$ satisfies those additional restrictions.

If to an isolated critical value c there corresponds not an isolated critical point, but some critical set P_c, then the discussion remains almost unchanged.

Indeed, by a displacement along orthogonal trajectories outside the neighborhood of the critical set, the set $(f \leqq c + \epsilon)$ may be isotopically deformed into the set $(f \leqq c - \epsilon) + \overline{U(P_c)}$, where $\overline{U(P_c)}$ is an arbitrarily small closed neighborhood of the critical set P_c. Accordingly,

$$n(f \leqslant c+\varepsilon) = n[(f \leqslant c-\varepsilon) + \overline{U(P_c)}] \leqslant n(f \leqslant c-\varepsilon) + n(\overline{U(P_c)}),$$

so that

$$n[\overline{U(P_c)}] \geqslant n(f \leqslant c+\varepsilon) - n(f \leqslant c-\varepsilon).$$

Suppose that the invariant $n(A)$ has the two following further properties:

(4) $n[\overline{U(P_c)}] = n(P_c)$, if $U(P_c)$ is a sufficiently small neighborhood of the critical set.

(5) $n(A) \geqq n(B)$, if $A \supset B$.

Note that from properties (5) and (1) for manifolds it follows that if the set A consists of a finite set of points or has a finite set of limit points, then $n(A) = 1$, since in a manifold such a point set can be enclosed in a cell.

The inequality $n[\overline{U(P_c)}] \geqq n(f \leqq c + \epsilon) - n(f \leqq c - \epsilon)$ may now be written in the form $n(P_c) \geqq n(f \leqq c + \epsilon) - n(f \leqq c - \epsilon)$, and if the right side of this last inequality is greater than unity, then on the level surface $(f = c)$ there is a continuum of critical points.[8]

Accordingly, if $n(A)$ satisfies conditions (1), (2), (3), (4), (5), then $n(M)$ bounds from below not only the number of geometrically distinct critical points, but also the number of different critical values, the confluence of which is possible only in the case that a continuum of critical points appears.

Let us now consider several of the most important invariants satisfying the conditions (1), (2), (3), or (1), (2), (3), (4) and (5), and thus giving lower estimates of the number of geometrically distinct critical points.

2. *Category.*[9] *The category of a closed set with respect to the space M*[10] is the minimal number k of sets A_i such that $A = \sum_{i=1}^{k} A_i$ and each set A_i contracts to a point in the set M under a continuous deformation. The category of a set A with respect to the space M will be denoted by $\operatorname{cat}_M A$. The category has the following properties:

(1) $\operatorname{cat}_M E = 1$ if E is a cell.

(2) $\operatorname{cat}_M E$ is an isotopic invariant.

(3) $\operatorname{cat}_M (A + B) \leqq \operatorname{cat}_M A + \operatorname{cat}_M B$,

and accordingly the category bounds from below the number of geometrically distinct critical points: $\operatorname{cat}_{M^n} M^n \leqq K(M^n)$.

We note some further essential properties of category:

(4) $\operatorname{cat}_M \overline{U(A, \epsilon)} = \operatorname{cat}_M A$, where $\overline{U(A, \epsilon)}$ is the closure of a sufficiently small ϵ-neighborhood of the closed set A, and M is a manifold.

(5) $\operatorname{cat}_M A \geqq \operatorname{cat}_M B$ if $A \supset B$. This property is obvious.

(6) Under a continuous deformation of the set its category can only increase.

---

[8] This last conclusion is valid even without the use of property (4).

[9] The concept of category was introduced by L. A. Ljusternik [148].

[10] In what follows M will usually be a manifold. However, the concept of category may be applied to any locally contractible space.

(7) Category is the largest of the invariants satisfying conditions (1), (3), (5), (6).

(8) $\mathrm{cat}_M A \leqq \dim A + 1$, where A is a closed set lying in the manifold M.[11]

(9) $\mathrm{cat}_M (A + B) \leqq \max [\mathrm{cat}_M A;\ \mathrm{cat}_M B] + \mathrm{cat}_M A \cdot B$, where M is a manifold.

Proof. If A and B do not intersect, the assertion is obvious. If the intersection $A \cdot B$ is not empty, then

$$\begin{aligned}
&\mathrm{cat}_M (A+B)\\
&= \mathrm{cat}_M \{[A - U(A\cdot B)] + [B - U(A\cdot B)] + \overline{U(A\cdot B)}\}\\
&\leqslant \mathrm{cat}_M \{[A - U(A\cdot B)] + [B - U(A\cdot B)]\} + \mathrm{cat}_M \overline{U(A\cdot B)}\\
&= \max \{\mathrm{cat}_M [A - U(A\cdot B)];\ \mathrm{cat}_M [B - U(A\cdot B)]\}\\
&\qquad + \mathrm{cat}_M A\cdot B \leqslant \max [\mathrm{cat}_M A;\ \mathrm{cat}_M B] + \mathrm{cat}_M A\cdot B.
\end{aligned}$$

Here we suppose that the neighborhood of the intersection, $U(A \cdot B)$, is sufficiently small.

(10) $\mathrm{cat}_{M_1 \times M_2} A \times B \leqq \mathrm{cat}_{M_1} A + \mathrm{cat}_{M_2} B - 1$, where $M_1 \times M_2$ is the topological product of the manifolds M_1 and M_2.[12]

Proof. Suppose $\mathrm{cat}_{M_1} A = k$ and $\mathrm{cat}_{M_2} B = l$. Then $A = \sum_{i=1}^{k} A_i$, where $\mathrm{cat}_{M_1} A_i = 1$ $(i = 1, 2, \cdots, k)$ and $B = \sum_{j=1}^{l} B_j$, where $\mathrm{cat}_{M_2} B_j = 1$ $(j = 1, 2, \cdots, l)$, $A \times B = \sum_{j=1}^{l} \sum_{i=1}^{k} c_{ij}$, where $c_{ij} = A_i \times B_j$, $\mathrm{cat}_{M_1 \times M_2} c_{ij} = 1$, since, continuously deforming the set A_i into a point in the manifold M_1 and the set B_j into a point in M_2, we realize a continuous deformation of the set c_{ij} into a point in the manifold $M_1 \times M_2$.

Consider the set $(A_1 \cdot A_2 \cdots A_k) \times (B_1 \cdot B_2 \cdots B_l) = d_{k+l}$, where the sign denotes the set-theoretical product, and the sign $\times$ denotes topological multiplication. $\mathrm{cat}_{M_1 \times M_2} d_{k+l} = 1$, since d_{k+l} is contained in some c_{ij}. Accordingly, for sufficiently small ϵ_0

$$\mathrm{cat}_{M_1 \times M_2} U(d_{k+l}, \varepsilon_0) = 1.$$

Denote by d_{k+l-1} the sum of all possible products of the form $(A_{\alpha_1} \cdot A_{\alpha_2} \cdots A_{\alpha_p}) \times (B_{\beta_1} \cdot B_{\beta_2} \cdots B_{\beta_q})$, while $\alpha_i \neq \alpha_j$ for $i \neq j$ and $\beta_s \neq \beta_r$ for $s \neq r$, $p + q = k + l - 1$.

$$\mathrm{cat}_{M_1 \times M_2} [d_{k+l-1} - U(d_{k+l}, \varepsilon_0)] = 1,$$

[11] Properties (3)-(8) of the category were established by L. A. Ljusternik and L. G. Šnirel'man [151].

[12] This theorem extends to a considerably wider class of spaces.

since the category of the separate terms of the form

$$(A_{\alpha_1}\cdot A_{\alpha_2}\ \ldots\ A_{\alpha_p})\times(B_{\beta_1}\cdot B_{\beta_2}\ \ldots\ B_{\beta_q})$$

is equal to unity and the terms do not have common points, so that from the intersection of sets of the form

$$(A_{\alpha_1}\cdot A_{\alpha_2}\ \ldots\ A_{\alpha_p})\times(B_{\beta_1}\cdot B_{\beta_2}\ \ldots\ B_{\beta_q})$$

and

$$\left(A_{\bar\alpha_1}\cdot A_{\bar\alpha_2}\ \ldots\ A_{\bar\alpha_{\bar p}}\right)\times\left(B_{\bar\beta_1}\cdot B_{\bar\beta_2}\ \ldots\ B_{\bar\beta_{\bar q}}\right),$$

where $\bar p+\bar q=k+l-1$ and at least one $\bar\alpha_i$ or $\bar\beta_j$ is distinct respectively from all α_s or from all β_r, it would follow that this intersection

$$\left(A_{\alpha_1}\cdot A_{\alpha_2}\ \ldots\ A_{\alpha_p}\cdot A_{\bar\alpha_1}\cdot A_{\bar\alpha_2}\ \ldots\ A_{\bar\alpha_{\bar p}}\right)$$
$$\times\left(B_{\beta_1}\cdot B_{\beta_2}\ \ldots\ B_{\beta_q}\cdot B_{\bar\beta_1}\cdot B_{\bar\beta_2}\ \ldots\ B_{\bar\beta_{\bar q}}\right)$$

would be contained in d_{k+l} and therefore not in

$$d_{k+l-1}-U(d_{k+l},\ \varepsilon_0).$$

The same considerations lead to the fact that the category of the set

$$d_{k+l-j-1}-\sum_{s=1}^{j}U(d_{k+l-s},\ \varepsilon_s)$$

is equal to unity, where d_{k+l-s} is the sum of all possible products of the form $(A_{\alpha_1}\cdot A_{\alpha_2}\cdots A_{\alpha_p})\times(B_{\beta_1}\cdot B_{\beta_2}\cdots B_{\beta_q})$, while $\alpha_i\neq\alpha_j$ for $i\neq j$ and $\beta_s\neq\beta_r$ for $s\neq r$ and $p+q=k+l-s$. Since $A\times B\subset\sum_{j=0}^{k+l-2}U(d_{k+l-j},\epsilon_j)$, for sufficiently small ϵ_j

$$\mathrm{cat}_{M_1\times M_2}(A\times B)\leqslant k+l-1.$$

(11) If an n-dimensional connected polyhedron has trivial homotopy groups up to the kth dimension inclusive, then its category with respect to itself does not exceed $[n/k+1]+1$ [**81**].

(12) If a polyhedron has a trivial Betti group and a nontrivial fundamental group, then its category is not less than three [**36**].

(13) If a closed manifold M^n contains a nonhomotopic proper subset, then $\mathrm{cat}_{M^n}M^n\geqq 3$.

Indeed, if the Betti groups are not trivial, then this assertion follows from part 4 of this section. If all the Betti groups are trivial but the fundamental group is nontrivial, then we can apply property (12). If now all the Betti groups B^r ($r = 1, 2, \cdots, n-1$) and the fundamental group are trivial, then according to the results of Hurewicz [84] there can be in the manifold M^n no nonhomotopic subset.

The properties of category mentioned above make it possible to estimate and, in certain cases, even to compute the category of various spaces. However, the direct computation of category is an extremely complicated problem, solvable in the present state of homotopy theory only in special cases. Therefore one usually must estimate the category from below by means of the more simply computable homotopy invariants. To this end Ljusternik and Šnirel'man introduced the concept of combinatorial category, which from now on we shall call homological category.

3. *Homological category. The homological category of a closed set $A \subset M$ with respect to the manifold M* is the minimal number k of sets A_i whose sum is A, while each set A_i contains no cycles which are not homologous to zero in M other than the zero-dimensional cycles. The homological category of a set A with respect to the space M will be denoted by $\operatorname{kat}_M A$. Evidently $\operatorname{kat}_M A \leqq \operatorname{cat}_M A$, so that $\operatorname{kat}_M A$ estimates from below the number of geometrically distinct critical points. Besides, it is not difficult to verify directly that $\operatorname{kat}_M A$ satisfies conditions (1), (2) and (3) of page 18. In spite of the fact that the definition of homological category makes use of homological concepts that have been thoroughly investigated, the computation of homological category nevertheless turns out to be very complicated, and therefore at present use is often made of the concept of length.

4. *Length. The cycle z has length l in the manifold M*, if in M there exist l cycles $\bar{z}_j$ such that the intersection $\bar{z} \cdot \bar{z}_1 \cdot \bar{z}_2 \cdots \bar{z}_l \underset{M}{\not\sim} 0 \bmod 2$,[13] but any $l+1$ cycles z_i of M have an intersection with the cycle z which is homologous to zero: $z \cdot z_1 \cdot z_2 \cdots z_{l+1} \underset{M}{\sim} 0$, while none of the cycles $\bar{z}_j$ and $\bar{z}_i$ coincides with M.

For example: (a) the projective plane in three-dimensional space has length 2, since its intersection with two other projective planes is not homologous to zero; (b) the projective line in three-dimensional space has length 1, since its intersection with the projective plane is not homologous to zero; (c) a meridian (or a parallel) on the torus, has length 1, since its intersection with a parallel (or a meridian) is not homologous to zero.

---

[13] Length may also be defined with respect to other moduli.

In all these examples it was obvious that it is impossible to enlarge the stated family of cycles whose intersections are not homologous to zero.

A closed set A has length l if it contains a cycle of length l and does not contain cycles of greater length.

The length of the empty set will be taken equal to -1.

The reader will easily calculate that: (a) the length of the sphere S^n with respect to S^n is equal to 1; (b) the length of the torus T^2 with respect to T^2 is equal to 2; (3) the length of the closed surface M_p^2 of any genus $p > 0$ with respect to M_p^2 is also equal to 2; (d) the length of n-dimensional projective space P^n with respect to P^n is equal to n; (e) the length of the product of two spheres $S^p \times S^q$ with respect to $W^p \times S^q$, where $p > 0$, $q > 0$ is equal to 2.

The length of a set A with respect to the space M will be denoted by $\operatorname{long}_M A$. The length has the following properties:

(1) $\operatorname{long}_M E = 0$ if E is a cell,

(2) $\operatorname{long}_M A$ is an isotopic invariant.

(3) $\operatorname{long}_M (A + B) \leqq \operatorname{long}_M A + \operatorname{long}_M B + 1$.

From properties (1), (2), and (3) it follows that $\operatorname{long}_M A + 1$ satisfies conditions (1), (2), and (3) of page 18, and accordingly estimates the number of critical points from below. The first two properties are obvious.

The proof of property (3) makes use of the following theorem of L. S. Pontrjagin,[14] known as the theorem on removal of cycles: *if the intersection of the cycle z with any cycle contained in the closed set A lying in the manifold M is homologous to zero, then there exists a cycle $\bar{z} \underset{M}{\sim} z$ and lying in $M - A$.*

PROOF OF PROPERTY 3. We shall show that

$$\operatorname{long}_M (A - B) \geqslant \operatorname{long}_M A - \operatorname{long}_M B - 1,$$

from which property (3) follows immediately. Suppose that $\operatorname{long}_M A = \alpha$, $\operatorname{long}_M B = \beta$, $\alpha > \beta$ (if $\alpha = \beta$ then the assertion of the theorem is trivial), so that there exists in A a cycle z whose intersection with the intersection of the cycles $z_1 \cdot z_2 \cdots z_\alpha$ is not homologous to zero.

Consider the cycle $x = z_1 \cdot z_2 \cdots z_{\beta+1}$. This cycle has a homologous to zero intersection with any cycle of the set B, since in the contrary case the length of the set B would be no less than $\beta + 1$. Applying the theorem on the removal of cycles, we may assert that there exists a cycle $\bar{x} \underset{M}{\sim} x$ lying in $M - B$. Then the intersection $\bar{x} \cdot z$ is a cycle lying in $A - B$, since $z \subset A$ and $\bar{x} \subset M - B$.

---

[14] A proof of this theorem may be found in [233].

$$\operatorname{long}_M(\bar{x}\cdot z) \geqslant \alpha-\beta-1,$$

since $(\bar{x}\cdot z)\cdot z_{\beta+2}, z_{\beta+3}\cdots z_\alpha \underset{M}{\not\sim} 0$. Accordingly,

$$\operatorname{long}_M(A-B) \geqslant \operatorname{long}_M A-\operatorname{long}_M B-1.$$

From properties (1), (2), (3) it follows that $\operatorname{long}_M M+1 \leqq K(M)$, where $K(M)$ is the lower bound of the number of geometrically distinct critical points of functions given on the manifold M.

(4) $\operatorname{long}_M A \geqq \operatorname{long}_M B$, if $A \supset B$. This property is obvious.

(5) $\operatorname{long}_M A+1 \leqq \operatorname{kat}_M A$.

PROOF. Obviously the assertions $\operatorname{long}_M A=0$ and $\operatorname{kat}_M A=1$ are equivalent. Suppose that $\operatorname{kat}_M A=k$. Then $A=\sum_{i=1}^k A_i$, where $\operatorname{kat}_M A_i=1$.

Applying property (3) and taking into account that $\operatorname{long}_M A_i=0$, we obtain:

$$\operatorname{long}_M A=\operatorname{long}_M \sum_{i=1}^{k} A_i \leqslant \sum_{i=1}^{k} \operatorname{long}_M A_i+k-1=k-1.$$

There is reason to suppose that $\operatorname{long}_M A+1=\operatorname{kat}_M A$, but so far a proof of this assertion has been found only for very special classes of manifolds.

(6) $\operatorname{long}_{M_1\times M_2} M_1\times M_2=\operatorname{long}_{M_1} M_1+\operatorname{long}_{M_2} M_2$, where $M_1\times M_2$ is the topological product of the manifolds M_1 and M_2.

PROOF. Suppose that $\operatorname{long}_{M_1} M_1=l_1$ and $\operatorname{long}_{M_2} M_2=l_2$. Then there exists in M_1 a cycle x_i distinct from M_1 and such that $x_1\cdot x_2\cdots x_{l_1} \underset{M_1}{\not\sim} 0$, and in M_2 there are cycles y_j distinct from M_2 and such that

$$y_1\cdot y_2\cdots y_{l_2} \underset{M_2}{\not\sim} 0.$$

It is easy to see that

$$(x_1\times M_2)\ldots(x_{l_1}\times M_2)\cdot(M_1\times y_1)\ldots(M_1\times y_{l_2}) \not\sim 0$$

$$\text{in } M_1\times M_2,$$

since this intersection is homologous to the product of the intersections $(x_1\cdot x_2\cdots x_{l_1})\times(y_1\cdot y_2\cdots y_{l_2})$, where the intersections inside the parentheses are taken in M_1 and M_2 respectively. Thus, $\operatorname{long}_{M_1\times M_2}(M_1\times M_2) \geqq \operatorname{long}_{M_1} M_1+\operatorname{long}_{M_2} M_2$. Suppose that $\operatorname{long}_{M_1\times M_2}(M_1\times M_2) > \operatorname{long}_{M_1} M_1+\operatorname{long}_{M_2} M_2$, so that there exist in $M_1\times M_2$ cycles z_i, distinct from $M_1\times M_2$ and such that $z_1\cdot z_2\cdots z_{l_1+l_2+1} \not\sim 0$ in $M_1\times M_2$, but each cycle

$$z_i \sim \sum_{k,j}(x_{ij}\times y_{ik}),$$

where x_{ij} and y_{ik} are cycles in M_1 and M_2 respectively, while $(x_{ij} \times y_{ik}) \neq M_1 \times M_2$ for all i, j, k, but separately either x_{ij} or y_{ik} may coincide with M_1 or M_2 respectively.

From this it follows that for some indices j_α and k_β

$$x_{1j_1} \cdot x_{2j_2} \cdots x_{l_1+l_2+1, j_{l_1+l_2+1}} \not\sim 0 \quad \text{in} \quad M_1$$

and

$$y_{1k_1} \cdot y_{2k_2} \cdots y_{l_1+l_2+1, k_{l_1+l_2+1}} \not\sim 0 \quad \text{in} \quad M_2.$$

Now since $\operatorname{long}_{M_1} M_1 = l_1$, the first of the two preceding intersections contains not more than l_1 cycles x_{ij_i} distinct from M_1. But then the second of these same intersections contains more than l_2 cycles distinct from M_2, which contradicts the supposition that $\operatorname{long}_{M_2} M_2 = l_2$.

(7) $\operatorname{long}_M (A + B) \leq \max [\operatorname{long}_M A; \operatorname{long}_M B] + \operatorname{long}_M A \cdot B + 1$.

PROOF. If $A \cdot B$ is empty, the theorem is obvious. If it is not empty, then

$$\begin{aligned}
&\operatorname{long}_M (A + B) \\
&\quad = \operatorname{long}_M \{[A - U(A \cdot B)] + [B - U(A \cdot B)] + \overline{U(A + B)}\} \\
&\quad \leqslant \operatorname{long}_M \{[A - U(A \cdot B)] + [B - U(A \cdot B)]\} + \operatorname{long}_M \overline{U(A \cdot B)} + 1 \\
&\quad\quad = \max \{\operatorname{long}_M [A - U(A \cdot B)]; \quad \operatorname{long}_M [B - U(A \cdot B)]\} \\
&\quad + \operatorname{long}_M A \cdot B + 1 \leqslant \max [\operatorname{long}_M A; \operatorname{long}_M B] + \operatorname{long}_M A \cdot B + 1,
\end{aligned}$$

where $U(A \cdot B)$ is a sufficiently small neighborhood of the set $A \cdot B$, so small that $\operatorname{long}_M \overline{U(A \cdot B)} = \operatorname{long}_M A \cdot B$.

The existence of a neighborhood of the closed set A sufficiently small that $\operatorname{long}_M \overline{U(A)} = \operatorname{long}_M A$, if M is a manifold, immediately follows from the definition of homology classes of a closed set with respect to the space M.

(8) $\operatorname{long}_M (A + B) = \max [\operatorname{long}_M A; \operatorname{long}_M B; \max \operatorname{long}_M z_i]$, where M is a manifold, z_i the so-called covering cycles, i.e., the cycles $z_i = C_{1i} + C_{2i}$, $C_{1i} \subset A$, $\Delta C_{1i} = y_i$, $C_{2i} \subset B$, $\Delta C_{2i} = y_i$, y_i a linking cycle ($y_i \subset A \cdot B$, $y_i \underset{A}{\sim} 0$ and $y_i \underset{B}{\sim} 0$), and $\max \operatorname{long}_M z_i$ is the maximal length of a covering cycle. The assertion of the theorem is obvious if one takes into account that aside from the classes of cycles generated by cycles contained in A and in B separately there appear in $A + B$ only classes generated by covering cycles.

At first glance it may seem that property (8) is difficult to use, since on the right side there also appear elements from $A + B$. However many properties of covering cycles are defined by the properties of linking cycles. Therefore in many cases one may succeed in calculating $\text{long}_M (A + B)$ from the homological properties of the sets A, B, and $A \cdot B$.

EXAMPLES:

(1) if the Betti numbers $p^i(A \cdot B) = 0$ $(i = 1, 2, \cdots)$, $p^0(A \cdot B) = 1$, then $\text{long}_M (A + B) = \max [\text{long}_M A; \text{long}_M B]$;

(2) if all the linking cycles in $A \cdot B$ are homologous to zero in M, then $\text{long}_M (A + B) = \max [\text{long}_M A; \text{long}_M B]$;

(3) if $A \cdot B$ does not contain linking cycles of dimension higher than $d - 1$, where $d \leqq \max [\text{long}_M A; \text{long}_M B]$, then $\text{long}_M (A + B) = \max [\text{long}_M A; \text{long}_M B]$;

(4) if the dimension of the linking cycles in $A \cdot B$ is less than d, then $\text{long}_M (A + B) \leqq \max [\text{long}_M A; \text{long}_M B; d]$.

Among the invariants estimating the number of geometrically distinct critical points, the length is a comparatively easy one to estimate, making it possible to estimate from below, and in some cases to compute, the category. For example:

(1) the length of n-dimensional projective space P^n with respect to P^n is equal to n, so that $\text{cat}_{P^n} P^n \geqq n + 1$, but since $\text{cat}_M A \leqq \dim A + 1$, therefore $\text{cat}_{P^n} P^n = n + 1$;

(2) the length of the n-dimensional hypertorus T^n with respect to T^n is equal to n, so that $\text{cat}_{T^n} T^n = n + 1$, since the category is not less than the length $+1$ and is not larger than the dimension $+1$.

In this way one can always compute the category, if the length coincides with the dimension. However, it is still possible to compute the category in some other cases. For example:

(3) the length of $2n$-dimensional complex projective space P^{2n} with respect to R^{2n} is equal to n (the intersection of $(2n - 2)$-dimensional complex projective spaces, taken n at a time, is not homologous to zero), so that $\text{cat}_{R^{2n}} R^{2n} \geqq n + 1$. Since the fundamental group of the space R^{2n} is trivial, we have $\text{cat}_{R^{2n}} R^{2n} \leqq [2n/2] + 1 = n + 1$ (see property (11), page 22), so that $\text{cat}_{R^{2n}} R^{2n} = n + 1$.

5. *The number of cells.* *The number of cells* of a closed set A with respect to a closed manifold M is the minimal number of cells of M whose sum contains the set A. The number of cells of the set A will be denoted by $\text{El}\, A$. Evidently $\text{cat}_M A \leqq \text{El}_M A$. It is also evident that $\text{El}_M A$ satisfies the three fundamental properties of invariants for estimating the number of geometrically distinct critical points:

(1) $\text{El}_M E = 1$ if E is a cell;

(2) $\text{El}_M A$ is an isotopic invariant;

(3) $\mathrm{El}_M(A+B) \leqq \mathrm{El}_M A + \mathrm{El}_M B$.

Accordingly, $\mathrm{El}_M M \leqq K(M)$, where $K(M)$ is the lower bound of the number of geometrically distinct critical points of twice continuously differentiable functions given on the manifold M. Thus, the number of cells formally gives a more precise estimate of the number of critical points than the category, and for subsets of a manifold one may give examples for which $\mathrm{cat}_M A < \mathrm{El}_M A$. However, up to the present there is no example of a manifold M such that for the whole manifold $\mathrm{cat}_M M < \mathrm{El}_M M$. An example of a set A for which $\mathrm{cat}_M A < \mathrm{El}_M A$, with M a manifold, is a chain of linked circumferences of circles of sufficiently small radii, situated along a one-dimensional nonbounding cycle in a three-dimensional space. The category of the set of points lying on these circumferences is equal to unity, since the category of each circumference is equal to unity and the circumferences do not intersect. The number of cells is equal to two, since if there existed a cell containing this entire chain of circumferences, it would have to contain a cycle homologous to the cycle z along which the circumferences lie, since they are linked, but $z \underset{M}{\not\sim} 0$.

If instead of circumferences one takes sufficiently small neighborhoods of circumferences then the linked chain of tori thus obtained (the whole torus in each case, not its surface) can be a region of smaller values for functions defined on the manifold, and in this case the desired estimate, namely of the number of critical points with different critical values lying in the given region of smaller values, that is obtained from the number of cells turns out to be more precise than the estimate obtained from the category.

We note some further properties, essential in what follows, of the number of cells.

(4) Among all invariants $n_M(A)$ satisfying the following conditions, the number of cells gives the best estimate of the number of critical points:

(a) $n_M(E) = 1$, where E is a cell;

(b) $n_M(A)$ is an isotopic invariant;

(c) $n_M(A+B) \leqq n_M(A) + n_M(B)$;

(d) $n_M(A) \geqq n_M(B)$, if $A \supset B$.

Proof. Suppose that $\mathrm{El}_M A = k$. Then $\sum_{i=1}^k E_i \supset A$, where E_i are cells of M. Applying successively conditions (d), (c), and (a), we obtain:

$$n_M(A) \leqslant n_M\Big(\sum_{i=1}^{k} E_i\Big) \leqslant \sum_{i=1}^{k} n_M(E_i) = k,$$

so that $\mathrm{El}_M(A) \geqq n_M(A)$.

Among the invariants satisfying only the first three properties but not satisfying property (d), there do exist some which exceed $\mathrm{El}_M A$, but these invariants are of another sort. They estimate the number of geometrically distinct critical points under the condition that the values of the functions at the critical points can coincide, whereas the invariants satisfying conditions (a), (b), (c), (d) estimate the number of geometrically distinct critical points *with distinct critical values*, the confluence of which leads to an infinite set of critical points.

(5) The number of cells of a polyhedron P with respect to a manifold M^n does not exceed the dimension of $P+1$:

$$\mathrm{El}_M P \leqslant \dim P + 1.$$ [15]

PROOF. Let $\dim P = d$. Triangulate P. Denote by R^{d-1} the complex consisting of the $(d-1)$-dimensional simplexes of the complex P. The set $E_0 = P - U(R^{d-1}, \epsilon_1)$,[16] where $U(R^{d-1}, \epsilon_1)$ is a sufficiently small ϵ_1-neighborhood of the set R^{d-1}, consists of a sum of nonintersecting cells and accordingly may be included in one element of M^n, i.e., $\mathrm{El}_M E_0 = 1$. Consider further the complex R^{d-2} consisting of the $(d-2)$-dimensional simplexes of the complex P, and set up the difference

$$E_1 = \overline{U_P(R^{d-1}, \epsilon_1)} - U(R^{d-2}, \epsilon_2),$$

where $U_P(R^{d-1}, \epsilon_1)$ is the closure of the ϵ_1-neighborhood of the set R^{d-1} in the polyhedron P, and $U(R^{d-2}, \epsilon_2)$ is the ϵ_2-neighborhood of the set R^{d-2}, with $\epsilon_2 > \epsilon_1$ and ϵ_2 sufficiently small. $\mathrm{El}_M E_1 = 1$, since E_1 consists of the sum of nonintersecting cells.

Proceeding in this way, we find that the set

$$E_i = U_P(R^{d-i}, \epsilon_i) - U(R^{d-i-1}, \epsilon_{i+1}) \quad (i = 1, 2, \ldots, (d-1)),$$

where $\epsilon_{i+1} > \epsilon_i$ and where ϵ_d is sufficiently small, consists of a sum of nonintersecting cells and accordingly $\mathrm{El}_M E_i = 1$. The set $E_d = \overline{U(R^0, \epsilon_d)}$ is also the sum of nonintersecting cells, so that $\mathrm{El}_M E_d = 1$. Thus, $P = \sum_{i=0}^{d} E_i$, where $\mathrm{El}_M E_i = 1$ $(i = 0, 1, \cdots, d)$. Hence $\mathrm{El}_M P \leqq \dim P + 1$.

6. *The number of cells with regular adjunction*. For the precise estimation of the number of distinct critical points of a function given on a manifold, one may use an isotopic invariant analogous to the number of cells with

[15] The theorem remains valid and the proof is unchanged if M^n is any n-dimensional polyhedron containing P.

[16] We have used the same symbol for a complex and the set of points lying in that complex.

regular adjunction, the so-called number of cells with proper adjunction.

By *the number of cells with proper adjunction* of a set A we shall mean the minimal number q of closed cells E_i $(i = 1, 2, \cdots, q)$ with the following properties:

(1) $\sum_{i=1}^{n} E_i = A$.

(2) The intersection $\sum_{i=1}^{s} E_i \cdot E_{s+1} = D_{s+1}$ $(s = 1, 2, \cdots, (q-1))$ is an $(n-1)$-dimensional bounded manifold or an $(n-1)$-dimensional sphere.

(3) If D_{s+1} is not an $(n-1)$-dimensional sphere, then there exists an arbitrarily small closed neighborhood $\overline{D}_{s+1}$ of the set D_{s+1}, taken with respect to the boundary of the set $\sum_{i=1}^{s} E_i$, which is monotonically isotopic to the set D_{s+1} on the boundary of the set $\sum_{i=1}^{s} E_i$ $(s = 1, 2, \cdots, (q-1))$.

(4) If D_{s+1} is not an $(n-1)$-dimensional sphere, then there exists a proper subset $\tilde{D}_{s+1}$ of the set D_{s+1} which is monotonically isotopic to the set D_{s+1} on the boundary $\sum_{i=1}^{s} E_i$ $(s = 1, 2, \cdots, (q-1))$.

The number of cells with proper adjunction of a set A will be denoted by $\operatorname{Pr El} A$. Evidently $\operatorname{Pr El} M \leqq K(M)$, where $K(M)$ is the number of geometrically distinct critical points of a twice continuously differentiable function defined on the manifold M, since the inclusion into the region of smaller values of one isolated critical point is equivalent to the attachment to the region of smaller values of a cell with proper adjunction, and thus the number of cells with proper adjunction of the region of smaller values under the inclusion of a critical point can go up by not more than unity.

Therefore it also follows that the number $\operatorname{Pr El} M$ is defined trivially for a manifold M.

THEOREM 3. $\operatorname{Pr El} M = K(M)$, *where M is a closed manifold, and $K(M)$ is the lower bound of the number of geometrically distinct critical points of continuous functions given on M.*

The proof does not differ from that of Theorem 1 (page 13), but since $\sum_{i=1}^{s} E_i \cdot E_{s+1} = D_{s+1}$ can have a still more complicated form, a degenerate critical point will in general appear at the center o_{s+1}.

Theorem 3 also remains valid in the case of bounded manifolds, if one considers functions which assume a constant maximum value on the boundary.

By adding superfluous cells of the decomposition of the manifold M and constructing a function in the same way as in Theorems 1 (page 13) and 3, we obtain a function defined on M and having any given number k of critical points, where $k \geqq \operatorname{Pr El} M$ while within certain limits the choice of the superfluous cells will influence the types of the additional critical points.

It is not difficult to verify that the number of elements with proper adjunction is:

(1) for projective space P^n equal to $n + 1$;

(2) for the n-dimensional hypertorus T^n equal to $n + 1$;

(3) for the product of spheres $S^k \times S^q$ equal to 3.

Concerning the third example one may say more. Let r and s be points of S^k and S^p respectively. As one cell we chose a spherical neighborhood U of some point (r, s) of the product $S^k \times S^p$, and as a second cell we chose the sum $(r \times S^p) + (S^k \times s) + V - U$, where V is the neighborhood of some path joining both cells of the sum $(r \times S^p) + (S^k \times s) - U$ and going to the boundary of U without selfintersection. The remaining part $(S^k \times S^p) - [(r \times S^p) + (S^k \times s) + V]$ forms a third cell. Proper adjunction, under an appropriate choice of V, is easily achieved.

Although in the first two cases one can easily obtain a decomposition into $n + 1$ cells with proper adjunction, there is no need to do so, in view of the following theorem of Gordon [**73**]: *on any n-dimensional manifold it is possible to define a function with* $(n + 1)$ *critical points.*

From this theorem and from the estimate of the number of critical points using length or category it immediately follows that

$$\operatorname{Pr}\operatorname{El} P^n = n + 1 \qquad \text{and} \qquad \operatorname{Pr}\operatorname{El} T^n = n + 1.$$

4. **Changes in the topological properties of level surfaces.**

1. *Changes in the Betti numbers.* As we remarked above, the topological and even the isotopic properties of level surfaces of twice continuously differentiable functions, defined on manifolds, can change only on the passage of the level surface through a critical point. Naturally the question arises as to how the fundamental topological invariants of a level surface $(f = x)$ change on the passage of x through a critical value corresponding to one critical point.

The investigation of the changes in the Betti numbers mod 2 of a level surface may be carried out with the aid of an addition theorem, in a manner similar to what we did above in investigating the change in the Betti numbers of the region of lower values.[17] However, this investigation can be carried out much more simply and completely by means of the Pontrjagin duality law [**233**]:

---

[17] Here we have to take into account the fact that on passing through a critical value corresponding to one nondegenerate critical point of type k the supercritical level surface will be formed from the subcritical level surface by subtracting the topological product of a $(k - 1)$-dimensional sphere with an $(n - k)$-dimensional cell and adding the topological product of an $(n - k - 1)$-dimensional sphere with a k-dimensional cell.

$$p^{n-r-1}(M^n - K)$$
$$= p^{n-r-1}(M^n) + q^r(K) + q^{r+1}(K) - p^{r+1}(K) \quad (r = 0, 1, 2, \ldots, n),$$

where M^n is a closed manifold, K is a complex, p^i is the i-dimensional Betti number mod 2, and $q^i(K)$ is the maximal number of homologically independent i-dimensional cycles of K, which are homologous to zero in M^n. Knowing the rule for the change of the Betti numbers mod 2 of the region $(f \leqq x)$, we can easily make use of the Pontrjagin duality law to study the changes of the Betti numbers mod 2 of the region $(f > x)$. Then, by again applying the same duality law, we may again determine the Betti numbers of the level surfaces. Indeed,

$$p^{n-r-1}[(f > x)]$$
$$= p^{n-r-1}(M^n) + q^r[(f \leqslant x)] + q^{r+1}[(f \leqslant x)] - p^{r+1}[(f \leqslant x)]$$
$$(r = 0, 1, 2, \ldots, n),$$

from which it follows, taking into account the results in §2, that if the point p_c is a nondegenerate critical point of type k of the twice continuously differentiable function f defined on the manifold M^n, while on the level surface $(f = c)$ there is only one critical point p_c, if we then have the following possibilities

(a) if p_c is a point of increasing type k for the function f, and if on passing through the isolated critical value the number $q^k[(f \leqq x)]$ increases (i.e., there is again included into $(f \leqq x)$ a k-dimensional cycle which is not homologous to zero in $(f \leqq x)$ but homologous to zero in the whole manifold), then the point p_c is a point of decreasing type $n-k$ for the function $-f$;

(b) if p_c is a point of increasing type k for the function f, and on the passage of x through the isolated critical value c the number $q^k[(f \leqq x)]$ does not change, then the point p_c is a point of increasing type $n-k$ for the function $-f$;

(c) if the point p_c is a point of decreasing type k for the function f, then the point p_c is a point of increasing type $n-k$ for the function $-f$.[18]

Again we apply the Pontrjagin duality law:

---

[18] If the manifold M^n is a homology sphere, then the Pontrjagin duality law goes over into the Alexander duality law

$$p^{n-r-1}[(f > x)] = p^r[(f \leqslant x)] \qquad (r = 1, 2, \ldots, (n-2))$$

and the investigation is very considerably eased. In particular, case (b) almost disappears.

$$
\begin{aligned}
p^{n-r-1}[(f=x)] = p^{n-r-1}(M^n) &+ q^r[(f \leqslant x-\varepsilon)+(f \geqslant x+\varepsilon)] \\
&+ q^{r+1}[(f \leqslant x-\varepsilon)+(f \geqslant x+\varepsilon)] \\
&\qquad - p^{r+1}[(f \leqslant x)] - p^{r+1}[(f \geqslant x)],
\end{aligned}
$$ [19]

where ϵ is sufficiently small.

In case (a), on the passage of x through a critical value c the right side changes only for $r = k$ or $r = n - k - 1$, and in both cases the second term goes up by unity.

In case (b), there are two possibilities as x passes through a critical value: (1) for $r = k$ only the second term on the right side goes up by unity; then for the remaining values of r, except for the dual value $n - r - 1$, the right side does not change, so that only $p^k[(f = x)]$ and $p^{n-k-1}[(f = x)]$ increase by unity. (2) For the case $r = k$ the right side does not change. Then for $r = k - 1$ the right side of the equation goes down by one in the fourth term, the other terms remaining unchanged. For other values of r, except for the dual value, the right side does not change. Accordingly, $p^{k-1}[(f = x)]$ and $p^{n-k}[(f = x)]$ go down by one.

In case (c) only for $r = n - k - 1$, and also, of course, for the dual value of r, does the right side change, the second term going down by one, while the remaining terms remain unchanged. Accordingly $p^{k-1}[(f = x)]$ and $p^{n-k}[(f = x)]$ go down by one.

It must be noted that if the dual values $k - 1$ and $n - k$ or k and $n - k - 1$ coincide, then although the separate terms on the right side change somewhat differently, the conclusions nevertheless remain valid, while the Betti numbers of the level surfaces in these dimensions may change up or down by two units, since $k - 1$ and $n - k$ or k and $n - k - 1$ coincide.

If now $n = 2k$, then there is a further possible exceptional case: for all values of r the right side does not change. This case is realized, for example, for a twice continuously differentiable function having only three critical points on the projective plane (see Figure 8, on which the diametrically opposed points of the boundary must be identified).

One can prove that a necessary and sufficient condition for this exceptional case is the presence in the manifold M^{2k} of a cycle z^k having the property $z^k \cdot z^k \underset{M^{2k}}{\nsim} 0$ [327].

The investigation of the changes of the Betti numbers of the level

---

[19] Instead of $p^{n-r-1}[(f \leqq x+\epsilon) \cdot (f \geqq x-\epsilon)]$ we write $p^{n-r-1}[(f=x)]$, since for noncritical values of x these quantities evidently coincide for sufficiently small ϵ.

surfaces can be supplemented, if necessary, by investigation of the Betti groups and the fundamental groups. However, we shall not dwell on these questions, referring the reader interested in this field to the author's paper [331].

2. *Changes of the invariants of category type.* We shall investigate the change of the category of level surfaces, although quite similar results may be obtained in the case of other invariants of category type, since in the proof of Theorem 1 we shall use only the following properties, intrinsic to all these invariants:

(1) $n(E) = 1$ if E is a cell;

(2) $n(A)$ is an isotopy invariant;

(3) $n(\overline{A+B}) \leqq n(A) + n(B)$;

(4) $n[\overline{U(A)}] = n(A)$, where $\overline{U(A)}$ is a sufficiently small closed neighborhood of the set A;

(5) $n(A) \geqq n(B)$, if $A \supset B$.

In the proof of Theorem 2, moreover, we used the property:

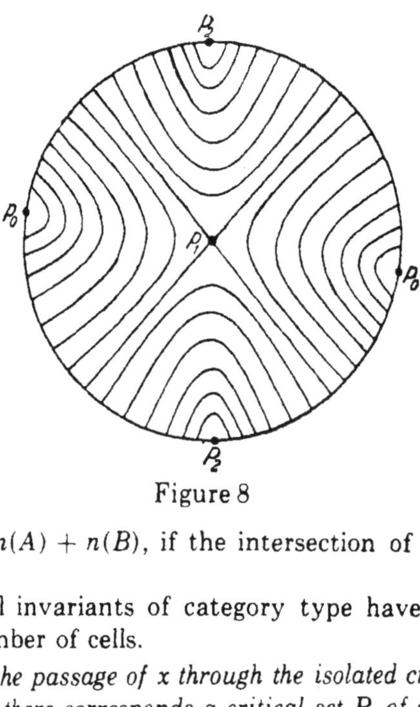

Figure 8

(6) $n(A+B) = n(A) + n(B)$, if the intersection of the closed sets A and B is empty.

The fundamental invariants of category type have this property, in addition to the number of cells.

THEOREM 1. *On the passage of x through the isolated critical value c of the function f, to which there corresponds a critical set P_c of category k, the category of the level surface $(f = c)$ changes by an integer s, where $-k \leqq s \leqq k$, while, as quite simple examples show, s may for various functions take on any value within the indicated limits.*

PROOF. The supercritical level surface $(f = x)$ for x sufficiently close to c, may be deformed isotopically, along trajectories orthogonal to the level surfaces, into a subcritical level surface $(f = c - \epsilon)$ everywhere outside a neighborhood of the critical set $U(P_c)$, so that, by properties (2) and (5) of the invariants $n(A)$ of category type, we obtain:

$$\operatorname{cat}_M [(f = c - \varepsilon) - U(P_c)] \leqslant \operatorname{cat}_M (f = x) \leqslant \operatorname{cat}_M [(f = c - \varepsilon) + U(P_c)],$$

where ϵ is so small that on the segment $[c - \epsilon, c + \epsilon]$, on which x varies, there is only one critical value c. By properties (3) and (4) we have:

$$\operatorname{cat}_M (f = c - \varepsilon) - k \leqslant \operatorname{cat}_M (f = x) \leqslant \operatorname{cat}_M (f = c - \varepsilon) + k,$$

so that the assertion of the theorem follows.

If P_c consists of one point or of a finite set of points, then on the passage of x through c the category $\operatorname{cat}_M (f = x)$ changes by $+1$ or by -1, or remains unchanged.

THEOREM 2. *If the twice continuously differentiable function f defined on the manifold M has only isolated critical points, then there exists a noncritical value x such that*

$$\operatorname{cat}_M (f = x) \geqslant \frac{\operatorname{cat}_M M - 1}{2}.$$

PROOF. We choose a finite set of noncritical values x_i $(i = 0, 1, 2, \cdots, m)$, monotonically increasing with increasing i, such that x_0 is less than the absolute minimum of the function f on M and x_m is larger than the absolute maximum of f on M and such that between any two of these values there lies not more than one critical value.

On the inclusion of one critical point into the region of smaller values, the category cannot go up by more than unity, so that

$$\operatorname{cat}_M [(f \geqslant x_i) \cdot (f \leqslant x_{i+1})] \leqslant \operatorname{cat}_M (f = x_i) + 1.$$

By properties (3) and (4) of invariants of category type, we obtain:

$$\operatorname{cat}_M \sum_{i=1}^{m-1} (f = x_i) \geqslant \operatorname{cat}_M M - \operatorname{cat}_M \sum_{i=0}^{m-1} (f \geqslant x_i + \eta_i) \cdot (f \leqslant x_{i+1} - \eta_{i+1}),$$

where $\eta_i > 0$ and is sufficiently small so that on the segments $[x_i - \eta_i; x_i + \eta_i]$ there are no critical values.

Using property (6) and the preceding inequality, we finally obtain:

$$\max_{1 \leqslant i \leqslant m-1} \operatorname{cat}_M (f = x_i) \geqslant \operatorname{cat}_M M - [\max_{1 \leqslant i \leqslant m-1} \operatorname{cat}_M (f = x_i) + 1],$$

or

$$\max_{1 \leqslant i \leqslant m-1} \operatorname{cat} (f = x_i) \geqslant \frac{\operatorname{cat}_M M - 1}{2}.$$

5. **Some applications.** Among the great number of applications of various kinds (see, for example, Chapter II and pages 90, 180), we shall for the time being consider only the four following problems:[20]

1. *Estimate of the number of equilibrium positions.* Consider a system of material points m_i $(i = 1, \cdots, n)$, each of which can move on some manifold M_i.

Let the points m_i influence each other according to any law, given only that the interaction force has a potential function. We shall estimate from below the number of equilibrium positions of such a system of points.

The equilibrium points will be critical points of the potential function, defined on the topological product

$$M = M_1 \times M_2 \times \cdots \times M_n,$$

so that the minimal number of equilibrium positions is not less than

$$\operatorname{long}_M (M_1 \times M_2 \times \cdots \times M_n) + 1 = \sum_{i=1}^{n} \operatorname{long}_{M_i} M_i + 1.$$

In particular, if all the M_i are topological circumferences, or spheres, then $\operatorname{long}_M M = n$, so that in this case there are no less than $n + 1$ equilibrium positions.

2. *Axes of a manifold.* Let M^n be a three times differentiable manifold, lying in Euclidean space E^p. By an axis of the manifold we shall mean a pair of points (a, b) of the manifold M^n such that the function $\rho(x, y)$ equal to the distance between the points x and y in E^p admits an extremum for $x = a$, $y = b$ under the conditions that x and y lie on M^n.

---

[20] The first three problems in a somewhat less general setting were considered by Ljusternik and Šnirel'man [151].

The distance between the points a and b will be called the length of the axis.

The problem comes down to the estimation of the number of critical points of the function $\rho(x,y)$, on the topological product $M^n \times M^n$. Here, however, in view of the symmetry of the functions $\rho(x,y)$, one must regard the points (x,y) and (y,x) as identified with each other and must also consider the points (x,x) as all identified to one point, or else consider the set of these points (x,x) as a given critical set of minimum type.

Thus, the problem reduces to the estimation of the number of critical points of a function defined on a space with singularities. This problem also is solved by using invariants of category type and their generalizations (see [**151**] and [**328**]).

If the manifold M^n is homeomorphic to the n-dimensional sphere, then, as L. A. Ljusternik calculated, the number of axes of different length is not less than $n+1$. A calculation of the analytically distinct axes of the 2-dimensional torus shows that their number is not less than 12.

3. *Eigenvalues*. Suppose that F is a twice continuously differentiable function of the variables $x_1, \cdots, x_n$. We shall seek the extremum of the function F under the condition $\sum_{i=1}^n x_i^2 = 1$.

The coordinates of the points in which the conditioned extremum is taken on are defined by the system of equations

$$\frac{1}{2}\frac{\partial F}{\partial x_i} - \lambda x_i = 0 \ (i = 1, 2, \ldots, n) \text{ and } \sum_{i=1}^{n} x_i^2 = 1.$$

The values λ for which this system has solutions are called the eigenvalues.

If F is a form of even degree, or *any* even function with respect to all the variables,

$$F(-x_1, -x_2, \ldots, -x_4) = F(x_1, x_2, \ldots, x_n),$$

then, as Ljusternik showed, the number of distinct eigenvalues is not less than $n+1$, while if two of these values merge, then there appears a continuum of solutions of the system of equations given above corresponding to this multiple eigenvalue λ.

PROOF. The twice continuously differentiable function F is defined on the $(n-1)$-dimensional sphere, but in view of its evenness it can be arbitrarily defined only on the hemisphere, and then at the diametrically opposed points of the bounding $(n-2)$-dimensional sphere the values of F must coincide. Therefore F may be given arbitrarily only on the hemisphere (or on the $(n-1)$-dimensional ball), in which the diametrically opposite points of the boundary of the $(n-2)$-dimensional sphere have been identified. Thus F is given arbitrarily on $(n-1)$-dimensional projective space. Since the length of $(n-1)$-dimensional projective space is equal to $n-1$, the function F has at least n critical points corresponding to various critical values, whose merging gives rise to the appearance of a continuum of critical points.

If we subject F to other restrictions, we can obtain analogous theorems for other classes of functions. For example, let us require that

$$F(x_1, x_2, x_3, x_4) \equiv F(x_2, -x_1, -x_3, -x_4).$$

Here F can be arbitrarily defined on a lens space $(4, 1)$[21] with four cells, so that the number of eigenvalues for such a function is not less than four.

4. *c-discriminant sets*. Let $z = f(x_i, c_j)$, where f is a twice continuously differentiable function of the local coordinates x_i $(i = 1, \cdots, p)$ and of the local values of the parameters c_j $(j = 1, \cdots, q)$. The point with coordinates x_i runs over some manifold M^p, and the point with coordinates c_j over a manifold N^q. A c-discriminant set D_c is a set defined by the equations

$$z = f(x_i, c_j) \quad \text{and} \quad \frac{\partial f}{\partial c_j} = 0 \qquad (j = 1, 2, \ldots, q).$$

For fixed x_i $(i = 1, \cdots, p)$ the function $f(x_i, c_j)$ is defined on the manifold N^q and the points of the set D_c are critical for $f(x_i, c_j)$, so that D_c contains at least $\operatorname{Pr}\operatorname{El} N^q$ geometrically distinct points. Changing x_i $(i = 1, \cdots, p)$ and taking into account that merging of geometrically distinct critical points is not possible, we can now assert that D_c contains at least $\operatorname{Pr}\operatorname{El} N^q$, generally speaking, nonintersecting[22] sets, homeomorphic to M^p.

The points of the set D_c at which the determinant $|\partial^2 f/\partial c_i \partial c_j| \neq 0$, can be classified into various types (see §1) and, depending on these types, one can make some inferences regarding their position relative to nearby level surfaces which makes it possible in turn to investigate certain properties of envelopes.

6. **The minimum-maximum principle and its generalization.** The present section contains a short exposition and generalization of the method of Ljusternik and Šnirel'man for estimating the number of geometrically distinct critical points.

---

[21] For lens spaces one may refer to the book of Seifert and Threlfall [**104**].

[22] Intersections can appear only in the case of the presence for some fixed x_i $(i = 1, \cdots, p)$ of a continuum of points belonging to D_c.

A collection of sets closed with respect to isotopic deformation in the space M^n will be called an *isotopy class.*

In other words, a collection of sets forming an isotopy class must, along with any set A, contain all isotopic deformations $\varphi_t(A)$ of the set A.

One can obtain an isotopy class by considering the collection of all the sets of a given space having in common some topological or isotopic property.

For example, the following sets form isotopy classes:

(a) the collection of sets of some manifold M containing a cycle homologous in M to a given cycle z;

(b) the collection of sets of a manifold M containing a cycle homotopic to a given cycle z;

(c) the collection of sets of a manifold M containing a cycle isotopic to a given cycle z;

(d) the collection of all sets of a manifold of a given length or of given category;

(e) the collection of sets whose category or length is not less than a given number p.

An isotopy class is said to be closed if it contains the topological limit of any sequence of sets in it.

As examples of closed topological classes we may take the classes consisting of sets A of the manifold M for which $n(A) \geqq i$, where $n(A)$ is some isotopically invariant integer function of sets, satisfying the following conditions:

(a) $n(U(A,\epsilon)) = n(A)$, where $U(A,\epsilon)$ is a sufficiently small ϵ-neighborhood of the set A;

(b) $n(A) \geqq n(B)$, if $A \supset B$.

For in fact, suppose that $\bar{A}$ is the topological limit of a sequence $A_1, A_2, \cdots, A_n, \cdots$, where $n(A_s) \geqq i$ $(s = 1, 2, \cdots)$. For any ϵ some sets of the sequence in question are contained in $U(\bar{A},\epsilon)$, so that $n(U(\bar{A},\epsilon)) \geqq i$, but for sufficiently small ϵ $n(U(\bar{A},\epsilon)) = n(\bar{A}) \geqq i$. As indicated above, the category and many invariants of category type satisfy conditions (a) and (b).

Suppose that there is defined on the manifold M a twice continuously differentiable function f, having only isolated critical values.

Choose an absolute maximum of the values of the function on each set of some isotopy class $|A|$ of closed sets, and then define a lower bound of these maxima over all sets of the isotopic class in question. The lower bound thus obtained will be called the *minimum-maximum* of the values of the function f on the class $|A|$ and denoted by $\operatorname{Inf}\max_{|A|} f$.

We shall prove that *the minimum-maximum is a critical value*. This assertion includes the principle of minimum-maximum proved under somewhat different definitions of classes by Ljusternik and Šnirel'man and used by them for the estimation of the number of geometrically distinct critical points [151].

PROOF. Suppose that $\operatorname{Inf}\max_{[A]} f = c$, and if on the level surface $f = c$ there are no critical points, let us deform the region of smaller values $(f \leqq c + \epsilon)$ along orthogonal trajectories into the set $(f \leqq c - \epsilon)$, where $\epsilon > 0$ and ϵ is chosen so that on the segment $[c - \epsilon, c + \epsilon]$ there are no critical values. Along with the set $(f \leqq c - \epsilon)$, there is at least one set A^0 of the class $[A]$ which is contained in the set $(f \leqq c - \epsilon)$ and is subjected to a monotonically isotopic deformation. This follows from the definition of the number c.

As a result of this isotopic deformation the set A^0 goes into some set $\varphi_t(A^0)$ of the same class, but lying in $(f \leqq c - \epsilon)$, which contradicts the definition of the number c, since on $\varphi_t(A^0)$ the maximum of the values of f does not exceed $c - \epsilon$, so that c could not be the minimum of the maxima of the values of the function f on the sets of the class $[A]$.

Thus, to each isotopy class there corresponds a critical value $c = \operatorname{Inf}\max_{[A]} f$. Analogously one may define the maximum-minimum $\sup\min_{[A]} f$ of the values of the functions on the sets of a topological class and prove that this value is also critical.

Consider a sequence of closed isotopy classes $[A_0], [A_1], \cdots, [A_n]$, where each set of the class $[A_j]$ contains a continuum of nonintersecting sets of the class $[A_i], j > i$. Such a sequence of classes will be called an increasing sequence of closed isotopy classes. Analogously one defines a decreasing sequence.

Evidently, $\min\max_{[Aj]} f \geqq \min\max_{[Ai]} f, j > i$, where instead of $\operatorname{Inf}\max_{[A]} f$ we have introduced the notation $\min\max_{[A]} f$, which emphasizes that in view of the closure of the isotopy class the value $\operatorname{Inf}\max_{[A]} f$ is taken on some set A^0 of the isotopy class $[A]$.

THEOREM 1. *If*

$$\min_{[A_i]}\max f = \min_{[A_j]}\max f = c \quad \text{for } j > i,$$

then on the level surface $(f = c)$ there is a continuum of critical points.[23]

[23] This theorem was also proved under somewhat different definitions of classes by Ljusternik and Šnirel'man [151].

Indeed, the first set A_j^0 of the class $[A_j]$ which is contained for $f = c$ in the region of smaller values $(f \leqq x)$ contains a continuum of nonintersecting sets $A_{i\alpha}^0$ of the class $[A_i]$ and each of these sets $A_{i\alpha}^0$ has at least one point on the level surface $(f = c)$, since in the contrary case $\min\max_{[A_i]} f$ would be less than $\min\max_{[A_j]} f$. Among the points of intersection $A_{i\alpha}^0 \cdot (f = c)$ there must be critical points, since in the contrary case there would be no critical points in some neighborhood of that intersection, and then one would be able, in that neighborhood to deform the level surface isotopically along the orthogonal trajectories in the direction of decrease of the function f and thus to deform the set $A_{i\alpha}^0$ monotonically and isotopically into a set $\varphi(A_{i\alpha}^0)$ on which the maximum of the function is less than c; but this contradicts the condition $c = \min\max_{[A_i]} f$.

Thus, on each of the continuum of nonintersecting sets $A_{i\alpha}^0$, there is not less than one critical point with the critical value c.

We note that for nonclosed isotopy classes, as is shown by the simplest examples, Theorem 1 is not true without additional assumptions.

The principle of the minimum-maximum and Theorem 1 make it possible to obtain many of the results of §3 on estimation of the number of geometrically distinct critical points.

For example, if all the closed sets of the manifold M whose homotopic category (or other invariant of category type) is not less than i $(i = 1, 2, \cdots, \mathrm{cat}_M M)$ are put into one isotopy class $[A_i]$, then to each isotopy class $[A_i]$ there will correspond at least one critical point with the critical value $\min\max_{[A_i]} f$, while the coinciding of two of these critical values leads, by Theorem 1, to the appearance of a continuum of critical points.

The method of estimation of the number of critical points by the principle of the minimum-maximum and Theorem 1 may be generalized. In the same way we have established the existence of critical values which are equal to the minimum-maximum or to the maximum-minimum on sets of some isotopy class, we can also set up iterated minimum-maxima with respect to the sets of some increasing sequence of classes.

Thus, for example, if we have two closed isotopy classes $[A_1]$ and $[A_2]$, while each set of the class $[A_2]$ contains a continuum of nonintersecting sets of the class $[A_1]$, then we can define the minimum-maximum of the values of the function on sets of the class $[A_1]$ contained in some definite set A_2 of the class $[A_2]$. This value one naturally denotes by $\mathrm{Inf}\max_{[A_1]A_2} f$.

Let us now consider all possible sets A_2 of the class $[A_2]$ and, after obtaining for each A_2 the value $\mathrm{Inf}\max_{[A_1]A_2} f$, choose the upper bound of these values over all sets of the class $[A_2]$. The value thus obtained is naturally denoted by $\sup \operatorname{Inf} \max_{|A_2|_M|A_1|_{A_2}} f$. If we have a decreasing sequence of isotopy classes $|A_k|, |A_{k-1}|, \cdots, |A_2|, |A_1|$, then by an analogous method we can construct the numbers $\sup \inf_{|A_k|_M|A_{k-1}|_{A_k}} \cdots \inf \max_{|A_1|_{A_2}} f$ (or $\operatorname{Inf} \sup_{|A_k|_M|A_{k-1}|_{A_k}} \cdots \inf \max_{|A_1|_{A_2}} f$ depending on whether the number k is odd or even), the definition of which is obvious from the notation or else they may also be defined inductively: $\sup \inf_{|A_k|_B|A_{k-1}|_{A_k}} \cdots \inf \max_{|A_1|_{A_2}} f$ is equal to the upper bound of the numbers $\inf_{|A_{k-1}|_{A_k}} \cdots \inf \max_{|A_1|_{A_2}} f$ over all the sets of the class $[A_k]$, contained in the set B. Analogously one defines the numbers $\operatorname{Inf} \sup_{|A_k|_M|A_{k-1}|_{A_k}} \cdots \sup \min_{|A_1|_{A_2}} f$ (or $\sup \inf_{|A_k|_M|A_{k-1}|_{A_k}} \cdots \sup \min_{|A_1|_{A_2}} f$).

Theorem. *The number* $\sup \inf \max_{|A_2|_M|A_1|_{A_2}} f = c$ *is a critical value.*

Proof. From the definition of the number c it follows that on each set of the isotopy class $[A_2]_M$ in the set $(f \leqq c + \epsilon)$, $\epsilon > 0$, there is a set of the class $[A_1]_M$.

Now if c were a noncritical value, we could monotonically and isotopically deform the set $(f \leqq c - \epsilon)$, by a shift along orthogonal trajectories, into the set $(f \leqq c + \epsilon)$ and extend this shift with attenuation to the entire manifold M (see page 4). Under this transformation any one of the sets $\overline{A}_2$ is transformed isotopically into a set of the same class, in which, however, no set of the class $[A_1]_M$ is contained in the set $(f \leqq c + \epsilon)$; but this contradicts the definition of the number c.

Theorem. *The number* $\inf \sup \inf \max_{|A_3|_M|A_2|_{A_3}|A_1|_{A_2}} f = c$ *is a critical value.*

Proof. From the definition of the number c it follows that there exists a set $\overline{A}_3$ of the class $[A_3]_M$ in which the set $(f \leqq c + \epsilon)$ contains sets of the class $[A_1]_M$ in each set of the class $[A_2]_M$, but the class $[A_3]_M$ does not contain any set $\overline{A}_3$ in which the set $(f \leqq c - \epsilon)$ contains a set of the class $[A_1]_M$ in each set of the class $[A_2]_M$.

If c were a noncritical value, then by a shift along orthogonal trajectories one could deform the set $(f \leqq c + \epsilon)$ monotonically and isotopically into the set $(f \leqq c - \epsilon)$ for a sufficiently small $\epsilon > 0$ and then extend this shift with attenuation to the entire manifold M. Here the set $\overline{A}_3$ is mapped into a set of the same class, in which, however, the set $(f \leqq c - \epsilon)$ already contains a set of the class $[A_1]_M$ in each set of the class $[A_2]_M$; but this contradicts the definition of the number c.

Analogously one proves that the numbers

$$\inf_{|A_k|_M} \sup_{|A_{k-1}|_{A_k}} \ldots \inf_{|A_1|_{A_2}} \max f = c_1,$$

$$\sup_{|A_p|_M} \inf_{|A_{p-1}|_{A_p}} \ldots \inf_{|A_1|_{A_2}} \max f = c_2,$$

$$\inf_{|A_p|_M} \sup_{|A_{p-1}|_{A_p}} \ldots \sup_{|A_1|_{A_2}} \min f = c_3,$$

$$\sup_{|A_k|_M} \inf_{|A_{k-1}|_{A_k}} \ldots \sup_{|A_1|_{A_2}} \min f = c_4$$

are critical values.

This assertion is the so-called principle of the iterated minimum-maximum.

Theorem. *If* $\min\max_{|A_1|_M} f = \sup\min\max_{|A_2|_M|A_1|_{A_2}} f$, *where* $|A_1|_M$ *is a closed topological class, then the level surface contains a set of critical points intersecting all the sets of the class* $|A_2|_M$, *from which it follows that there will be a continuum of critical points on the level surface* $(f = c)$, *if there is a continuum of nonintersecting sets of class* $|A_2|_M$ *in the manifold* M.

Proof. From the definition of the number $\min\max_{|A_1|_M} f = c$ it follows that for $f = c$ the set $(f \leqq x)$ includes sets of the class $|A_1|_M$ for the first time.

From the definition of the number $\sup\min\max_{|A_2|_M|A_1|_{A_2}} f = c$ it follows that on the passage of x through the value c the set $(f \leqq x)$ will contain sets of the class $|A_1|_M$ in all sets of the class $|A_2|_M$.

Accordingly, for $f = c$ each set of the class $|A_2|_M$ will for the first time include a set A_1^0 of the class $|A_1|_M$ and for the proof of the theorem it is sufficient to show that every set A_1^0 contains critical points.

Suppose that the intersection $A_1^0 \cdot (f = c)$ contains no critical points. Then in the neighborhood of this intersection we carry out a shift along orthogonal trajectories in the direction of decreasing values of the function f. Under this deformation the set A_1^0 goes into a set of the same class but lying in $(f \leqq c - \epsilon)$, $\epsilon > 0$, which contradicts the definition of the number $c = \min\max_{|A_1|_M} f$.

In almost the same way one proves that under analogous conditions the coincidence of any of the numbers

$$c_k = \sup_{|A_k|_M} \inf_{|A_{k-1}|_{A_k}} \ldots \inf_{|A_1|_{A_2}} \max f \quad \left(\text{or } \inf_{|A_k|_M} \sup_{|A_{k-1}|_{A_k}} \ldots \inf_{|A_1|_{A_2}} \max f\right)$$

$$(k = 0, 1, \ldots, p;\ c_0 = \max f \text{ on } M)$$

implies the appearance of a continuum of critical points.

We note that from the definition of the numbers c_k it follows that for an odd k $c_0 \geqq c_2 \geqq c_n \geqq \cdots \geqq c_{k-3} \geqq c_{k-1} \geqq c_k \geqq c_{k-2} \geqq \cdots \geqq c_3 \geqq c_1$. Analogous inequalities may be obtained for even k, and also for

$$c_k = \sup_{\{A_k\}_M} \inf_{\{A_{k-1}\}_{A_k}} \ldots \sup_{\{A_1\}_{A_2}} \min f,$$

or

$$c_k = \operatorname{Inf}_{\{A_k\}_M} \sup_{\{A_{k-1}\}_{A_k}} \ldots \sup_{\{A_1\}_{A_2}} \min f.$$

Interesting relations may also be obtained for the numbers c_k calculated for different decreasing sequences of isotopy classes.

Defining the closed isotopy classes $[A_i]$ $(i = 1, 2, \cdots, n_M(M))$ by the condition $n(A_i) \geqq i$ if $A_i \subset [A_i]$, where n is any invariant of category type (see page 18) satisfying conditions (a) and (b) of page 39, we obtain a decreasing sequence of closed isotopy classes. On each level surface $f = c_k$, where

$$\operatorname{Inf}_{\{A_k\}_M} \sup_{\{A_{k-1}\}_{A_k}} \ldots \inf_{\{A_1\}_{A_2}} \max f = c_k,$$

there lies at least one critical point, while if $c_k = c_l$ for $k \neq l$, then on the level surfaces $(f = c_k)$ there will appear continua of critical points, so that $n_M(M) \leqq K(M)$, where $K(M)$ is the number of geometrically distinct critical points of the function f, defined on M.

A decreasing sequence of isotopy classes may be obtained in another way. For example, suppose that the manifold M contains cycles z_i such that $z_1 \cdot z_2 \cdots z_l \not\sim 0$ in M. Then one may consider the isotopy class $[A_l]$ of closed sets containing cycles homologous to the cycle z_1, and then the isotopy class of sets containing cycles homologous to the cycle $z_1 \cdot z_2$ and so forth. The class $[A_{l-i+1}]$ consists of closed subsets of the manifold M containing cycles homologous to the cycle $z_1 \cdot z_2 \cdots z_i$.

Obviously each successive class of the sequence is contained in the preceding one, since the cycle $z_1 \cdot z_2 \cdots z_i$ excises the cycle z_i from the cycle $z_1 \cdot z_2 \cdots z_{i-1}$ and accordingly is contained in sets of the class $[A_{l-i}]$.

But in this case also, when we apply principle of iteration of the minimum-maximum, we obtain only the result, already known in §3, that the number of critical points of a twice continuously differentiable function defined on a manifold is larger than or equal to the length of the manifold plus unity.

An essentially new result is obtained if in the analogous way we construct the classes $[A_i]$ of closed sets containing cycles which are homologous or isotopic to a given cycle z_i, where each cycle z_{i+1} of the class $[A_{i+1}]$ contains a continuum of nonintersecting cycles $z_i \subset [A_i]$, but the cycles z_i are not intersected by any of the z_{i+1} cycles; that is, there is no cycle y in M such that $z_{i+1} \cdot y \underset{M}{\sim} z_i$.

As shown by the example given below, such sequences of classes exist on certain manifolds. Let T^3 be the three-dimensional torus, which we shall represent as a cube with identification of points at opposite ends of a common perpendicular to two faces. As carriers of the representatives of a homology basis of the one-dimensional cycles we take three mutually perpendicular edges z_1^1, z_2^1, z_3^1 of the cube, and as carriers of the representatives of the two-dimensional basis cycles we choose three mutually perpendicular faces z_{12}^2, z_{23}^2, z_{31}^2 of the cube, where the subscripts indicate the indices of the edges z_1^1, z_2^1, z_3^1 which lie on a given face. We remove from T^3 a sufficiently small ϵ-neighborhood $U(z_3^1, \epsilon)$ of the edge z_3^1. In $T^3 - U(z_3^1, \epsilon)$ all cycles are destroyed which are homologous to the cycle z_{12}^2, since $z_{12}^2 \cdot z_3^1 \nsim 0$ in T^3. The boundary of the bounded manifold $T^3 - U(z_3^1, \epsilon)$ is a torus, which we denote by T_1^2. This boundary T_1^2 we identify with the boundary T_2^2 of the body of some torus N^3 in such a way that the meridians of one of T_1^2 and T_2^2 are identified with the parallels of the other. As a result of such an identification we obtain a manifold M^3 in which cycles homologous in T^3 to the cycles z_3^1 become homologous to zero, since these cycles are homologous, as a result of the identification, to meridians of the torus T_2^2 which are homologous to zero in the body of the torus N^3.

Cycles homologous in T^3 to cycles z_{12}^2 are destroyed after identification. Cycles homologous in T^3 to cycles z_1^1 (or z_2^1) which are excised from z_{13}^2 (or z_{23}^2) by cycles homologous to $z_{12}^2, z_1^1 \sim z_{13}^2 \cdot z_{12}^2$ in T^3, already in M^3 are not excised by any cycle, but the isotopy class of closed sets containing cycles isotopic to z_{13}^2 (or z_{23}^2) and the isotopy class of closed sets containing cycles isotopic to the cycles z_1^1 (or to z_2^1 respectively) form two terms of a decreasing sequence of classes. Application of the principle of iteration of the minimum-maximum to the decreasing sequences containing these two classes, gives an estimate of the number of critical points which cannot be obtained from use of the concept of length by the methods of this section or of §3.

7. **Some generalizations in finite-dimensional space.** If a function f is defined on the manifold M^n, a mapping of the manifold onto a segment is thereby defined, and the critical points of the function f are the critical points of such a mapping.

A natural generalization of the problem of the estimation of the number of critical points of a function is the problem of the estimation of the number of critical elements of the mapping of a manifold onto another closed or bounded manifold or onto a space of a more extended class.

Of course, in such a wide setting the problem is very complicated and there has been no attempt to solve it in this form.

We consider the simplest special cases of this problem: the mapping of a manifold M^n onto a one-dimensional closed manifold, namely onto a circumference, and the mapping of M^n into n-dimensional Euclidean space.

1. *Mapping of a closed manifold onto a circumference.* Let f be a twice continuously differentiable mapping of the closed manifold M^n onto a circumference S^1. The mapping f may be defined by twice continuously differentiable functions of the local coordinates. Locally this case at hand does not differ from the question of defining a function, since a closed neighborhood of any point of a circumference is a segment, so that, as in the definition of a function, we have a mapping onto a segment. Therefore, the concept of critical point is analytic or topological and its classification may be left without change (see §§1 and 2). But points of types k and $n-k$ are indistinguishable unless one defines a positive direction on the circumference.

The image of a critical point under the mapping f is naturally called a critical value. The complete preimage under the mapping f of any value is called the level surface corresponding to that value. If the value is critical, then the corresponding level surface is said to be critical.

We shall suppose that the mapping f has only isolated critical values.

Consider a manifold M^{n-1}, which is contained in M^n and is a noncritical level surface under the mapping f. The manifold M^{n-1} may consist of one or several components. Suppose that M^{n-1} is the complete preimage under the mapping f of the point $p \in S^1$.

The complete preimage of an ϵ-neighborhood $U(p,\epsilon)$ of p which does not contain a critical value, will be the topological product of the manifold M^{n-1} with the interval E^1, so that the difference $M^n - M^{n-1} \times E^1$ maps onto the segment $S^1 - U(p,\epsilon)$. Hence, for the purpose of estimating the number of critical points contained in $M^n - M^{n-1} \times E^1$ one may apply the usual methods for estimating the number of critical points of a function defined on a bounded manifold $M^n - M^{n-1} \times E^1$, while on each of the two copies M_1^{n-1} and M_2^{n-1} of the manifold M^{n-1} which constitute the boundary of the set $M^n - M^{n-1} \times E^1$, the function has a constant value of maximum type of one and of minimum type on the other.

Accordingly, the problem comes down to the estimation of the number of critical points of a function defined on a bounded manifold $M^n - M^{n-1} \times E^1$ for different choices of the manifold $M^{n-1} \subset M^n$ among those $(n-1)$-dimensional manifolds which can be noncritical level surfaces for some mapping.

Evidently, it is far from true that any manifold $M^{n-1} \subset M^n$ can be a noncritical level surface. For this it is necessary in any case that on M^{n-1} there should be no cycles z_1 and z_2 whose intersection in M^n is not homologous to zero, since cycles of the same class lie on nearby level surfaces, and level surfaces do not intersect.

In many cases these remarks are already sufficient to obtain a good, and sometimes a precise, estimate of the number of critical points of the mapping of a manifold onto a circumference. For example:

(1) *The sphere S^n under mapping onto the sphere S^1 has at least two geometrically distinct critical points.*

Proof. The noncritical level surface $M^{n-1} = \sum_{i=1}^{p} M_i^{n-1}$, where the M_i^{n-1} are connected manifolds without common points, are homologous to zero in S^n, so that $S^n - \sum_{i=1}^{p} M_i^{n-1} \times E^1$ consists of $p+1$ components, while at least two of these components have boundaries consisting of one component on which the function takes on a constant value and consequently on each of these two components there further exists at least one interior critical point of minimum or maximum type.

The simplest examples show that the estimate just found is exact.

(2) *Projective space P^n under a mapping onto a circumference has at least $n+1$ geometrically distinct critical points.*

Proof. Suppose that the noncritical level surface M^{n-1} consists of one component. Suppose that P^k is a cycle homologous to the projective plane of maximal dimension contained in M^{n-1}. Then on the basis of the theorem on removing a cycle one may assert that at least one of the two components of the set $M^n - M^{n-1} \times E^1$ contains a cycle homologous to the projective plane P^{n-k-1}, since the intersection of the cycle P^{n-k-1} with cycles lying in M^{n-1} is homologous to zero. In the other of the two components of the set $M^n - M^{n-1} \times E^1$, there is always a cycle homologous to P^k, since this component contains level surfaces isotopic to M^{n-1}. Accordingly, from §3, the one component contains at least $n-k$ geometrically distinct critical points, since $\text{long}_{P_n} P^{n-k-1} = n-k-1$, and the other contains at least $k+1$ critical points, since $\text{long}_{P_n} P^k = k$.

If M^{n-1} consists of several components, then the idea of the proof remains the same but the discussion becomes rather more complicated.

(3) *The n-dimensional torus or any other topological product, one of whose factors is a circumference, cannot have critical points under a mapping into a circumference.* This assertion is obvious.

Turning to the general case of the mapping of a manifold M^n onto a circumference, we obtain estimates analogous to the results of Ljusternik and Morse presented in §§2-4.

Suppose as before that f is a twice continuously differentiable mapping of the manifold M^n onto a circumference S^1, and φ is a twice differentiable mapping of the circumference S^1 onto a segment, with only one minimum point and one maximum point and no other critical points. The existence of such a mapping φ is evident, while the points p_1 and p_2 may be chosen arbitrarily among the noncritical values of the function f. For convenience in what follows one may choose p_1 and p_2 such that on one of the arcs $\widehat{p_1p_2}$ there is no critical value of the mapping f. This may be realized by taking the point p_2 sufficiently close to p_1. Under such a choice of the points p_1 and p_2, the level surfaces corresponding to them, $M_1^{n-1} = f^{-1}(p_1)$ and $M_2^{n-1} = f^{-1}(p_2)$, will be isotopic in M^n. The mapping φf has two critical sets: M_1^{n-1} of minimum type and M_2^{n-1} of maximum type, and the remaining critical points of the mapping φf of the manifold M^n onto the segment coincide with the critical points of the mapping f. The types of the isolated critical points of the mappings φf and f coincide if on the circumference we choose a positive direction of passing from p_1 to p_2 along the arc, on which there may lie critical values of the mapping f (if we do choose a positive direction on the circumference, then points of type k and $n-k$ will be indistinguishable).

All these assertions, if one starts from the analytic definition of critical point, immediately follow from the chain rule of differentiation.

Now we apply the results presented in §§2 and 3 to the mapping φf:

$$p^i = m_+^i + m_+^i(M_1^{n-1}) + m_+^i(M_2^{n-1}) - m_-^{i+1} - m_-^{i+1}(M_1^{n-1}) - m_-^{i+1}(M_2^{n-1}), \qquad (2)$$

where p^i $(i = 0, 1, \cdots, n)$ is the i-dimensional Betti number mod 2 of the manifold M^n, m_+^i is the number of critical points of increasing type i, m_-^{i+1} is the number of critical points of decreasing type $i+1$, $m_-^i(M_j^{n-1})$ and $m_+^i(M_j^{n-1})$ $(j = 1, 2)$ are the type numbers of the critical set M_j^{n-1} of decreasing and increasing types respectively (see page 7 and [**190**; **191**]). From (2), as in §2, there follows the set of inequalities

$$\left.\begin{aligned}
&p^0 \leqslant m^0 + m^0(M_1^{n-1}) + m^0(M_2^{n-1}),\\
&p^0 - p^1 \geqslant [m^0 + m\ (M_1^{n-1}) + m^0(M_2^{n-1})]\\
&\qquad\qquad - [m^1 + m^1(M_1^{n-1}) + m^1(M_2^{n-1})],\\
&\cdots\cdots\cdots\cdots\cdots\cdots\cdots\cdots\\
&\sum_{i=0}^{n}(-1)^i p^i = \sum_{i=0}^{n}(-1)^i m^i + \sum_{i=0}^{n}(-1)^i m^i(M_1^{n-1})\\
&\qquad\qquad + \sum_{i=0}^{n}(-1)^i m^i(M_2^{n-1}),
\end{aligned}\right\} \quad (3)$$

where

$$m^i = m_-^i + m_+^i, \quad m^i(M_j^{n-1}) = m_-^i(M_j^{n-1}) + m_+^i(M_j^{n-1}) \quad (j = 1,\ 2).$$

Equation (2) and inequality (3) lead to a convenient estimate of the number of distinct critical points, if the numbers $m_-^i(M_j^{n-1})$ and $m_+^i(M_j^{n-1})$ are expressed in terms of the Betti numbers and the number of cells of a null-basis of the manifolds M_1^{n-1} and M_2^{n-1}. Readers who are interested in this question are referred to the author's paper [**324**].

For the estimation of the number of geometrically distinct critical points we make use of invariants of category type. With the previous notations, let us identify in $M^n - M^{n-1} \times E^1$ all the points of the manifold M_1^{n-1} and denote the point obtained as a result of this identification by the symbol q_1. Analogously identifying the points of the manifold M_2^{n-1}, we obtain a point q_2. After this identification the bounded manifold $M^n - M^{n-1} \times E^1$ is turned into a space $\overline{M}^n$ in which two points q_1 and q_2 have been distinguished, corresponding to the minimum and maximum points of the function φf. We decompose $\overline{M}^n$ into the smallest possible number k of sets $A_0, B_1, B_2, \cdots, B_{k-1} A_k$, where A_0 and A_k are isotopic to neighborhoods of the points q_1 and q_2 respectively, and the sets $B_1, B_2, \cdots, B_{k-1}$ have category 1 in $\overline{M}^n$. The number k is called the category of the space $\overline{M}^n$ with respect to itself and to the singular set Q, consisting in the case at hand of the two points q_1 and q_2, and it is denoted by ${}_Q\mathrm{cat}_{\overline{M}^n}\overline{M}^n$.

The number ${}_Q\mathrm{cat}_{\overline{M}^n}\overline{M}^n$ evidently estimates the lower limit of the number of critical points of twice continuously differentiable functions defined on $\overline{M}^n$ with given minimum and maximum points q_1 and q_2, and consequently the number of critical points distinct from q_1 and q_2 is not less than ${}_Q\mathrm{cat}_{\overline{M}^n}\overline{M}^n - 2$.

This assertion follows immediately from the fact that the category with respect to the point q_1 of the region of smaller values ($\varphi f \leqq x$) changes no more than by unity on the inclusion of each critical point.

Analogous estimates also hold for other invariants of category type.

2. *Mapping of an n-dimensional manifold into n-dimensional Euclidean space.* Suppose that on the closed manifold M^n there is given a system of twice continuously differentiable functions $f_1, \cdots, f_n$. The critical points of such a system of functions are those points in which the rank of the matrix

$$\left\| \begin{matrix} \frac{\partial f_1}{\partial x_1} & \frac{\partial f_1}{\partial x_2} & \cdots & \frac{\partial f_1}{\partial x_n} \\ \cdot & \cdot & \cdot & \cdot \\ \cdot & \cdot & \cdot & \cdot \\ \frac{\partial f_n}{\partial x_1} & \frac{\partial f_n}{\partial x_2} & \cdots & \frac{\partial f_n}{\partial x_n} \end{matrix} \right\|$$

is less than n, where the x_j are the local coordinates of the manifold M^n. If the rank of this matrix at a critical point is equal to $n - \rho$, then ρ is called the genus of the critical point, $1 \leqq \rho \leqq n$.

This classification for the case at hand does not have an essential meaning, since for a given problem critical points of genus $\rho \geqq 2$ will exist only as rare exceptions, but in the study of critical points of a mapping of an n-dimensional manifold into an m-dimensional Euclidean space points of genus $\rho \geqq 2$ must of necessity occur [232].

A system of functions $f_1, \cdots, f_n$ defines a mapping φ of the manifold M^n into Euclidean space R^n. The set of critical points of genus 1 has, generally speaking, dimension $n - 1$, since the coordinates of these points satisfy one equation

$$\left| \begin{matrix} \frac{\partial f_1}{\partial x_1} & \frac{\partial f_1}{\partial x_2} & \cdots & \frac{\partial f_1}{\partial x_n} \\ \frac{\partial f_2}{\partial x_1} & \frac{\partial f_2}{\partial x_2} & \cdots & \frac{\partial f_2}{\partial x_n} \\ \cdot & \cdot & \cdot & \cdot \\ \cdot & \cdot & \cdot & \cdot \\ \frac{\partial f_n}{\partial x_1} & \frac{\partial f_n}{\partial x_2} & \cdots & \frac{\partial f_n}{\partial x_n} \end{matrix} \right| = 0.$$

Suppose that P is the set of all critical points, and Q the set of ordinary points, i.e., noncritical points of the manifold M^n. Represent Q in the form $Q = \sum_{i=1}^s Q_i^*$, where Q_i^* are the components of the set Q, $M^n = P + \sum_{i=1}^s Q_i^*$, or, in order to deal with closed sets, represent M^n in the form $M^n = \overline{U(P)} + \sum_{i=1}^s Q$, where $\overline{U(P)}$ is the closure of a sufficiently small

ϵ-neighborhood of the set P and $Q_i = Q_i^* - U(P)$. Applying the theorem on the length of a sum, we obtain

$$\operatorname{long}_{M^n} M^n \leqslant \operatorname{long}_{M^n} P + \max_{1 \leqslant i = s} \operatorname{long} Q_i + 1. \tag{4}$$

Since the rank of the matrix $\|\partial f_i/\partial x_j\|$ is equal to n, we can prove that each cycle of Q is homologous to some cycle of the boundary Q_i. Hence

$$\max_{1 \leqslant i \leqslant s} \operatorname{long}_{M^n} Q_i \leqslant \operatorname{long}_{M^n} \overline{U(P)} = \operatorname{long}_{M^n} P$$

and from (4) we obtain $\operatorname{long}_{M^n} M^n \leqq 2 \operatorname{long}_{M^n} P + 1$, so that $\operatorname{long}_{M^n} P \geqq (\operatorname{long}_{M^n} M^n - 1)/2 = [(\operatorname{long}_{M^n} M^n)/2]$, where we have denoted the integer part of $(\operatorname{long}_{M^n} M^n)/2$ by $[(\operatorname{long}_{M^n} M^n)/2]$. In the same way one proves an analogous theorem for homological category:

$$\operatorname{kat}_{M^n} P \geqslant \frac{\operatorname{kat}_{M^n} M^n}{2}.$$

8. **Generalization to the infinite case.** The methods used above for estimating the number of critical points of functions may also be used for the estimation of the number of solutions of variational problems.

In this connection there arise considerable analytic difficulties of a function-theoretic nature, analogous to those which appear in the proof of the existence of functions realizing the minimum of a given functional. Moreover, the deformations analogous to shifts along the orthogonal trajectories turn out to be more complicated. We shall not deal with these questions, since in this section we intend only to give a very short survey of the basic papers on qualitative methods of estimation of the number of solutions of variational problems, referring the reader who wishes more detailed information to the work of Ljusternik [**146**; **147**; **151**], and to M. Morse [**191**].

So under certain restrictions on the functional v defined on a functional space Ω, which also must satisfy certain conditions, the following assertion is valid:

The region of smaller values ($v \leqq x$) changes its topological properties only when critical elements, that is, elements of the space Ω at which the variation of the functional is equal to zero, are included in the region ($v \leqq x$).

M. Morse [**191**] carried over to the functional case the concept of nondegenerate critical point of type k and established that when one nondegenerate critical point of type k is included in the region of smaller values, there are only the two possibilities: either the k-dimensional Betti number mod 2 of the region of smaller values goes up by 1, or the $(k-1)$-dimensional Betti number mod 2 of the same region goes down by one.

By finding the Betti number of the space of rectifiable curves passing through two given points on a manifold homeomorphic to the sphere S^2, and the Betti number of the space of closed rectifiable curves on spaces homeomorphic to the sphere S^n, M. Morse obtained an estimate of the number of solutions of two fundamental variational problems for functions defined on the spaces indicated. In particular, one obtains an estimate of the number of geodesic curves passing through two given points of a manifold homeomorphic to the sphere S^n, and the number of closed geodesic curves on manifolds homeomorphic to the sphere S^n.

It turns out that here and in other problems there is an infinite set of solutions. However, since the question is one of estimating the number of analytically distinct elements, the possibility is not excluded that, at least in the second of the indicated problems, these geodesics may coincide geometrically and one and the same closed geodesic may be repeated several times.

In order to estimate the number of geometrically distinct critical elements and, in particular, the number of geometrically distinct closed geodesic curves, L. A. Ljusternik and L. G. Šnirel'man applied the concept of category, which without essential changes carried over to the functional case. They proved that the number of geometrically distinct critical elements of a functional v defined on a space Ω is not less than the category of that space.

Ljusternik and Šnirel'man computed the category of spaces of closed nonselfintersecting curves on the sphere S^2 and thus in particular estimated the number of geometrically distinct geodesic curves on manifolds homeomorphic to the sphere.

It turned out that on such manifolds there exist not less than three closed nonselfintersecting distinct geodesic curves, whose lengths can coincide only in the case of a continuum of geodesic curves of equal length. For example, the ellipsoid $(x^2/a^2) + (y^2/b^2) + (z^2/c^2) = 1$, $a \neq b$, $b \neq c$, $c \neq a$, has coordinate planes closed geodesics of different lengths consisting of its sections by the planes $x = 0$, $y = 0$, $z = 0$. But if one considers an ellipsoid of revolution $a = b \neq c$ or the sphere $a = b = c$, there is a continuum of geodesics of different lengths.

The case of a functional defined on a manifold homeomorphic to the sphere is very complicated in the two-dimensional case, since on the sphere all closed curves can be continuously deformed into a point, and the curves passing through two given points can be deformed into some such fixed curve. In other words, in both cases the functional space consists of one component.

The situation is much simpler when the functional is defined on closed curves, or on curves passing through two given points lying on other closed two-dimensional manifolds, for example on the torus.

In this last case the functional space in question consists of an infinite set of components.

For example, on the torus one component of the space of closed curves is formed by the closed curves continuously deformable into a meridian of the torus, denoted by z_1, and another by the curves continuously deformable into a parallel z_2 of the torus. Also, one component is formed by the closed spirals wrapped k times around the torus (in the meridian direction), and p times around the torus (in the direction of a parallel). These spirals are homologous to $kz_1 + pz_2$.

In each of these components, it is possible under certain conditions to establish the existence of an absolute minimum (or maximum) of the functional, and thus to prove the existence of an infinite set of solutions of the variational problem in question.

Further efforts were directed towards the estimation of the number of solutions of variational problems in which the functional was defined on manifolds of arbitrary numbers of dimensions. If in such a manifold the fundamental group has infinite order, then the functional space Ω in the two fundamental variational problems considered above has an infinite set of components and the estimation of the number of solutions is simpler. It is sufficient to prove the existence of one critical element in each component.

It is substantially more difficult to estimate the number of solutions of variational problems for functionals defined on spaces of curves on manifolds with trivial fundamental groups, for example on spheres or on the products of spheres.

The fundamental difficulty consists here in calculating the necessary topological invariants. It turned out to be particularly difficult to calculate the category. Therefore, it was natural to attempt to carry over to the infinite-dimensional case the concept of length, whose computation in the finite-dimensional case was very much simpler than that of category. In order to define length in infinite-dimensional spaces it was first necessary to develop in them a theory of intersections, which also ran into fundamental difficulties, since finite-dimensional cycles in general position in infinite-dimensional spaces do not intersect one another.

In 1939 Ljusternik [138-140] and [142-147], using upper homologies, carried over to functional spaces the fundamental concepts of the theory of intersections and the concept of length.

Using some results of Pontrjagin, Ljusternik succeeded in estimating the length of spaces of curves on manifolds homeomorphic to the n-dimensional sphere (such spaces are of fundamental importance in the calculus of variations), and thus in estimating the number of solutions of the corresponding variational problems.

In the same direction interesting results were obtained by A. I. Fet [**270-273**] and by S. I. Al′ber, who calculated the Betti groups, the ring of intersections and the length of some functional spaces and at the same time obtained lower estimates for the number of solutions of variational problems in which the functional was defined on these spaces.

In recent years deep results concerning estimation of the number of solutions of variational problems have been obtained by Serre [**254**], using a quite new and rather complicated topological apparatus. In conclusion we note that the methods of estimation in the examples on pages 37-38 also generalize to the functional case and in addition make it possible to estimate the number of eigenvalues of many operators (see Ljusternik [**137**], V. I. Sobolev [**255**], E. S. Citlanadze [**298-302**], M. A. Krasnosel′skiĭ [**105**],[24] M. M. Vaĭnberg [**48-50**]).

[24] There is an extensive bibliography in this paper.

CHAPTER II

Qualitative Methods in the Theory of Functions of Complex Variables

1. **Fundamental concepts.** Qualitative methods in the theory of analytic functions of one complex variable have found wide application in the study of the structure of Riemann spaces and in the estimation of the number of singular and critical points of various types. We shall deal here only with the second of these questions, but shall treat it more generally, extending it to include the estimation of the number of singular and critical points of analytic functions of several complex variables.[1]

The estimation of the number of critical points of an analytic function $f(z) = u(x,y) + iv(x,y)$, i.e., of the number of points at which the derivative $f'(z) = 0$, may be reduced to the estimation of the number of critical points of $u(x,y)$ or of $v(x,y)$, since in view of the Cauchy-Riemann conditions

$$\frac{\partial u}{\partial x} = -\frac{\partial v}{\partial y}; \quad \frac{\partial u}{\partial y} = \frac{\partial v}{\partial x}$$

the critical points of the functions $u(x,y)$ and $v(x,y)$ coincide.

Nondegenerate interior critical points of the function $u(x,y)$ or $v(x,y)$ can only be of the first type, since a harmonic function cannot take on a maximum or minimum value at an interior point. As the result of the confluence of m^1 critical points of type 1 there may arise a critical point of type $(0, m^1, 0)$. Critical points of the function $f(z)$ will be singular points of the vector field of the gradient of the function $u(x,y)$ (or $v(x,y)$).

An important characteristic of a singular point of a vector field, with which we shall have more than once to deal in what follows, is the index of the singular point. For the case of a plane vector field $\vec{v}(p)$ the index of an isolated singular point O may be defined as follows: around the point O we draw a circle S of radius small enough that inside the circle there is no other singular point. The angle through which the vector of the field rotates on one turn around the circumference S in the positive direction is denoted by $2\pi j$, and the integer j is called the index of the singular point O.

[1] A. T. Cvetkov has investigated this question.

Figure 9 illustrates singular points of vector fields of index 1 and -1.

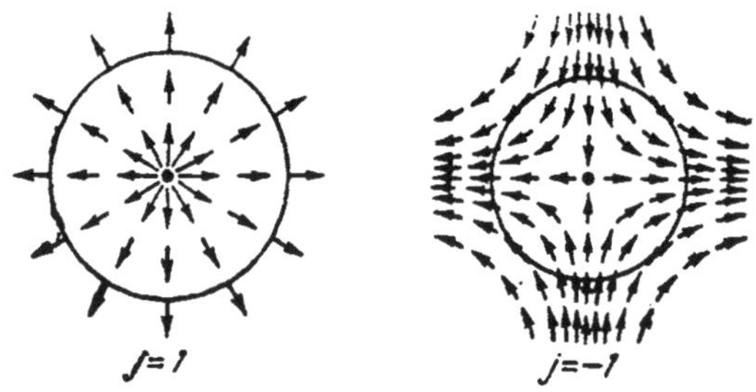

Figure 9

Evidently the index is independent of the choice of the circle S. Moreover, S may be replaced by any other closed curve into which it can be continuously deformed without passing through singular points of the vector field. Indeed, under such a deformation the rotation angle of the vector field under one transit of S can change only in a continuous way. Since that angle is equal to $2\pi j$ and j is an integer, the angle can only change by not less than 2π and hence cannot change at all.

For n-dimensional Euclidean space, including the case $n = 2$, one can give the following definition of a singular point of the vector field $\vec{v}(p)$. The index of the isolated singular point O of a vector field $\vec{v}(p)$ is the degree of the mapping of a sphere S^{n-1} with center at the point O and with sufficiently small radius onto an appropriately oriented Euclidean sphere, under which the point p is put into correspondence with the point $\vec{v}(p)/|\vec{v}(p)|$.

The definition of the index of a singular point may be generalized to completely continuous infinite-dimensional vector fields. For such fields we construct an approximating n-dimensional space, we define the index j_n of a singular point in this latter space and prove that for a sufficiently large n the index j_n becomes stable, i.e., it ceases to be dependent on n, $j_n = j$ for $n \geqq N$. The number j is called the index of the singular point in the given completely continuous vector field.

The index of a nondegenerate critical point of type 1 of the function $u(x,y)$ which is a singular point of the gradient field and a singular point of saddle type for the family of curves $u(x,y) = C$ (or $v(x,y) = C$), is equal to -1. The index of an m^1-multiple critical point of type 1 (a point of multiplicity $(0, m^1, 0)$) is equal to $-m^1$.

As is well known,[2] the sum of the indices of all the singular points of a

[2] Hopf [72]. For the case $n = 2$ this theorem was proved by Poincaré.

continuous vector field which has only a finite number of singular points and is defined on a closed n-dimensional manifold is equal to the Euler characteristic of that manifold.

Since the Euler characteristic of a sphere is equal to 2, and the index of a simple or multiple critical point of an analytic function is negative, it already follows that it is impossible to define a nonconstant analytic function on the whole complex sphere.

The behavior of the functions $u(x,y)$ and $v(x,y)$ in the neighborhood of singular points of the analytic function $f(z) = u(x,y) + iv(x,y)$ is rather complicated. Thus it is more convenient to investigate another function, which makes it possible to estimate the number of critical points of the function $f(z)$. This is the function $U(x,y) = \ln|f(z)|$, where $f(z)$ is a meromorphic function.

In the neighborhood of a zero or pole the function $f(z)$ may be represented in the form $(z-a)^m\varphi(z)$, where m is a nonzero integer and $\varphi(z)$ has neither a zero nor a pole at the point a. Accordingly, in the neighborhood of a zero or a pole of the function $f(z)$

$$U(x, y) = \ln|f(z)| = m\ln|z-a| + \omega(x, y),$$

where $\omega(x,y)$ is a harmonic function continuous at the point z. The critical points of the function $U(x,y)$, because of the Cauchy-Riemann conditions, are also critical points of the function $\ln f(z)$, and therefore also critical points of the function $f(z)$, since

$$\frac{d}{dz}\ln f(z) = \frac{f'(z)}{f(z)}.$$

Thus the zeros and poles of the function $f(z)$ are logarithmic poles of the functions $U(x,y)$, and critical points of the function $f(z)$, which are not zeros of that function, are at the same time critical points of the function $U(x,y)$. The logarithmic poles of $U(x,y)$ which correspond to the zeros of the function $f(z)$ play the role of minimum points, since on approaching the point a from any direction, $m\ln|z-a|$ decreases unboundedly. Analogously, the logarithmic poles corresponding to the poles of the function $f(z)$ play the role of maximum points since on approaching the point a from any direction the quantity $m\ln|z-a|$ increases unboundedly. Since the level lines of the function $U(x,y)$ in the neighborhood of the point a in both cases are closed curves enclosing the point a, we may excise an arbitrarily small neighborhood of the point a along a level curve and identify the points of that level curve. Thus instead of the logarithmic pole one obtains an ordinary maximum or minimum point. Thus for meromorphic functions defined on the entire complex sphere the estimates of the numbers of zeros, critical points not coinciding with zeros, and poles are valid, which are obtained by means of the Betti numbers, category, the number of cells and other invariants used in the estimation of the number of critical points of continuous functions defined on manifolds.

For example, applying the Morse inequalities we obtain:

$$m^0 \geqslant 1,$$
$$m^0 - m^1 \leqslant 1,$$
$$m^2 - m^1 + m^0 = 2,$$

where m^0 is the number of geometrically distinct zeros of the meromorphic function $f(z)$, m^1 is the sum of the multiplicities of the critical points which do not coincide with zeros of the function $f(z)$, and m^2 is the number of geometrically distinct poles of the function $f(z)$. Instead of harmonic functions $U(x,y)$ with logarithmic poles, one may consider functions topologically equivalent to them, i.e., obtained from $U(x,y)$ by isotopic transformations T of the domain of definition.

Denoting the function $U(x,y)$ by $U(p)$ for brevity, we consider the function $U_1(p) = U(T(p))$.

Such a function $U(T(p))$ will be called *pseudologarithmic* with logarithmic poles. Evidently, in the study of topological or isotopic properties, one may replace harmonic functions with logarithmic poles by pseudoharmonic functions with logarithmic poles.

2. **Interdependences among the zeros, critical points, and poles of a meromorphic function.** In this section we give a short review of the results of Morse [188], using a different method of investigation.

Suppose that $U(x,y) = \ln|f(z)|$, where $f(z)$ is a meromorphic function (it is sufficient to take $U(x,y)$ to be pseudoharmonic with logarithmic poles) defined in a closed region R. A sufficiently small ϵ-neighborhood of an interior ordinary point p of the function $U(x,y)$ is divided by the level surface of that function into two parts, in one of which $U(x,y) < U(p)$, and in the other $U(x,y) > U(p)$, where $U(p)$ is the value of the function $U(x,y)$ at the point p. This property is frequently taken as the definition of an ordinary point.

A sufficiently small ϵ-neighborhood of an interior nondegenerate point p of type 1 intersects the region of smaller values ($U \leqq U(p) - \eta$) where η is sufficiently small, in a neighborhood of the zero-dimensional sphere, i.e., the level surface $U = U(p)$ divides a sufficiently small ϵ-neighborhood of the critical point in question into four parts, in two of which $U < U(p)$ and in the two others $U > U(p)$ (Figure 10). In the case of a m^1-tuple critical point p of type 1 (i.e., a point of type $(0, m^1, 0)$) it is also easy to show that a sufficiently small ϵ-neighborhood of the point p is divided by the level surface $U = U(p)$ into $2m^1 + 2$ parts in $m^1 + 1$ of which $U < U(p)$ and in $m^1 + 1$ of which $U > U(p)$. Such a structure of the neighborhood of a simple or multiple critical point of type 1 may also be taken as the topological definition of an isolated critical point of type 1 of the given multiplicity. The region R_l of definition of the function $U(x, y)$ will be taken to be a connected portion of the complex sphere bounded by l topological circumferences. We note that the Euler characteristic $\chi(R_l)$ of this region is equal to $2 - l$, since the Betti number of this region $p^0 = 1$

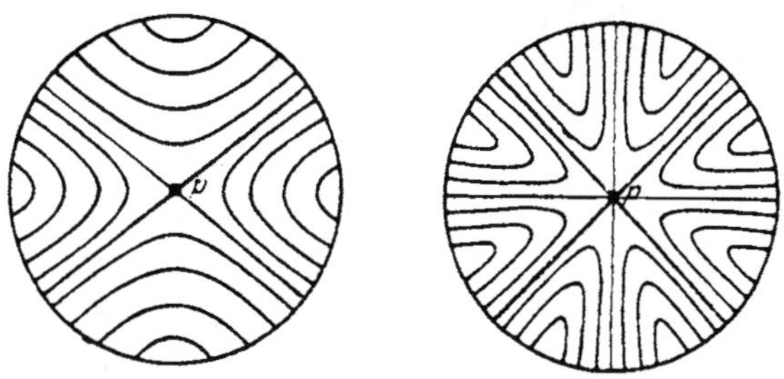

Point of type 1 Point of type (0.3.0)

Figure 10

and $p^1 = l - 1$. The sum of the multiplicities of the interior critical points, which, as we mentioned above, can be only simple or multiple points of type 1, is denoted by m^1. We replace the logarithmic poles by minimum and maximum points and denote the number of such minimum points by m^0 and the number of maximum points by m^2. Suppose that no segment of the boundary is a portion of a single level curve of the function $U(x, y)$. We classify the boundary points by the topological structure of their neighborhoods, as follows:

(1) If a sufficiently small ϵ-neighborhood of the point p is divided by the level curve $U = U(p)$ into only two parts, in which $\Delta U = U(x, y) - U(p)$ has different signs, then the point p is said to be ordinary.

(2) If in a sufficiently small neighborhood of the point p the difference ΔU does not change sign, then for $\Delta U \geqq 0$ the point p is said to be a minimum point and for $\Delta U \leqq 0$ a maximum point.

(3) If the neighborhood of the given point p has a region of smaller values (i.e., $\Delta U < 0$) consisting of m components, then the point p is said

to be a point of type 1 of multiplicity $m-1$ or a point of type $(0, m-1, 0)$. These definitions are special cases of the following more general definition, which we shall use in the following section: if the region of smaller values of the function $U(x_1, x_2, \cdots, x_n)$ in a sufficiently small ϵ-neighborhood of the point p has the Betti numbers $p^0, p^1, \cdots, p^{n-1}$, then the point p is said to be a critical point of type $(0, p^0-1, p^1, p^2, \cdots, p^{n-1})$. A point of type $(0, 0, \cdots, 0)$ is an ordinary point, and a point of type $(0, -1, 0, \cdots, 0)$ by definition coincides with $(1, 0, 0, \cdots, 0)$ and is called a minimum point. We call attention to the disparity between the roles played by the regions of smaller and greater values in the definition of critical points.

The whole classification scheme is carried out in terms of the topological structure of the region of smaller values. For example,

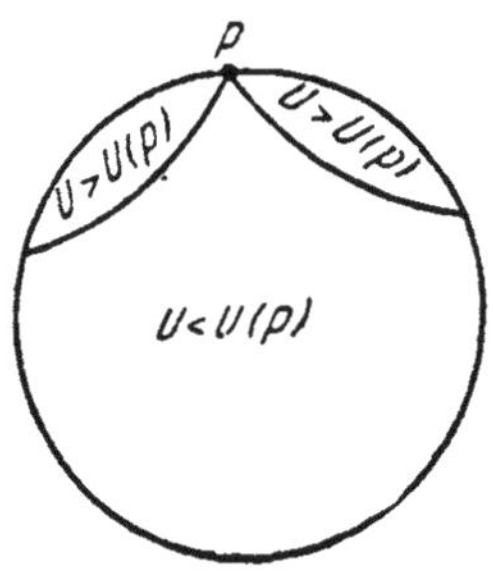

Figure 11

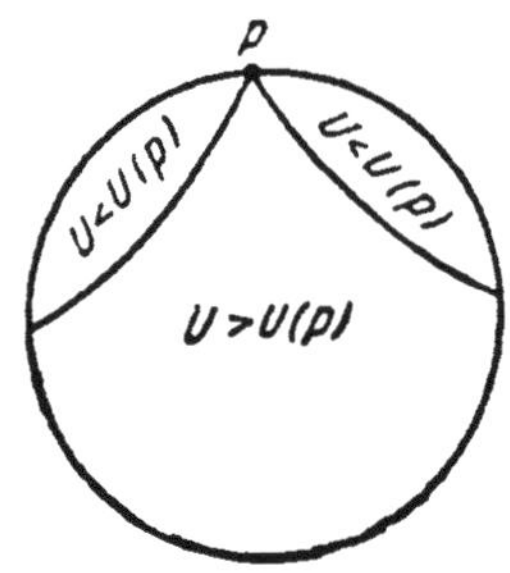

Figure 12

the boundary point p shown in Figure 11 is ordinary, and the boundary point shown in Figure 12 is a critical point of type 1.

Supposing that the function $U(x, y)$ has only a finite set of critical points and applying the results of §§ 1 and 2 of Chapter I, we obtain the following theorem: *as the level surface $U(x, y) = t$ passes through a nondegenerate critical point p of type k $(k = 0, 1, 2)$, the following cases are possible*: (a) *the* kth *Betti number of the region of smaller values of the function U goes up by unity and the other Betti numbers do not change*; (b) *the* $(k-1)$st *Betti numbers of the region of smaller values goes down by unity and the other Betti numbers do not change.*

This theorem is valid both for interior and for boundary critical points, except for boundary critical points of maximum type, the inclusion of which does not change the topological properties of the region of smaller values. Indeed, applying the addition theorem for Betti numbers and retaining the notations of § 2 of Chapter I, we will have:

$$\Delta p^2 = p^2(E) - p^2(S^{k-1}) + n^r + n^{r-1} \qquad (r = 0, 1, 2).$$

In all cases, except the inclusion of boundary points of maximum type, the separate terms change in the same way as in §2 of Chapter I. But for boundary maxima S is no longer a circumference, but rather a one-dimensional cell, and therefore on the inclusion of a boundary maximum point all the terms of the right side for $r = 1, 2$ are equal to zero, and for $r = 0$ the first two terms cancel each other out. The others are equal to zero, i.e., $\Delta p^r = 0$ $(r = 0, 1, 2)$. Accordingly, the Morse inequalities

$$m^0 \geqslant p^0,$$
$$m^0 - m^1 \leqslant p^0 - p^1,$$
$$m^0 - m^1 + m^2 = p^0 - p^1 + p^2,$$

and also the estimate of the number of analytically distinct critical points

$$\sum_{i=0}^{2} p^i \leqslant K(R_l)$$

remain valid, if one takes m^2 to be the number of interior poles of the function $f(z)$ (or the number of interior maximum points for the function $U(x, y)$).

From the fact that $m^0 - m^1 + m^2 = p^0 - p^1 + p^2$ it follows that:

(1) if the function $U(x, y) = \ln |f(z)|$ has no logarithmic poles, then $\chi(R_l) = m - s$, where $\chi(R_l)$ is the Euler characteristic of the region R_l, m the number of minima and s the sum of the multiplicities of points of type 1 (saddle points). Or since $\chi(R_l) = 2 - l$, therefore $2 - l = m - s$.

(2) $\chi(R_l) = m - S - s + M$, or $2 - l = m - S - s + M$, where m is the number of minimum points of $U(x, y)$, S the sum of the multiplicities of the interior points of type 1, s the sum of the multiplicities of the boundary points of type 1, M the number of logarithmic poles of the function $U(x, y) = \ln |f(z)|$.

(3) $\chi(R_l) = m_1 + m_2 - S - s + M_1$ or $2 - l = m_1 + m_2 - S - s + M_1$, where m_1 is the number of boundary minimum points of $U(x, y)$, m_2 is the number of zeros of the meromorphic function $f(z)$ without regard to their multiplicities, S and s have their previous meanings, and M_1 is the number of poles of the function $f(z)$, again without regard to their multiplicity.

The inequality $\sum_{i=0}^{2} p^i \leqq \sum_{i=0}^{2} m^i$ may be written in the following form:

(1) $\sum_{i=0}^{2} p^i \leqq m_1 + m_2 + S + s + M_1$, where $\sum_{i=0}^{2} p^i = l$ for $l \neq 0$ and $\sum_{i=1}^{2} p^i = 2$, if $l = 0$.

(2) $\sum_{i=0}^{2} p^i \leqq m + S + s + M$.

We note that the use of other topological invariants for the estimation of the number of singular and critical points in the case at hand will not lead to new results, because of the extremely simple topological properties of the set R_l.

Up to this point we have been supposing that no segment of the boundary coincides with a segment of a level curve of $U(x, y)$. Now we suppose that $U(x, y)$ has a constant value on

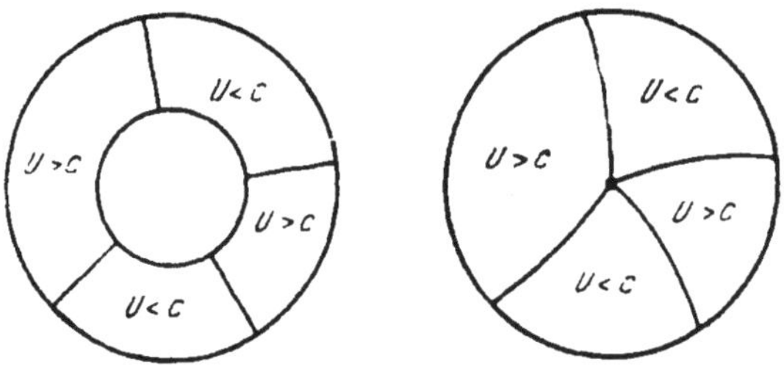

Figure 13

certain components of the boundary of the region R_l. In this case it is convenient to identify all the points of each of those components as one point. If the component of the boundary is at the same time a component of the level curve $U(x, y) = c$, then the identified point thus obtained will be a maximum or minimum point, since in the neighborhood of it ΔU preserves its sign. If the boundary component on which $U(x, y)$ is constant does not coincide with some component of the level surface (Figure 13) the identified point will be of another type. After the identification R_l goes into some new region R_{l_1}, where $l_1 \leqq l$, and if no segment of the boundary of R_{l_1} now coincides with a segment of a level curve of the function $U(x, y)$, then the above estimates may be applied.

3. **Functions of several complex variables.** The analytic function

$$f(z_1, z_2, \ldots, z_n) = u(x_1, y_1, x_2, y_2, \ldots, x_n, y_n) + iv(x_1, y_1, x_2, y_2, \ldots, x_n, y_n),$$

where

$$z_j = x_j + iy_j \qquad (j = 1, 2, \ldots, m),$$

is usually defined on the so-called function-theoretic space M^{2n}, which from the topological point of view is the product of n two-dimensional spheres S^2, or on the complex projective space P^{2n}, or on portions of these spaces.

In what follows we shall need some information on the fundamental topological invariants of these spaces. For the function-theoretic space $M^{2n} = S^2 \times S^2 \times \cdots \times S^2$ the even-dimensional Betti numbers $p^{2r} = C^r_n$, since p^r is equal to the coefficient of x^r in the polynomial $(x^2 + 1)^n$, and the odd-dimensional Betti numbers $p^{2r+1} = 0$ $(r = 0,1,2, \cdots, n)$. The fundamental group is trivial. The length of the manifold M^{2n} is equal to n (see page 27), the category of M^{2n} and the number of cells is equal to $n + 1$, the number of cells with regular adjunction is equal to 2^n. The Betti number of the complex projective space $p^{2r} = 1$, $p^{2r+1} = 0$ $(r = 0, 1, \cdots, n)$; the category (see page 27), the number of cells, the length $+1$, and the number of cells with regular adjunction, and the Euler characteristic of the space p^{2n} are all equal to $n + 1$.

At regular points of the analytic function

$$f(z_1, z_2, \ldots, z_n) = u(x_1, y_1, x_2, y_2, \ldots, x_n, y_n) + iv(x_1, y_1, x_2, y_2, \ldots, x_n, y_n)$$

conditions analogous to the Cauchy-Riemann conditions must be satisfied, i.e., u and v satisfy the equations

$$\left.\begin{array}{c} \dfrac{\partial^2 u}{\partial x_k \partial x_s} + \dfrac{\partial^2 u}{\partial y_k \partial y_s} = 0, \quad \dfrac{\partial^2 u}{\partial x_k \partial y_s} - \dfrac{\partial^2 u}{\partial x_s \partial y_k} = 0 \\ (k, s = 1, 2, \ldots, n)^3 \end{array}\right\} \tag{1}$$

and, in particular, u and v are harmonic functions in each pair of variables x_k, y_k $(k = 1, 2, \cdots, n)$:

$$\frac{\partial^2 u}{\partial x_k^2} + \frac{\partial^2 u}{\partial y_k^2} \equiv 0$$

and

$$\frac{\partial^2 v}{\partial x_k^2} + \frac{\partial^2 v}{\partial y_k^2} \equiv 0.$$

Accordingly, the functions u and v cannot have maxima or minima at interior points of their regions of definition. Moreover, they cannot have maxima or minima at interior points even with respect to one pair of variables x_k, y_k with the other variables fixed. Thus interior critical points of the functions u and v are points of type 1 (minimax points) in each pair of variables x_k, y_k for fixed values of the remaining variables. Hence it easily follows that they are of type n with respect to the collection of all variables $x_1, y_1, x_2, y_2, \cdots, x_n, y_n$. This assertion may also be proved directly by bringing the second differential into canonical form in the neighborhood of a nondegenerate interior critical point. Indeed,

[3] Functions satisfying these equations are often said to be semiharmonic. But since this term is required also for the solution of the equation $\Delta^n u = 0$ $(n \geqq 2)$, we shall call them n-harmonic to avoid ambiguity.

$$d^2u = \sum_{i=1}^{n}\sum_{j=1}^{n}\left(\frac{\partial^2 u}{\partial x_i\,\partial x_j}\right)_0 \Delta x_i\,\Delta x_j + \sum_{i=1}^{n}\sum_{j=1}^{n}\left(\frac{\partial^2 u}{\partial y_i\,\partial y_j}\right)_0 \Delta y_i\,\Delta y_j + 2\sum_{i=1}^{n}\sum_{j=1}^{n}\left(\frac{\partial^2 u}{\partial x_i\,\partial y_j}\right)_0 \Delta x_i\,\Delta y_j.$$

The last double sum is equal to zero from conditions (1). In the two other double sums, from conditions (1) again, the coefficients on $\Delta x_i\,\Delta x_j$ and on $\Delta y_i\,\Delta y_j$ are equal and opposite. Thus if the first double sum, when brought into canonical form, contains k negative terms, the second double sum in canonical form will contain $n-k$ negative terms.

Since isolated degenerate critical points may be considered a result of the confluence of nondegenerate critical points, they can only be multiple critical points of type n. Hence it follows at once that there cannot exist a nonconstant harmonic function without singularities on all of M^{2n} or P^{2n}, since such a function must have not less than $p^s(M^{2n})$ (or $p^s(P^{2n})$) points of type s.

Suppose that u is the real part of the analytic function $f(z_1, z_2, \cdots, z_n)$, defined and regular on a bounded subset N^{2m} of the manifold M^{2n} (or P^{2n}). Suppose that on the boundary of N^{2n} there are only a finite number of critical points (as to the classification of these critical points see page 59).

We suppose also that in a neighborhood of an ordinary critical point the region of smaller values ($u < c$) may be isotopically deformed into the region ($u < c - \epsilon$), $\epsilon > 0$ (the possibility of such a deformation can be proved without serious restrictions on the boundary of the manifold N^{2m}, and the level surfaces in its neighborhood, but the proof is very awkward (see [**191**; **316**; **328**]).

In this case one can apply the method of estimation of the number of critical points of §2 of Chapter I, so that the following inequalities are valid:

$$p^i(N^{2n}) \leqslant m^i \qquad (i = 0, 1, 2, \ldots, n),$$

where $p^i(N^{2n})$ is the ith Betti number of the bounded manifold N^{2m} and m^i is the number of analytically distinct points of type i:

$$\mathrm{cat}_{N^{2n}} N^{2n} \leqslant K(N^{2n}),$$

where $K(N^{2n})$ is the number of geometrically distinct critical points. One obtains further estimates by means of other invariants of category type.

Now suppose that the function u has N^{2n} singular points of pole type, but no singular points of other types (in particular, no points of nondefinition), and, as before, satisfies the other conditions imposed on that function.

In this case it is convenient to consider the function $U = \ln |u|$ in place of u. Its logarithmic poles correspond to the poles and zeros of the function u, and critical points which are not at the same time zeros of the function u coincide with critical points of u. The logarithmic poles corresponding to the zeros of the function u play the role of the minimum points. The logarithmic poles corresponding to the poles of the function $f(z_1, \cdots, z_n)$ play the role of the maximum points.

As is well known, the singular points and zeros of the function $f(z_1, \cdots, z_n)$ are distributed continuously, filling out $2n-2$-dimensional analytic surfaces. Assigning to the set P of logarithmic poles of minimum type and to the set R of logarithmic poles of maximum type the corresponding multiplicities (see §2 of Chapter I), we may now rewrite the estimates obtained above as follows:

$$p^i(N^{2n}) \leqslant m^i + m^i(P) + m^i(Q) \qquad (i = 0, 1, 2, \ldots, 2n),$$
$$\operatorname{cat}_{N^{2n}} N^{2n} \leqslant K(N^{2n}) + \operatorname{cat}_{N^{2n}} P + \operatorname{cat}_{N^{2n}} Q.$$

We note that aside from trivial cases, the theory just presented cannot be applied to analytic functions defined on the entire projective space or on the entire function-theoretic space, since in these cases there almost necessarily appear singular points of other types (points of nondefinition), which very much complicate the investigation.

CHAPTER III

The Fixed Point Method

At the end of the first quarter of the twentieth century, and also in recent years, in the papers of P. S. Aleksandrov, S. Banach, R. Cacciopoli, V. V. Nemyckiĭ, A. N. Tihonov, L. Brouwer, J. Schauder and others, there began the application of new and very powerful methods of proof of the existence of solutions to various equations and systems of equations, namely the fixed point method.

In theorems on fixed points it is asserted that under one or another set of conditions a mapping of some set onto itself will have a fixed point.

One then studies equations or systems of equations written in the form $p = A(p)$, where A is a finite, differential, integral or other kind of operator operating on the points p of some space, for example, $x = f(x)$, or $y = y_0 + \int_{x_0}^{x} f(x, y)dx$, or $y = f(x) + \int_a^b K(x, s)y(s)ds$.

To find the solution of the equation $p = A(p)$ one finds a point p which is fixed under the operation of the operator A.

Accordingly, in order to prove the existence of a solution of such an equation one must consider the mapping which is defined by the operator $A(p)$. If then on the basis of already proved theorems on fixed points it is possible to assert that under such a mapping there always exists a fixed point, then the equation $p = A(p)$ has a solution.

1. **Theorems on fixed points.** The simplest theorem on fixed points is the principle of contraction mappings, which gives the conditions under which the method of successive approximations may be used.

1. *Principle of contraction mappings. If in a space M which is metric*[1] *and complete*[2] *there is given an operator A satisfying the following conditions:*

---

[1] A space is said to be metric if there is defined in it a function $\rho(y,z)$ on the pairs of points of that space, satisfying for any points of that space the following:

(1) $\rho(x,y) \geqq 0$, while $\rho(y,y) = 0$ and $\rho(y,z) = 0$ implies that $y = z$;

(2) $\rho(y,z) = \rho(z,y)$;

(3) $\rho(y,z) \leqq \rho(y,u) + \rho(u,z)$ (the triangle inequality).

[2] A metric space is said to be complete if every fundamental sequence of points in the space (i.e., a sequence satisfying the condition $\rho(y_n, y_{n+m}) < \epsilon$ for $n \geqq N(\epsilon)$ for arbitrary $m > 0$, $\epsilon > 0$) converges.

(1) *The operator A carries elements of the space M into elements of the same space*:

if $y \subset M$, *then* $A(y) \subset M$,

(2) $\rho(A(y), A(z)) \leqq \alpha\rho(y, z)$, *where y and z are any points of the space M*, $\alpha < 1$ *and* α *does not depend on the choice of y and z*, $\rho(y, z)$ *being the distance between y and z in the space M*,

then there exists a unique fixed point $\bar{y}$ *of the space M*: $A(\bar{y}) = \bar{y}$. *This point may be found by the method of successive approximations, i.e.*,

$$\bar{y} = \lim_{n \to \infty} y_n, \quad where \quad y_n = A(y_{n-1}) \qquad (n = 1, 2, \ldots),$$

with the point y_0 *chosen arbitrarily in the space M*.

Proof. (1) The limit $\bar{y}$ of the sequence $y_0, y_1, \cdots, y_n, \cdots$ exists since this sequence is fundamental. Indeed, applying the triangle inequality several times, we find

$$\rho(y_n, y_{n+m}) \leqslant \rho(y_n, y_{n+1}) + \rho(y_{n+1}, y_{n+2}) + \ldots + \rho(y_{n+m-1}, y_{n+m}),$$

so that, taking into account the inequalities

$$\rho(y_2, y_1) = \rho[A(y_1), A(y_0)] \leqslant \alpha\rho(y_1, y_0),$$

$$\rho(y_3, y_2) = \rho[A(y_2), A(y_1)] \leqslant \alpha\rho(y_2, y_1) \leqslant \alpha^2\rho(y_1, y_0),$$

$$\cdots\cdots\cdots\cdots\cdots\cdots\cdots\cdots$$

$$\rho(y_{n+1}, y_n) = \rho[A(y_n), A(y_{n-1})] \leqslant \alpha(y_n, y_{n-1}) \leqslant \alpha^n\rho(y_1, y_0),$$

$$\cdots\cdots\cdots\cdots\cdots\cdots\cdots\cdots$$

we will have

$$\rho(y_n, y_{n+m}) \leqslant (\alpha^n + \alpha^{n+1} + \ldots + \alpha^{n+m-1})\rho(y_1, y_0) = \frac{\alpha^n - \alpha^{n+m}}{1-\alpha}\rho(y_1, y_0) < \frac{\alpha^n}{1-\alpha}\rho(y_1, y_0) < \varepsilon,$$

for n sufficiently large.

(2) The limit $\bar{y}$ of the sequence y_n $(n = 0, 1, \cdots, n, \cdots)$ is a fixed point of the space M: $A(\bar{y}) = \bar{y}$. Indeed, suppose that $A(\bar{y}) = \bar{\bar{y}}$. Then for sufficiently large n

$$\rho(\bar{y}, \bar{\bar{y}}) \leqslant \rho(\bar{y}, y_n) + \rho(y_n, y_{n+1}) + \rho(y_{n+1}, \bar{\bar{y}}) < \varepsilon,$$

since for sufficiently large n we have $\rho(\bar{y}, y_n) < \epsilon/3$, in view of the fact that $\bar{y} = \lim_{n\to\infty} y_n = \bar{y}$; the second term $\rho(y_n, y_{n+1}) < \epsilon/3$ since the sequence

y_n is fundamental, and finally

$$\rho(y_{n+1}, \bar{\bar{y}}) = \rho[A(y_n), A(\bar{y})] \leqslant \alpha\rho(y_n, \bar{y}) < \frac{\varepsilon}{3}.$$

Thus the distance between the two fixed points $\bar{y}$ and $\bar{\bar{y}}$ can be made arbitrarily small, so that $\bar{\bar{y}} = \bar{y}$ and $A(\bar{y}) = \bar{y}$.

(3) The fixed point is unique. Indeed, under the action of the operator A the distance between any two points decreases:

$$\rho[A(\bar{y}), A(\bar{z})] \leqslant \alpha\rho(\bar{y}, \bar{z}), \quad \alpha < 1.$$

Thus not both the points $\bar{y}$ and $\bar{z}$ can be fixed.

Examples of the application of the principle of contraction mappings are given in the following section.

2. *Brouwer's theorem. Under a continuous mapping φ of an n-dimensional cell into itself there exists at least one fixed point $\bar{p}$:*

$$\varphi(\bar{p}) = \bar{p}.$$

Evidently it is sufficient to prove the theorem for mappings of n-dimensional simplexes into themselves.

We begin by proving two lemmas.

LEMMA I. *If all the vertices of a simplicial subdivision of a simplex E^n whose vertices are labelled $0, 1, \cdots, n$, are labelled with the same numbers in an arbitrary fashion, subject to the single condition that the vertices of the subdivision lying on any given k-dimensional face T^k $(k = 0, 1, \cdots, n)$ of the basic simplex E^n are labelled with numbers appearing in the enumeration of the vertices of the face T^k, then there exists an odd number σ of n-dimensional simplexes (and therefore at least one) of the subdivision all of whose vertices are labelled with distinct numbers.*

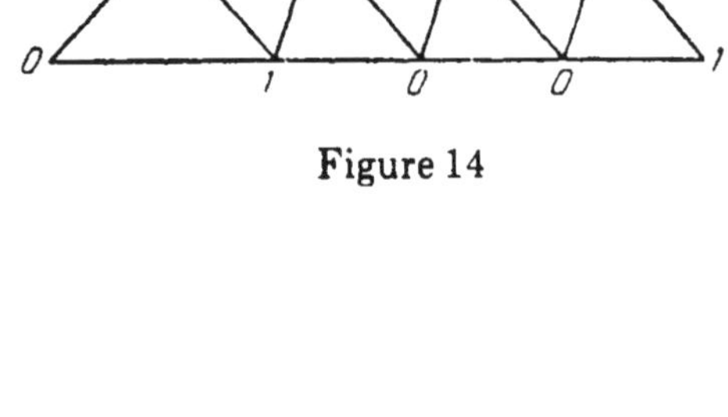

Figure 14

An r-dimensional simplex of the subdivision whose vertices are labelled with distinct numbers and moreover exactly with the numbers $0, 1, \cdots, r$, will be called an *α_r-simplex*. The lemma asserts that under the conditions indicated above there exists an odd number σ_n of n-simplexes. We shall carry out the proof by the method of complete induction. For $n = 0$ the lemma is true. Suppose that it is valid for the $(n-1)$-dimensional simplex.

We shall count in two ways the number s of α_{n-1}-simplexes entering into the subdivision of E^n. Any α_{n-1}-simplex which is a face of two simplexes of the subdivision is counted twice.

Now s is equal to the sum of the number σ_{n-1} of α_{n-1}-simplexes lying on the $(n-1)$-dimensional face of the basic simplex E^n whose vertices are labelled by the numbers $0, 1, \cdots, (n-1)$ (no α_{n-1}-simplex can lie on any other $(n-1)$-dimensional face of the simplex E^n, since in the labelling of the vertices of the simplexes lying on these faces at least one of the numbers $0, 1, \cdots, (n-1)$ cannot appear) and an even number $2q$ of α_{n-1}-simplexes lying inside the simplex E^n, each of which is the common boundary of two n-dimensional simplexes of the subdivision. Since by the induction hypothesis σ_{n-1} is odd, the number $s = \sigma_{n-1} + 2q$ is also odd. On the other hand, s is equal to the sum of the number σ_n of α_n-simplexes (since each α_n-simplex contains one face which is an α_{n-1}-simplex) and an even number $2l$ of α_{n-1}-simplexes which are faces of the n-dimensional simplexes whose vertices are labelled by the numbers $(0, 1, 2, \cdots, n-1, d)$, where d is equal to one of the numbers $0, 1, 2, \cdots, (n-1)$. Each of those n-dimensional simplexes has exactly two faces which are α_{n-1}-simplexes. Accordingly, the number $\sigma_n = s - 2l$ is odd.

Lemma II. *If the simplex E^n is entirely covered by $n+1$ closed sets M_i $(i = 0, 1, 2, \cdots, n)$, while each k-dimensional face $(k = 0, 1, \cdots, (n-1))$ whose vertices are labelled by the indices $\beta_0, \beta_1, \cdots, \beta_k$ is contained in the sum of the sets $\sum_{i=0}^{k} M_{\beta_i}$, then the sets M_i have at least one common point* (*Figure* 15).

Consider the following system of finer and finer simplicial ϵ_j-subdivisions[3] of the simplex E^n $(j = 1, 2, \cdots)$, $\lim_{j\to\infty} \epsilon_j = 0$. The vertices of the simplexes of each subdivision are labelled by the indices $0, 1, \cdots, n$, under the following conditions:

(a) a vertex is arbitrarily labelled with one of the indices i of the sets M_i in which it is contained;

(b) the vertices lying on a k-dimensional face $(k = 0, 1, \cdots, (n-1))$ of

[3] The ϵ-subdivisions contain simplexes of diameter no greater than ϵ.

the simplex E^n can be labelled, in accordance with condition (a), only with the indices which label the vertices of that k-dimensional face.

If conditions (a) and (b) are satisfied the labelling of the vertices will evidently satisfy the conditions of Lemma I, so that in each ϵ_j-subdivision there will be an α_n-simplex. Any point p_j of an α_n-simplex of the ϵ_j-subdivision will be distant by not more than ϵ_j from all the sets M_i $(i = 0, 1, \cdots, n)$. The sequence of points p_j $(j = 1, 2, \cdots)$, each of which is chosen arbitrarily from the α_n-simplex of the ϵ_j-subdivision $(j = 1, 2, \cdots)$, has at least one limit point $\bar{p}$ in E^n, whose distance from all the sets M_i $(i = 0, 1, \cdots, n)$ is equal to zero, so that, the sets M_i being closed, the point $\bar{p}$ belongs to all of these sets.

Proof of the theorem. Consider the closed set M_i of the points of E^n which are not brought nearer to the vertex i by the mapping φ, i.e., $p \subset M_i$ if the distance of the point p from the vertex of E^n labelled with the index i is not greater than the distance of the point $\varphi(p)$ from the same vertex. Evidently the conditions of Lemma II are satisfied and so the sets M_i have a common point $\bar{p}$. This point does not approach any of the vertices of the simplex E^n under the mapping φ, from which it follows that it is a fixed point of the mapping φ.

It will be more convenient, using the barycentric coordinates $m_0, m_1, \cdots, m_n, \sum_{i=0}^{n} m_i = 1$ of the points of the simplex E^n, to define the set M_i as the set of points for which the coordinate m_i does not increase under the mapping φ. Under such a mapping the sets M_i also satisfy the conditions of Lemma II and accordingly have a common point $\bar{p}$. None of the barycentric coordinates of this point increase under the mapping φ. But since $\sum_{i=0}^{n} m_i = 1$, none of these barycentric coordinates can decrease either, since then some other coordinate would have to increase. Accordingly, the barycentric coordinates of $\bar{p}$ do not change under the mapping φ, and thus $\bar{p}$ is fixed under φ.

3. *Schauder's Principle. Let F be a continuous mapping into itself of a*

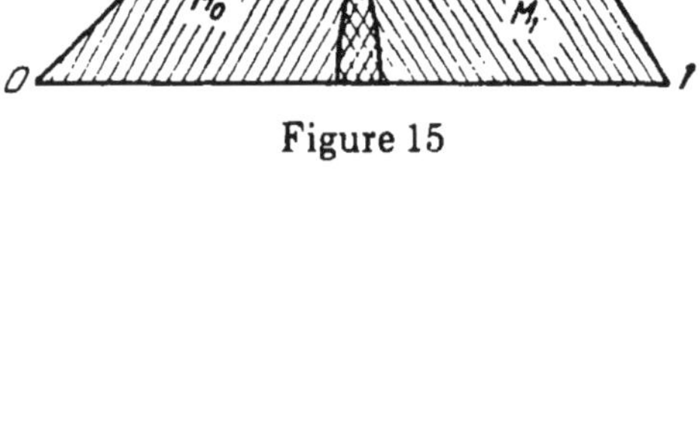

Figure 15

closed compact convex set H lying in a complete linear normed space; then the mapping F has at least one fixed point.

Proof. Since H is compact, it contains a finite ϵ_n-net[4] consisting of points $p_1, p_2, \cdots, p_s$. Let $H^{(n)}$ be the smallest convex body of the space M which contains all the points p_i $(i = 1, 2, \cdots, s)$. The dimension of the body $H^{(n)}$ is not greater than $s - 1$. Let us consider an ϵ_n-subdivision of the cell $H^{(n)}$ which includes all the points p_i $(i = 1, 2, \cdots, s)$ among its vertices. The mapping F of the cell $H^{(n)}$ is in general not simplicial. We replace it by an approximating simplicial mapping $F^{(n)}$ which is defined at the vertices of the simplicial decomposition of the cell $H^{(n)}$ as follows: we obtain $F^{(n)}(a_j)$ by shifting the point $F(a_j)$ to the closest of the points p_i $(i = 1, 2, \cdots, s)$. If $F(a_j)$ is at the same distance from several points p_i of the ϵ_n-net, we shift it to any one of these points. Thus the points $F^{(n)}(a_j)$ and $F(a_j)$ are not distant from one another by more than ϵ_n. We extend this mapping from the vertices of the simplicial subdivision to the entire cell $H^{(n)}$ by taking it to be linear on the entire simplex. Thus the mapping $F^{(n)}$, for a sufficiently small ϵ_n-net, is arbitrarily close to the mapping F at all the points of $H^{(n)}$; more precisely, $\rho[F(x), F^{(n)}(x)] \leqq 3\epsilon_n$ for $x \subset H^{(n)}$, where ρ is the distance in the space M. By Brouwer's theorem $F^{(n)}$ has a fixed point $\bar{x}_n$:

$$F^{(n)}(\bar{x}_n) = \bar{x}_n,$$

and since $\rho(F^{(n)}(x), F(x)) \leqq 3\epsilon$ for any $x \subset H^{(n)}$, the point $\bar{x}_n$ is shifted under the mapping F by not more than $3\epsilon_n$. Let us now take a sequence $\{\epsilon_n\}$ tending to zero and for each n construct the corresponding ϵ_n-net and subdivision of the body $H^{(n)}$ as indicated above; in this way we obtain simplicial subdivisions $F^{(n)}$ of the $H^{(n)}$ which approximate the mapping F to within $3\epsilon_n$. In addition, we obtain a sequence of points $\bar{x}_n$ which are fixed under the corresponding mappings $F^{(n)}$ and do not shift under the mapping F by more than $3\epsilon_n$. The sequence $\bar{x}_n$ has at least one limit point $\bar{x}$ on the compact set H, which because F is continuous must be a fixed point of that mapping: $F(\bar{x}) = \bar{x}$.

Remark. By Mazur's lemma[5] which states that the smallest closed convex set containing a given compact set B in a complete linear normed space is compact, one may formulate Schauder's principle as follows: *if in*

---

[4] An ϵ-net of a set H lying in a metric space is a set A of points with the property that for any point q of the set H there is at least one point of A which is distant by less than ϵ from q.

[5] Mazur, *Über konvexe Mengen in linearen nomierten Räumen*, Studia Math. 4 (1933), 70-84 and also Studia Math. 2 (1930).

a complete linear normed space a continuous operator carries a closed convex set into a compact part of itself, then there is a fixed point.

An application of Schauder's principle is given in the following section.

We list here without proof several other theorems on fixed points which are not directly used in the remainder of this book.

TIHONOV'S THEOREM [264]. *Under every continuous mapping of a convex bicompact set in a linear locally-convex space into a subset of itself there exists at least one fixed point.*

This very general fixed point principle makes it possible, for example, to prove under very wide hypotheses that there exist solutions of systems of differential equations with a continuum of unknown functions.

SECOND THEOREM OF SCHAUDER.[6] *A weakly continuous operator defined in a complete normed separable space, which maps a closed weakly compact set A into a part of itself, leaves at least one point of the set A fixed.*

THEOREM OF HOPF-LEFSCHETZ.[7] *If the number $l_\varphi = \sum_{r=0}^{n}(-1)^r S_P B_0^r$ of the continuous mapping φ of the polyhedron P^n into itself is not equal to zero, then there exists at least one fixed point of the mapping φ. Here $S_P B_0^r$ is the trace of the endomorphism of the r-dimensional group of weak homologies B_{00}^r of the polyhedron P^n induced by the mapping φ.*

From this theorem, for example, one may obtain the following strengthening of Brouwer's theorem: every continuous mapping of a connected polyhedron all of whose Betti numbers, with the exception of the zero-dimensional one, are equal to zero, has at least one fixed point.

Indeed, in this case $S_P B_{00}^0 = 1$ and $S_P B_{00}^r = 0$ for $r > 0$, so that $l_\varphi = 1$. Among the polyhedra P^n which satisfy the above conditions but are not cells one may mention, for example, the even-dimensional projective spaces. We note that the Hopf-Lefschetz condition $l_\varphi \neq 0$ is not necessary for the existence of a fixed point. Moreover, necessary and sufficient conditions cannot in general be given in terms of homology invariants. However, since the complicated homotopy invariants and homotopy theorems on fixed points have not yet been sufficiently studied, the Hopf-Lefschetz condition remains useful at the present day and is the fixed point theorem most frequently applied in the study of mappings of polyhedra.

---

[6] Schauder, Math. Z 26 (1927).

[7] A detailed exposition of homology theorems on fixed points may be found in the book [1] of P. S. Aleksandrov.

2. **Some applications of fixed point theorems.** We consider in this section some applications of the principle of contraction mappings and the principle of Schauder. Here we shall dwell rather in detail on the less known applications of these principles to the theory of differential equations with retarded argument.

SOME APPLICATIONS OF THE PRINCIPLE OF COMPRESSION MAPPINGS.

1. *The finite equation $x = f(x)$, where the real function f of the real variable x is defined and continuously differentiable for all values of x, has a unique real solution if $|f'(x)| \leqq \alpha < 1$, and this solution may be found by the method of successive approximations.*

Indeed, the operator $f(x)$ and the space of real numbers with the usual metric $\rho(x_1, x_2) = |x_1 - x_2|$ satisfy the conditions of the principle of contraction mappings.

We shall verify that the second condition is satisfied by the operator A, since it is obvious that the other conditions of the principle of contraction mappings are satisfied:

$$|f(x_1) - f(x_2)| = |f'(\xi)|\,|x_1 - x_2| \leqslant \alpha\,|x_1 - x_2|,$$

where $x_1 < \xi < x_2$.

2. *The differential equation $dy/dx = f(x,y)$ with the initial condition $y(x_0) = y_0$, or the equivalent integral equation $y(x) = y_0 + \int_{x_0}^{x} f(x, y(x))dx$, in which the function f is continuous in some neighborhood D of the initial point (x_0, y_0) and satisfies in this neighborhood a Lipschitz condition in y, has a unique continuous solution in some neighborhood $D_1 \subset D$ of the initial point, and that solution may be found by the method of successive approximations.*

Indeed, the operator $A(y) = y_0 + \int_{x_0}^{x} f(x, y(x))dx$ and the space of continuous functions C_0 with the uniform convergence topology[8] on the segment $x_0 - h \leqq x \leqq x_0 + h$ satisfy, for sufficiently small h, all the conditions of the principle of contraction mappings.

We shall verify that condition 2 applied to the operator A is true, the other conditions being obvious:

$$\rho\,(A(y),\, A(z)) = \max\left|\int_{x_0}^{x} f(x,\, y)\,dx - \int_{x_0}^{x} f(x,\, z)\,dx\right|$$

$$\leqslant N \max\left|\int_{x_0}^{x} |y - z|\,dx\right| \leqslant Nh \max|y - z| \leqslant \alpha\rho\,(y,\, z),$$

[8] *Translator's note*: I.e., the distance defined by $\rho(y(x), z(x)) = \max\limits_{x_0 - h \leqq x \leqq x_0 + h} |y(x) - z(x)|$.

if one takes $h \leqq \alpha/N$, where $0 < \alpha < 1$.

An analogous theorem may easily be verified for systems of n differential equations.

3. *The integral equation*

$$y(x) = f(x) + \lambda \int_a^b K(x, s, y(s))\, ds,$$

in which K and f are continuous functions and $K(x,s,y)$ satisfies a Lipschitz condition in the third argument, has for sufficiently small $|\lambda|$ a unique continuous solution, and that solution may be found by the method of successive approximations (see [**203**]).

Proof. Taking the operator $A(y) = f(x) + \lambda \int_a^b K(x,s,y(s))ds$ to be defined in the space of continuous functions on $[a,b]$ with the uniform convergence topology and noting that the first condition of the principle of contraction mappings is satisfied, we verify condition (2):

$$\begin{aligned}
&\rho(A(y), A(z)) \\
&= \max \left| \lambda \int_a^b [K(x, s, y(s)) - K(x, s, z(s))]\, ds \right| \\
&\leqslant |\lambda| N \left| \int_a^b |y(s) - z(s)|\, ds \right| \\
&\leqslant |\lambda| N \max |y - z| \cdot |b - a| \\
&= |\lambda| \cdot |b - a| N \rho(y, z) \leqslant \alpha \rho(y, z),
\end{aligned}$$

where $\alpha < 1$ if $|\lambda| \cdot |b - a| \cdot N \leqq \alpha$.

4. *The differential equation with retarded argument*[9]

$$\dot{x}(t) = f(t, x(t), x(t - \tau(t))),$$

where the continuous function $\tau(t) \geqq 0$ with the initial condition $x(t) = \varphi(t)$ on the initial set E_{t_0} consisting of the point $t = t_0$ and the values of $t - \tau(t)$ less than t_0 for $t \geqq t_0$, and $\varphi(t)$ is a given continuous function, has a unique continuous solution for $t_0 \leqq t \leqq t_0 + h$, if the function f is continuous in the neighborhood of the values $(t, \varphi(t))$ for $t \subset E_{t_0}$ and satisfies in this neighborhood a Lipschitz condition in the second and third arguments and h is

[9] For a more detailed discussion of equations with retarded argument see Chapter V.

sufficiently small. This unique solution may be found by the method of successive approximations.

Proof. We replace the differential equation by the following equivalent integral equation with the same initial conditions:

$$x(t) = \varphi(t_0) + \int_{t_0}^{t} f(t, x(t), x(t - \tau(t)))\, dt;$$

$$x(t) = \varphi(t) \text{ on } E_{t_0}.$$

We shall verify that the principle of contraction mappings is verified by the operator

$$A(x(t)) = \varphi(t_0) + \int_{t_0}^{t} f(t, x(t), x(t - \tau(t)))\, dt,$$

defined in the space C_0 of continuous functions given on E_{t_0} and on the segment $t_0 \leqq t \leqq t_0 + h$ with the uniform topology and with all the functions coinciding with $\varphi(t)$ on E_{t_0}. It is necessary to verify condition 2 of the principle of contraction mappings, since condition 1 and the restrictions imposed on the space on which the operator A is defined are obviously satisfied:

$$\begin{aligned}
&\rho\left[A(x(t)),\ A(y(t))\right] \\
&= \sup\left|\int_{t_0}^{t} \left[f(t, x(t), x(t-\tau(t))) - f(t, y(t), y(t-\tau(t)))\right] dt\right| \\
&\leqslant N \sup\left|\int_{t_0}^{t} \left[|x(t) - y(t)| + |x(t-\tau(t)) - y(t-\tau(t))|\right] dt\right| \\
&\qquad \leqslant Nh\Big[\sup_{t_0 \leqslant t \leqslant t_0+h} |x(t) - y(t)| \\
&\qquad + \sup_{t_0 \leqslant t \leqslant t_0+h} |x(t-\tau(t)) - y(t-\tau(t))|\Big] \\
&\qquad\qquad \leqslant 2Nh\rho(x(t),\ y(t)).
\end{aligned}$$

For $h \leqq \alpha/2N$, where $0 < \alpha < 1$, condition (2) of the principle of contraction mappings will be satisfied.

In a quite analogous way one formulates and proves the existence and uniqueness of the solution of one differential equation with several delays:

$$\dot{x}(t) = f(t, x(t - \tau_1(t)), x(t - \tau_2(t)), \ldots, x(t - \tau_n(t)))$$

and the same for systems of m such equations.

5. If the delay is continuously distributed, then instead of the differential-difference equation one obtains integro-differential equations, with the following form in the simplest case:

$$\dot{x}(t)=\int_0^{\infty} K(t, s, x(t-s))\, ds, \tag{1}$$

or in integral form:

$$x(t)=\varphi(t_0)+\int_{t_0}^{t} dt \int_0^{\infty} K(t, s, x(t-s))\, ds$$

with the initial conditions $x(t-s)=\varphi(t-s)$ for $t-s \leqq t_0$, where the continuous function φ can differ from zero only on a finite segment.

If in equation (1) *the functions f and K are continuous, $\int_0^\infty K(t,s,0)ds$ converges uniformly for $t_0 \leqq t \leqq t_0+h_1$ where $h_1>0$, and K satisfies a Lipschitz condition in the third argument, then there exists a unique continuous solution of* (1) *for $t_0 \leqq t \leqq t_0+h$, where h is sufficiently small, and this solution may be found by the method of successive approximations.*

PROOF. The operator

$$A(x(t))=\int_{t_0}^{t} dt \int_0^{\infty} K(t, s, x(t-s))\, ds,$$

defined in the space C_0 of continuous functions with the uniform topology, given for values $t \leqq t_0+h$, and coinciding with $\varphi(t)$ for $t \leqq t_0$, evidently satisfies the first condition of the principle of contraction mappings. We shall verify that A also satisfies the second condition of that principle:

$$\begin{aligned} &\rho[A(y), A(z)] \\ &\quad=\sup\left|\int_{t_0}^{t} dt \int_0^{\infty}[K(t, s, y(t-s))-K(t, s, z(t-s))]\, ds\right| \\ &\quad\leqslant N\sup\left|\int_{t_0}^{t} dt \int_0^{t-t_0} |y(t-s)-z(t-s)|\, ds\right| \\ &\qquad\qquad\leqslant Nh^2 \max|y(t)-z(t)| \leqslant \alpha\rho(y(t), z(t)), \end{aligned}$$

where $0<\alpha<1$, if $h^2 \leqq \alpha/N$, where N is the Lipschitz constant.

SOME APPLICATIONS OF SCHAUDER'S PRINCIPLE.

1. *The differential equation $\dot{x}(t)=f(t,x(t))$, where f is continuous in the*

neighborhood of (t_0, x_0), *has a continuous solution satisfying the initial condition* $x(t_0) = x_0$.

PROOF. Without loss of generality we may suppose that $t_0 = 0$ and $x_0 = 0$. We replace the differential equation by the following equivalent integral equation:

$$x(t) = \int_0^t f(t, x(t))\, dt, \quad -h \leqslant t \leqslant h, \quad |f(t, x)| \leqslant M.$$

The operator $A(x(t)) = \int_0^t f(t, x(t))dt$, defined in the linear function-space of continuous functions C_0, satisfies the conditions of Schauder's principle. Indeed, the closed convex set R of this space consisting of the functions $x(t)$ satisfying $|x(t)| \leqq M_1$ is mapped by the operator A for sufficiently small h into a compact part of itself. This is because $\int_0^t f(t, x(t))dt$ is a continuous function, and the collection of functions of the form $\int^t f(t, x(t))dt$, where $x(t) \subset R$, is uniformly bounded: $|\int_0^t f(t, x(t))dt| \leqq Mh \leqq M_1$ for $h \leqq M_1/M$, and equicontinuous:

$$\left| \int_0^{t+\Delta t} f(u, x(u))\, dt - \int_0^t f(u, x(u))\, du \right|$$
$$= \left| \int_t^{t+\Delta t} f(u, x(u))\, du \right| \leqslant M|\Delta t| < \varepsilon$$

for $|\Delta t| < \epsilon/M$, and thus, from Arzelà's theorem, this collection of functions is compact.

Evidently the proof can easily be carried over to systems of finite numbers of differential equations.

2. An analogous theorem is valid also for differential equations with retarded argument:

$$\dot{x}(t) = f(t, x(t), x(t - \tau(t))), \tag{2}$$

with $x(t) = \varphi(t)$ on the initial set E_{t_0} (see page 74), where the functions f, φ, and τ are continuous and $\tau(t) \geqq 0$ for $t \geqq t_0$. *In this case also there exists a solution of equation* (2).

Without loss of generality we may suppose that $t_0 = 0$ and $\varphi(t_0) = 0$, and, replacing the differential equation by its integral analogue, we obtain

$$x(t) = \int_0^t f(t, x(t)), x(t - \tau(t)))\, dt, \quad x(t) = \varphi(t) \text{ on } E_0,$$
$$0 \leqslant t \leqslant h, \quad |f(t, x(t), x(t - \tau(t))| \leqslant M.$$

Consider the operator

$$A(x(t)) = \int_0^t f(t, x(t), x(t-\tau(t)))\,dt,$$

defined on the set of continuous functions coinciding with $\varphi(t)$ on the set E_0, with the metric of the space C_0. The operator A carries the closed convex set R of continuous functions satisfying the condition $|x(t)| \leqq M_1 \neq 0$ for $0 \leqq t \leqq h$ into a compact part of itself. In fact, if $x(t) \subset R$, then $A(x(t)) \subset R$ for sufficiently small h: $|A(x(t))| \leqq Mh \leqq M_1$, if $h \leqq M_1/M$, and moreover the set of functions $A(x(t))$ is uniformly bounded for $x(t) \subset R$:

$$|A(x(t))| \leqslant M_1$$

and equicontinuous:

$$\Big| \int_0^{t+\Delta t} f(u, x(u), x(u-\tau(u)))\,du$$

$$- \int_0^t f(u, x(u), x(u-\tau(u)))\,du \Big|$$

$$= \Big| \int_t^{t+\Delta t} f(u, x(u), x(u-\tau(u)))\,du \Big| \leqslant M|\Delta t| < \varepsilon$$

$$\text{for } |\Delta t| < \frac{\varepsilon}{M}$$

and therefore compact.

This theorem may be easily generalized to differential equations with retarded arguments with several delays, and to systems of a finite number of such equations, and also to many cases of continuously distributed arguments.

3. **Theorems on fixed points using invariants of category type.** Theorems estimating the number of critical cells of a function or of a functional, presented in Chapter I, are at the same time theorems on fixed points of several sufficiently restricted classes of mappings.

We shall say that a mapping φ of the manifold M^n into itself is of gradient type if $\varphi(p) = p$ for $df(p) = 0$, where $f(p)$ is a twice continuously differentiable function defined on the manifold M^n.

An example of mappings of gradient type is given by the transformations which carry out displacements along the trajectories of a dynamical system:

$$\frac{dx_i}{dt} = \frac{\partial u(x_1, x_2, \ldots, x_n)}{\partial x_i} \quad (i = 1, 2, \ldots, n)$$

or

$$\frac{dx_i}{dt} = M_i(x_1, x_2, \ldots, x_n) \frac{\partial u(x_1, x_2, \ldots, x_n)}{\partial x_i} \quad (i = 1, 2, \ldots, n),$$

or

$$\frac{dx_i}{dt} = -\frac{\partial u(x_1, x_2, \ldots, x_n, y_1, y_2, \ldots, y_n)}{\partial y_i},$$
$$\frac{dy_i}{dt} = \frac{\partial u(x_1, x_2, \ldots, x_n, y_1, y_2, \ldots, y_n)}{\partial x_i} \quad (i = 1, 2, \ldots, n),$$

where M_i are continuous functions of the local coordinates and u is a twice continuously differentiable function of the local coordinates $x_1, x_2, \cdots, x_n$, and in the latter case of the coordinates $x_1, x_2, \cdots, x_n, y_1, y_2, \cdots, y_n$. Under mappings of gradient type of M^n into itself the number of geometrically distinct stationary points is estimated from below by invariants of category type:

$$\operatorname{kat}_{M^n} M^n, \ \operatorname{cat}_{M^n} M^n,$$
$$\operatorname{long}_{M_n} M^n + 1, \ \operatorname{El}_{M^n} M^n, \operatorname{Pr} El\, M^n$$

(see Chapter I, §3).

We shall call a fixed point of the mapping φ of gradient type nondegenerate, if there corresponds to it a nondegenerate critical point of the function f mentioned above. A stationary point p of the mapping φ is of type k if the point p is a critical point of type "k" of the function f. To a degenerate isolated fixed point p one assigns a multiplicity equal to the multiplicity of the critical point p of the function f.

If the fixed point p is ordinary (i.e., noncritical), then it is assigned the multiplicity $(0, 0, \cdots, 0)$.

From the results of Chapter I it follows that the number of analytically distinct fixed points (the sum of the multiplicities of the fixed points) of a mapping of gradient type may be estimated from below by the Betti numbers mod 2 (or with another modulus), the number of cells with regular adjunction of M^n, and by other invariants of this type.

We note that the definition of nondegeneracy of a stationary point, its type, and its multiplicity, depend on the choice of the function f, and in this sense this concept is not an intrinsic characteristic of fixed points of the mapping φ.

One may generalize somewhat further the class to which the above estimates of the number of geometrically distinct fixed points may be applied, by taking along with the previous class those mappings which may be continuously deformed, without decreasing the number of stationary points into mappings of gradient type. Even then the class of mappings is rather small, but given this restriction we can guarantee for these mappings the existence of not one, but rather several fixed points.

CHAPTER IV

Qualitative Methods in the Theory of Differential Equations

Qualitative methods of the investigation of differential equations began to develop around 75 years ago. In the course of the last 30-40 years they have developed to such a degree that it would be quite impossible to present even the basic results of this theory within the confines of a single chapter.

Therefore we will restrict ourselves to the presentation only of some selected questions, more or less connected with the other chapters of this book, asking the reader who wishes to become better acquainted with the qualitative theory of differential equations to turn to the monograph *Qualitative theory of differential equations* by Nemyckiĭ and Stepanov [**210**].

1. Estimates of the number of stationary points. A stationary point of the dynamical system

$$\frac{dx_i}{dt} = X_i(x_1, x_2, \ldots, x_n) \qquad (i = 1, 2, \ldots, n) \tag{1}$$

is a solution of that system $x_i = x_i^0$ $(i = 1, 2, \cdots, n)$ where the x_i^0 are constants. Thus trajectories which correspond to stationary points consist of single points in the phase space with coordinates $x_1, x_2, \cdots, x_n$. The coordinates of stationary points are obviously defined by the system of equations

$$X_i(x_1, x_2, \ldots, x_n) = 0 \quad (i = 1, 2, \ldots, n).$$

The distribution of stationary points and the behavior of the integral curves in the neighborhood of these points strongly influences the entire qualitative map of the distribution of trajectories of the dynamical system, so that the investigation of the stationary points of the dynamical system is an essential part of the qualitative analysis of the system.

Suppose that the right sides of the system (1) are continuous and satisfy Lipschitz conditions or some other conditions guaranteeing the uniqueness of the solution of the fundamental initial value problem and the continuous dependence of the solutions on the initial values. Suppose that also the system (1) has only isolated stationary points, i.e., in a sufficiently small neighborhood of a stationary point there are no other stationary points.

If one takes a stationary point as the origin of coordinates and supposes that in the neighborhood of this point the right sides of the system (1) are differentiable, then in that neighborhood the system may be represented in the form

$$\frac{dx_i}{dt} = \sum_{j=1}^{n} a_{ij}x_j + R_i \qquad (i = 1, 2, \ldots, n), \tag{2}$$

where $a_{ij} = (\partial X_i(0, 0, \cdots, 0))/\partial x_j$, and R_i has an order higher than unity with respect to $\sqrt{\sum_{i=1}^{n} x_i^2}$.

In the qualitative analysis of the distribution of the trajectories in the neighborhood of the stationary point $x_1 = x_2 = \cdots = x_n = 0$, the system (2) can often be replaced by the linear system

$$\frac{dx_i}{dt} = \sum_{j=1}^{n} a_{ij}x_j \qquad (i = 1, 2, \ldots, n). \tag{3}$$

Such a replacement does not change the qualitative nature of the map in the neighborhood of the stationary point (still taken at the origin of coordinates) if the orders of the terms R_i are at least $1 + \alpha$, where $\alpha > 0$, with respect to $\sqrt{\sum_{i=1}^{n} x_i^2}$, and all the roots of the characteristic equation

$$\begin{vmatrix} a_{11} - k & a_{12} & \ldots & a_{1n} \\ a_{21} & a_{22} - k & \ldots & a_{2n} \\ \cdot & \cdot & \cdot & \cdot \\ \cdot & \cdot & \cdot & \cdot \\ a_{n1} & a_{n2} & \ldots & a_{nn} - k \end{vmatrix} = 0 \tag{4}$$

have nonzero real parts.

The signs of the real parts of the roots of the characteristic equation strongly influence the qualitative map of the distribution of trajectories in the neighborhood of a stationary point, so that it is appropriate to classify the stationary points according to the number of roots of the characteristic equation (4) with negative (or with positive) real parts.

We shall call a stationary point for which equation (4) has k roots with negative real parts and $n - k$ roots with positive real parts *a stationary point of type k* or a *k-saddle*. Points of type 0 and points of type n will also be called *nodes* if the roots of (4) are all real, or *foci* if not all roots of (4) are real. Figures 16, 17, and 18 show the distribution of trajectories in the neighborhoods of points of types 0, 1, 2, for $n = 2$, in these cases called:

an unstable node (Figure 16 left); an unstable focus (Figure 16 right); a saddle (Figure 17); a stable node (Figure 18 left), and a stable focus (Figure 18 right), respectively. The arrows indicate the direction of motion along the trajectories as t increases.

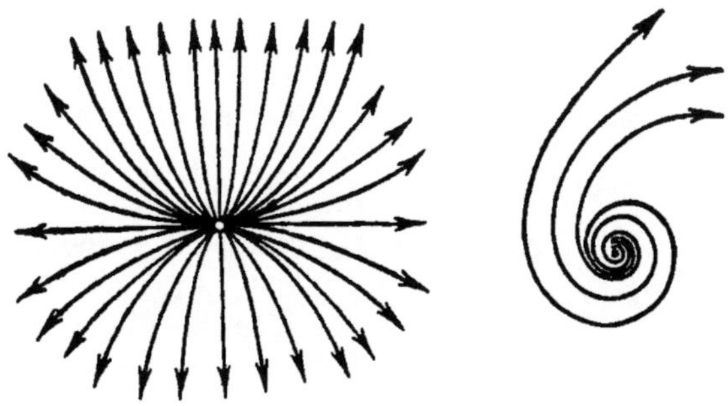

Figure 16

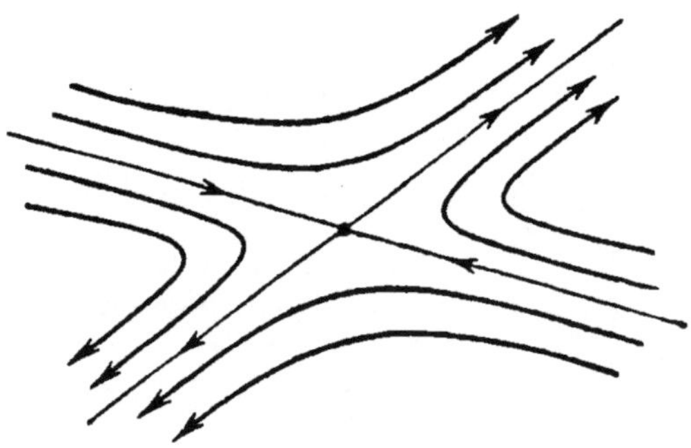

Figure 17

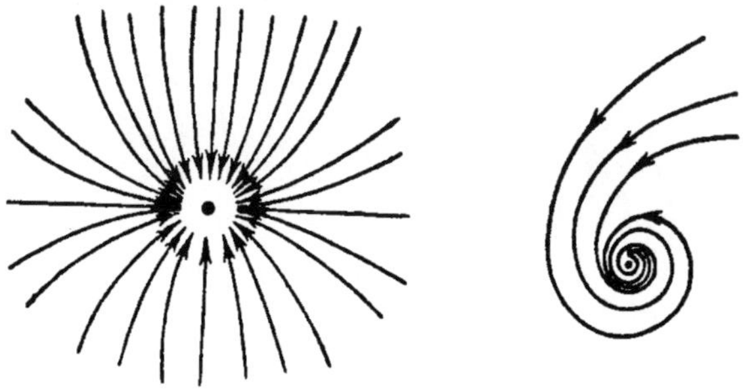

Figure 18

If the real parts of several roots of equation (4) are equal to zero, then terms of higher order begin to have an influence, and stationary points may be qualitatively of a great many different sorts.

Several examples for the case $n = 2$ are presented in Figures 19, 20 and 21. Figure 19 depicts a center, that is a stationary point corresponding, for example, to the case of pure imaginary roots of the characteristic equation of the system (3) for $n = 2$. Figure 20 illustrates a nodal region, which can be formed from the confluence of stable and unstable nodes. Figure 21 shows a quadruple saddle, which can be formed as the result of the confluence of four simple saddles.

In many problems it is useful to consider the notion of the index of a stationary point, i.e., the index of that point with respect to the field of directions defined by system (1).

Although the concept of the index of a singular point of a vector field was already utilized in Chapter II (see page 55), we nevertheless consider it worthwhile to discuss this important concept again. One can introduce the concept of the index of a stationary point in a particularly simple graphic form in the two-dimensional case. Suppose that the dynamical system

$$\left.\begin{aligned} \frac{dx_1}{dt} &= f_1(x_1, x_2); \\ \frac{dx_2}{dt} &= f_2(x_1, x_2), \end{aligned}\right\} \qquad (5)$$

satisfies the conditions formulated above on the Euclidean plane or on some portion of that plane.

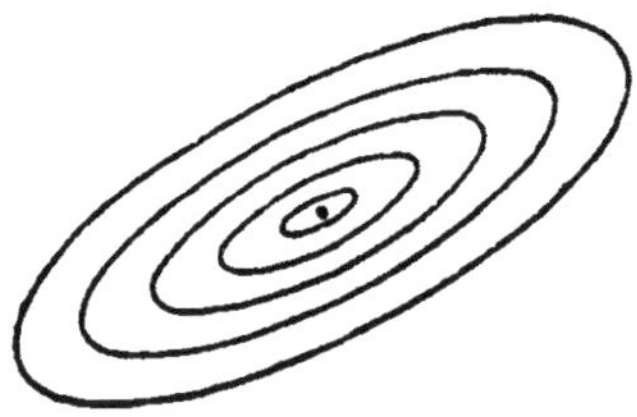

Figure 19

System (5) defines on the (x_1, x_2) plane an

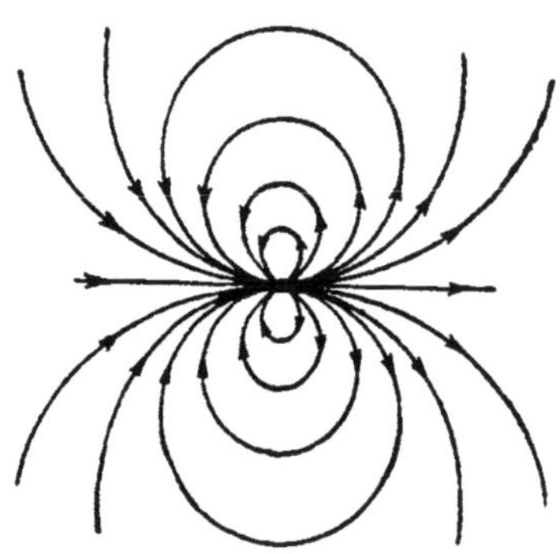

Figure 20

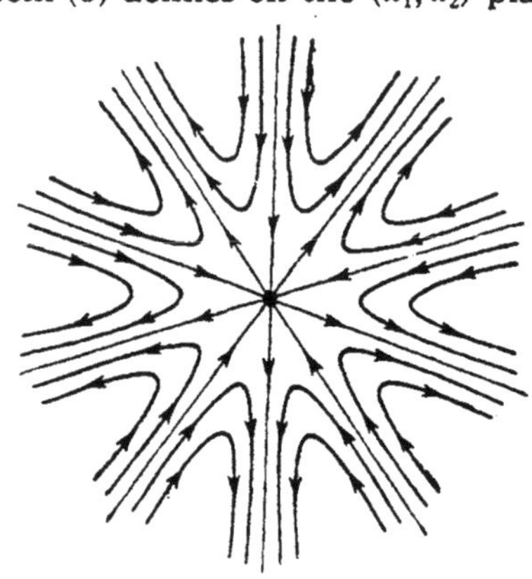

Figure 21

everywhere (except for the stationary points) continuous vector field of tangents to the trajectories. At the stationary points the direction of the vector field is not defined. On each arc AB not passing through the stationary point one can define the angle through which the vector field rotates as the motion proceeds from A to B (Figure 22).

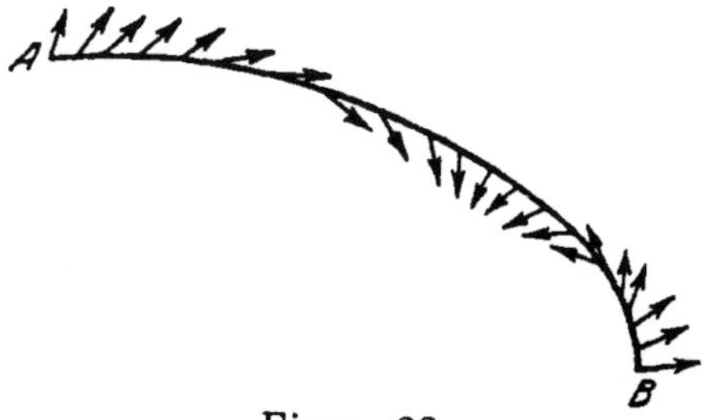

Figure 22

In particular, one may consider the angle of rotation φ_C of the vector field on traversing a closed curve C, which does not pass through a stationary point of the system (5). The rotation angle φ_C of the vector field on the closed curve C is obviously equal to $2\pi j$, where j is an integer (Figure 23).

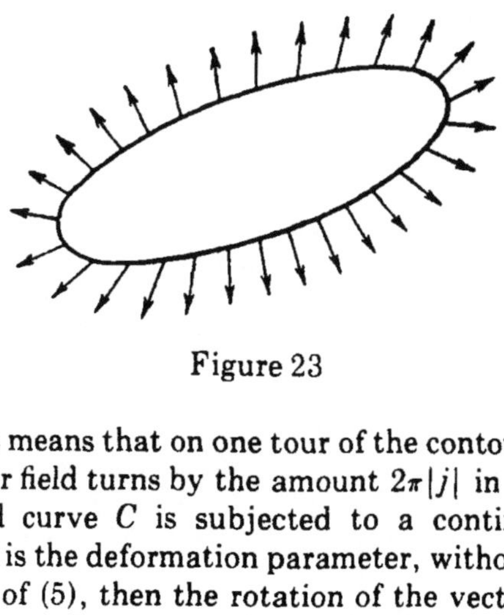

Figure 23

If j is negative this means that on one tour of the contour C in the positive direction the vector field turns by the amount $2\pi|j|$ in the negative direction. If the closed curve C is subjected to a continuous deformation $\Phi_t(C) = C_t$, where t is the deformation parameter, without passing through a stationary point of (5), then the rotation of the vector field φ_{C_t} remains unchanged. This is true because on the one hand, the rotation φ_C can only change in a continuous way because of the continuity of the field, and on the other hand $\varphi_{C_t} = 2\pi j$ can only change by discrete amounts at least 2π in magnitude. Hence φ_{C_t} does not change.

Suppose that the curve C bounds a portion of the plane in which there are no stationary points of the system (5). Then C can be continuously deformed into a point distinct from a stationary point without passing through a stationary point of the system (5). Since under such a continuous deformation the rotation of the vector field on the curve C does not change, and on a curve degenerating to a point the rotation is equal to zero, the rotation of the vector field φ_C will be zero also on the curve C.

Accordingly, if on some closed curve C the rotation of the vector field is different from zero, then inside the curve there must be a stationary point.

In particular, if C is a trajectory of the system (5), then the rotation φ_C on a single tour in the positive direction of the curve C is equal to 2π, and therefore inside a closed trajectory there must always be at least one stationary point. Of course, as in the preceding theorems, this is true under the hypothesis that on the curve C and inside the region of the plane bounded by this curve the functions f_1 and f_2 are defined and satisfy the conditions indicated above. The rotation of a vector field on a contour C which goes once around a stationary point p in the positive direction does not depend on the choice of the contour C.

Indeed, any other contour C_1 satisfying the conditions indicated above can be continuously deformed without passing through a stationary point into the contour C, and under such a deformation, as we already noted above, the rotation of the vector field does not change. Hence $\varphi_C = \varphi_{C_1}$.

Thus, the integer $j = \varphi_C/2\pi$ does not depend on the choice of the contour C satisfying the conditions indicated above, but rather only on the type of the stationary point p. This number is called the index of the stationary point p.

We shall illustrate with several examples: the index of a node, of a focus, and a center are equal to 1 (see Figure 9, page 56), the index of a saddle is equal to -1 (see Figure 9), the index of a nodal region is equal to $+2$ (see Figure 20), the index of a double saddle is equal to -2 (Figure 24). We

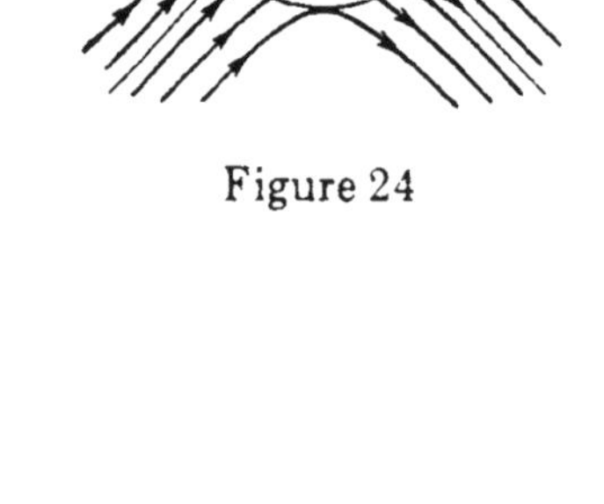

Figure 24

note that the indicated properties of the rotation of the vector field are consequences of well-known properties of curvilinear integrals of complete differentials (the fact that the integral around a closed curve is zero if inside the contour there are no discontinuities of the integrand function, the independence of the integral around a closed curve from the form of the contour, if the latter goes around the point of discontinuity once, and so forth). Indeed, the rotation of the vector field

$$\varphi_C = \int_C d\varphi = \int_C d \operatorname{arctg} y'_x = \int_C \frac{dy'}{1+y_x'^2} = \int_C \frac{f_1 df_2 - f_2 df_1}{f_1^2 + f_2^2}.$$

The integrand is discontinuous only at the points where $f_1 = 0$ and $f_2 = 0$, i.e., at the stationary points of the system (5).

Making use of the known properties of curvilinear integrals of total differentials, or proceeding directly from the definition of the rotation of a field, it is easy to prove the theorem that *the rotation of the vector field defined by* (5) *on one passage around the closed nonselfintersecting contour C which does not pass through a stationary point is equal to* $2\pi \sum_k j_k$, *where* $\sum_k j_k$ *is the sum of the indices of all the stationary points which are included inside the contour C.*

Indeed, dividing the portion of the plane bounded by the contour C into curvilinear triangles (or other closed curves), so that every stationary point p_i lies inside these triangles, while in each of them there is not more than one stationary point (Figure 25), and orienting all of these triangles

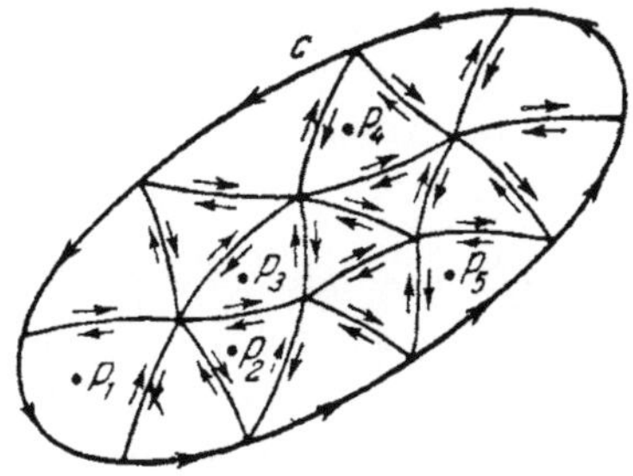

Figure 25

according to the orientation of the curve C, we find that the rotation of the field on the contour C is equal to the sum of the rotations of all of the triangles C_i, $\varphi_C = \sum_i \varphi_{C_i}$, since on the common boundaries of the triangles opposite orientations will have been assigned, so that their rotations in the sum $\sum_i \varphi_{C_i}$ will have cancelled out.

Thus, $\varphi_C = \sum_i \varphi_{C_i}$, but $\sum_i \varphi_{C_i} = 2\pi \sum_k j_k$, so that $\varphi_{C_i} = 0$, if inside the triangle C_i there are no stationary points, and $\varphi_{C_i} = 2\pi j_i$ if there is a stationary point p_i of index j_i.

If the right side of (5) is varied continuously, keeping to the restrictions imposed, then the stationary points also move continuously, while the index of an isolated stationary point can only change after it comes into confluence with another stationary point (in this connection it follows directly from what was said above that the indices of the confluent points preserve their sum), or when a stationary point separates into several points, the sum of whose indices must be equal to the index of the original point.

In particular, a point of nonzero index cannot be annihilated without first turning into a point with zero index. Points of nonzero index can disappear only after coming into confluence with one or several points the sum of whose indices is opposite to the index of the point in question, so that as a result of confluence a point of zero index appears, which may then disappear. In the same way, new stationary points arise either as points of zero index, after which they can break into several points the sum of whose indices is equal to zero; or by the breakup of a point of index j into a collection of points, the sum of whose indices is equal to j. All these assertions follow immediately from the fact that the rotation of the vector field around a closed contour not passing through a stationary point cannot change under a continuous variation of the right sides of (5), since φ_C can change under such a variation only in a continuous way, whereas φ_C can only take on the values $2\pi j$, where j is an integer, and thus cannot change continuously. Thus the sum of the indices of the stationary points lying inside the contour C cannot change at all.

Analogous theorems can be proved also in the n-dimensional case.

The index of an isolated stationary point of the system

$$\frac{dx_i}{dt} = X_i(x_1, x_2, \ldots, x_n) \qquad (i = 1, 2, \ldots, n) \tag{6}$$

is the degree of the mapping of a sphere σ, with center at the stationary point and of sufficiently small radius that inside this sphere there is only one stationary point, onto an identically oriented sphere S of unit radius with center at the origin of coordinates, the mapping being defined as follows: to each point p of the sphere σ there is assigned a point of the sphere S defined by the unit vector $(\vec{v}(p))/|\vec{v}(p)|$, where $\vec{v}(p)$ is a vector field of tangents to the trajectories of the system (6) at the point p. It is easy to see that this definition of index coincides in the two-dimensional case with the definition given on page 84.

Although the concept of index of a stationary point has been introduced here for Euclidean spaces, it is valid also for dynamical systems defined on manifolds, since it is a local concept and therefore applies equally well to neighborhoods of points of the manifold, which are topologically equivalent to neighborhoods of points of Euclidean space.

For closed manifolds it is possible to prove the following theorem ([72]): *the sum of the indices of all the stationary points of the dynamical system* (6), *where* $x_1, x_2, \cdots, x_n$ *are the local coordinates of the manifold* M^n, *is equal to the Euler characteristic of the manifold* M^n.

This theorem provides an estimate of the number of critical points. For example, it enables us to assert that on every closed manifold with an Euler characteristic different from zero any dynamical system has at least one stationary point.

One can prove that this estimate is exact, namely that on any closed manifold it is possible to define a dynamical system having not more than one stationary point whose index, from the previous theorem, is equal to the Euler characteristic of the manifold.

We present several examples:

Of all the oriented closed surfaces, only on the torus is it possible to define a dynamical system without stationary points, since the Euler characteristic of a surface of genus p is equal to $2 - 2p$. Only for $p = 1$, i.e., only for the torus, does this vanish.

A uniform motion along a meridian or along a parallel of the torus gives an example of a system without stationary points. The Euler characteristic of the sphere is equal to 2, so that the sum of the indices of the stationary points of a dynamical system defined on the sphere is equal to two. If the trajectories on the sphere are a family of parallels, then there are two stationary points of index 1, namely two centers at the poles of the sphere. If the trajectories are the family of meridians, then there are also two stationary points of index 1, namely two nodes.

If the trajectories are sections of the sphere by a bundle of planes passing through a straight line tangent to the sphere, then we obtain a single stationary point of index 2, a nodal region (Figure 26).

Thus, on a manifold with an Euler characteristic different from zero it is possible to guarantee the existence of at least one stationary point. However, for several special classes of dynamical systems this general estimate may be significantly improved.

We shall call the dynamical system

$$\frac{dx_i}{dt} = X_i(x_1, x_2, \ldots, x_n) \qquad (i = 1, 2, \ldots, n) \tag{7}$$

potential, if there exists a twice continuously differentiable function $u(x_1, x_2, \cdots, x_n)$ satisfying the condition

$$\frac{\partial u}{\partial x_i} = X_i(x_1, x_2, \ldots, x_n) \qquad (i = 1, 2, \ldots, n).$$

We shall call the dynamical system (7) a system of potential type if there exists a twice continuously differentiable function $u(x_1, x_2, \cdots, x_n)$ such that if $\partial u/\partial x_i = 0$ $(i = 1, 2, \cdots, n)$ holds then $X_i(x_1, x_2, \cdots, x_n) = 0$, i.e.,

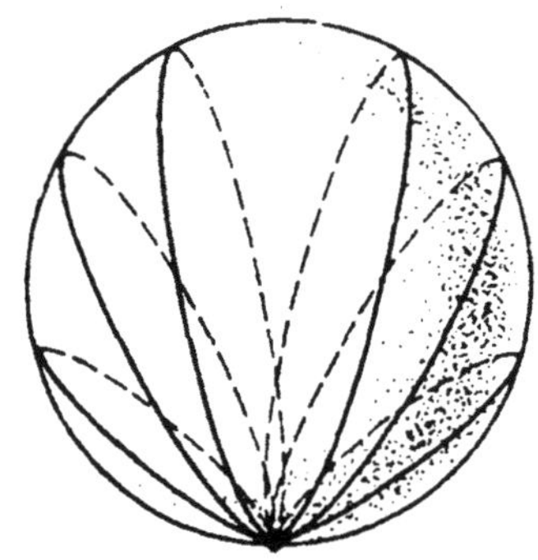

Figure 26

the critical points of the function u are stationary points of the system (7). As an example of such a system one can take the canonical system of differential equations

$$\frac{dx_i}{dt} = \frac{\partial u}{\partial y_i}; \quad \frac{dy_i}{dt} = -\frac{\partial u}{\partial x_i} \qquad (i = 1, 2, \ldots, n),$$

given that the function u is twice continuously differentiable. If the system (7) can be replaced by a system of potential type without changing the number of stationary points, then the system will be called *quasipotential*.

Evidently the number of stationary points of quasipotential systems defined on the manifolds can be estimated by means of invariants estimating the number of critical points. I.e., it may be asserted that the length $+1$ of the manifold M^n, the homological category of M^n, the homotopy category of M^n, the number of cells of M^n, the number of cells with proper adjunction and other invariants of category type estimate from below the number of stationary points of a quasipotential dynamical system defined on the manifold M^n (see Chapter I, §3).

If one regards the stationary points of a dynamical system of potential type as being nondegenerate under the condition that the corresponding critical points of the function u are nondegenerate, then in the estimation of the number of nondegenerate stationary points of the dynamical system one can use the invariants estimating the number of nondegenerate critical points: the sum of the Betti numbers mod 2, the number of cells with regular adjunction and so forth (see Chapter I, §2).[1]

For example, every dynamical system of potential type defined on the two-dimensional torus has at least three geometrically distinct or four analytically distinct stationary points. These theorems on the number of stationary points of quasipotential systems can be also generalized to certain other dynamical systems (see [332]).

2. **Dependence of the solutions on small coefficients of the highest derivatives.** The well-known theorem on the continuous dependence of the solutions of a system of differential equations on a parameter makes it possible to assert that the solution of the system

$$\frac{dx_i}{dt} = f_i(t, x_1, x_2, \ldots, x_n, \mu); \quad x_i(t_0) = x_{i0}$$
$$(i = 1, 2, \ldots, n)$$

will depend continuously on the parameter μ if all the functions f_i are continuous in the set of their arguments, μ included, in the region of their variation, and satisfy a Lipschitz condition in the variables $x_1, x_2, \cdots, x_n$.[2]

This theorem is of basic significance, since in the actual systems described by differential equations the parameters of the system can almost never be measured exactly, and moreover, as a rule, they vary slightly under the influence of various perturbing factors. Thus if a small variation of the parameters were to lead to a large distortion in the solutions, the solution would have no value and would not even approximately describe the phenomena under investigation.

Frequently the differential equation, or system of equations, has a small coefficient of the highest derivative.

The conditions of the theorem on the continuous dependence of the solutions on a parameter are usually satisfied, with the exception of one frequently occurring case, which we shall study in this section.

In the simplest case the equation has the form:

$$\mu \frac{dx}{dt} = f(t, x), \tag{8}$$

---

[1] In addition, it is possible to estimate the critical points of the various types k, to which for a potential system, k-saddles will correspond.

[2] See for example I. G. Petrovskiĭ, *Lectures on the theory of ordinary differential equations*, 3rd ed., GITTL, Moscow, 1949. (Russian)

where μ is a small positive parameter and f is continuous and satisfies a Lipschitz condition in x.

The right side of the equation $dx/dt = f(t,x)/\mu$ is discontinuous for $\mu = 0$, and therefore one cannot assert without further investigation that the solution of the equation $\mu(dx/dt) = f(t,x)$ for small μ will be close to the solution of the so-called degenerate equation $f(t,x) = 0$.

Suppose that $x = \varphi_i(t)$ is one of the roots of the degenerate equation. As the parameter μ approaches zero the derivative dx/dt of the solutions of the equation $dx/dt = f(t,x)/\mu$ at points where $f(t,x) \neq 0$ will increase unboundedly in absolute value, and its sign will coincide with the sign of the function $f(t,x)$. Therefore the tangents to the integral curves at every point at which $f(t,x) \neq 0$ will tend as $\mu \to \infty$ to a direction parallel to the Ox axis. If $f(t,x) > 0$ the solution $x(t,\mu)$ increases with increasing t, since $dx/dt > 0$. If $f(t,x) < 0$, then $x(t,\mu)$ decreases with increasing t.

Thus, if with increasing x but with t fixed the sign of the function $f(t,x)$ changes from plus to minus on passing through the graph $x = \varphi_i(t)$ of the root of the degenerate equation, then integral curves which are close to $x = \varphi_i(t)$ in initial values will for small μ approach the graph of $x = \varphi_i(t)$ almost along straight lines parallel to the Ox axis, and from then on cannot

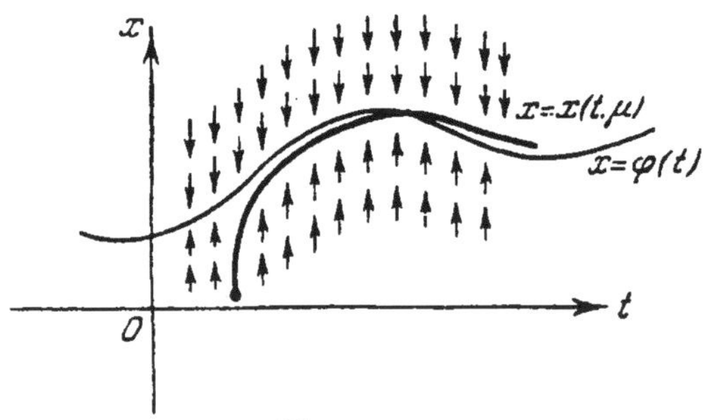

Figure 27

recede from it (Figure 27). In this case the root $x = \varphi_i(t)$ of the degenerate equation is called stable.

A sufficient condition for stability of the root is obviously that $\partial f(t,x)/\partial x < 0$ on the root $\varphi_i(t)$. On the other hand, if on passing through that root with x increasing the sign of the function f changes from minus to plus, then the integral curves for sufficiently small μ are almost along lines parallel to the Ox axis and receding from the graph of the root as t increases (Figure 28).

In this case the root $x = \varphi_i(t)$ is said to be unstable. A sufficient condition for the instability of the root is that $\partial f(t,x)/\partial x > 0$ on the curve $x = \varphi_i(t)$.

Also possible is the case of semistability, occurring when the root $x = \varphi_i(t)$ is of even multiplicity and thus the sign of the function $f(t,x)$ does not change on crossing it. Then for small μ the integral curves on one

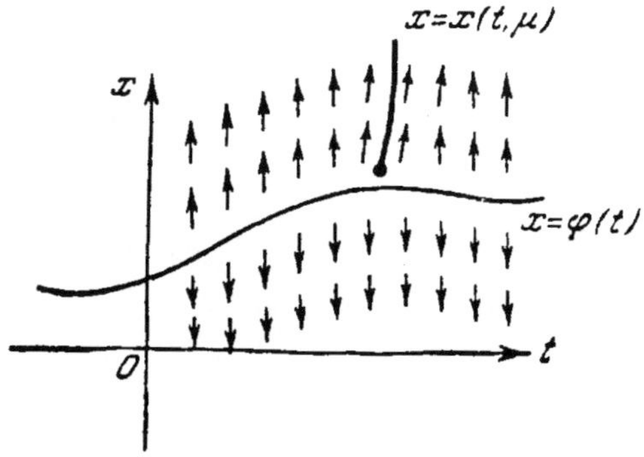

Figure 28

side of the root tend to the graph of the root, and on the other side for small μ rapidly recede from it (Figure 29).

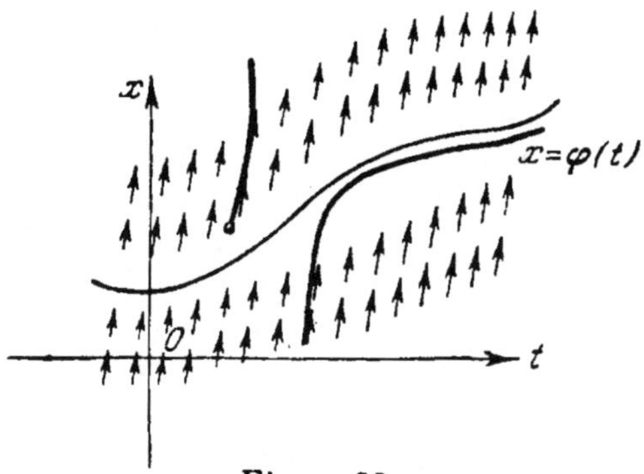

Figure 29

We suppose for definiteness that in the neighborhood of the semistable root $x = \varphi_i(t)$, $f(t,x) \geqq 0$.

If $\varphi_i' > 0$ here, then the integral curves which approach the root $x = \varphi_i(t)$ cannot intersect the graph of $\varphi_i(t)$ since at the points of that graph the tangent to the integral curves is parallel to the Ot axis.

Accordingly, in this case the integral curves approaching the graph of the root $x = \varphi_i(t)$ remain in its neighborhood for $t > t_0$ (see Figure 29).

If on the other hand $\varphi_i'(t) < 0$, then the integral curves approaching the graph of the function $\varphi_i(t)$ intersect it and on the other side of the graph rapidly recede from it.

If $\varphi_i'(t) > 0$ for $t_0 \leqq t < t_1$ and $\varphi_i'(t) < 0$ for $t > t_1$, then for sufficiently small μ the integral curves approaching the graph of $\varphi_i(t)$ remain close to this graph for $t_0 + \delta < t < t_1$, where $\delta > 0$, and in the neighborhood of the point t_1 intersect the root $\varphi_i(t)$ and then rapidly recede from it (Figure 30).

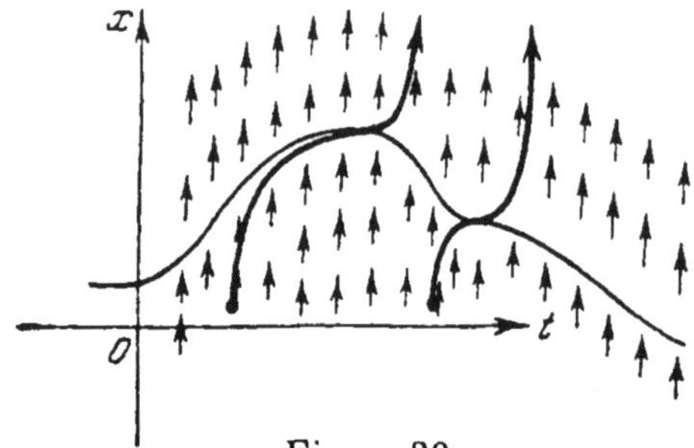

Figure 30

If in the neighborhood of the semistable root we have $f \leqq 0$, then for the assertions above to be correct the sign of the derivative $\varphi_i'(t)$ has to be changed.

It must also be noted that, although in the case of semistability, for $f \geqq 0$ and $\varphi_i'(t) > 0$ or $f \leqq 0$ and $\varphi_i'(t) < 0$ and with initial values lying in the neighborhood of the root $x = \varphi_i(t)$ on one side of this graph the solutions of the original equation theoretically remain also for $t > t_0 + \delta$, $\delta > 0$ in the neighborhood of the root of the degenerate equation, nevertheless this is valid only in the absence of any perturbations. If perturbations are not excluded, then one must deal with the possibility of a passage, because of these perturbations, of the integral curve $x = x(t, \mu)$, on approaching the root $x = \varphi_i(t)$, through the graph of this root onto the unstable side, and thereupon the integral curve will recede from the graph of $x = \varphi_i(t)$, and the smaller we take the values of μ the smaller the perturbation that can cause this effect.

If the degenerate equation $f(t, x) = 0$ has several roots, then the question arises which of the stable or semistable roots is approached, for sufficiently small μ, by the solution of the original equation $x = x(t, \mu)$ defined by the initial conditions $x(t_0) = x_0$.

This question has an obvious answer. For sufficiently small μ the integral curve $x = x(t, \mu)$ almost coincides with the line $x = x_0$. As t increases this integral curve, increasing (if $f(t_0, x_0) > 0$) or decreasing (if $f(t_0, x_0) < 0$), approaches the closest stable (or in the cases discussed above, unstable) root $x = \varphi_j(t)$ in the direction of increasing x and then remains in its neighborhood. In this case we say that the initial point lies in a region of influence of the root $x = \varphi_j(t)$.

Thus the integral curve $x = x(t, \mu)$ of the equation (8), for $\mu \to 0$ and $t > t_0$ tends to a stable (sometimes semistable) root of the degenerate equation whose region of influence includes the initial point (t_0, x_0). If no such roots exist, then $x(t, \mu) \to \infty$ or $x(t, \mu) \to -\infty$. In these two latter cases one says that the initial point (t_0, x_0) lies in a region of influence of plus or minus infinity.

If stable (or semistable) branches of the degenerate equation $f(t, x) = 0$ intersect, then we can determine precisely which stable (or semistable) branch is approached as $\mu \to 0$ by the integral curve $x = x(t, \mu)$ defined by the initial conditions $x(t_0) = x_0$. For this purpose we need only note the sign of the derivative and take account of the fact that dx/dt vanishes only on the curve $f(t, x) = 0$. The analysis from a general point of view of the various possibilities arising in the case of intersection of the graphs of the roots of the degenerate equation would be rather tedious and furthermore useless in practice, since the small perturbations inevitable in actual systems would make our results illusory.

Therefore we shall restrict ourselves to a single example:

$$\mu \dot{x} = (x^2 - t^2)(x^2 - 4t^2); \quad \mu > 0,$$

and investigate as $\mu \to 0$ the behavior of the solutions defined by the initial values (a) $x(-3) = -4$; (b) $x(-3) = 0$.

The distribution of signs of the function $f(t, x) = (x^2 - t^2)(x^2 - 4t^2)$ in the various regions is illustrated in Figure 31. Thus, the stable roots of the equation $f(t, x) = 0$ are $x = 2t$, $x = -t$ for $t < 0$ and $x = -2t$, $x = t$ for $t > 0$. The solution determined by the initial point $M_1(-3, -4)$ for sufficiently small μ approaches the stable root $x = 2t$ for $t < 0$ and remains in its neighborhood up to the point $t = 0$. For $t > 0$ the integral curve will lie close to the graph of the stable root $x = -2t$.

The solution determined by the initial point $M_2(-3, 0)$ for sufficiently small $\mu > 0$ will approach the root $x = -t$, which is stable for $t < 0$, and remain in the neighborhood of this root up to the point $t = 0$. In the neighborhood of the point $t = 0$ the solution begins to grow with ever increasing velocity (see Figure 31).

If now in the neighborhood of the point $t = 0$ various small perturbations are possible, then one may state, concerning the integral curves of the equation in question defined by the same initial conditions, only that for $t > 0$ and for small μ they either remain close to one of the graphs of the stable roots, or rapidly go off to infinity.

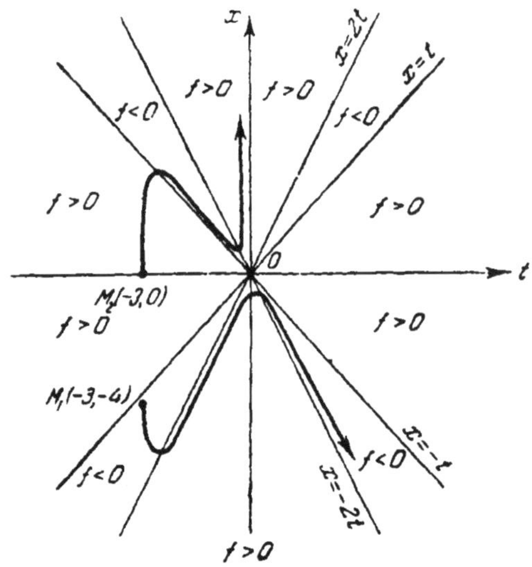

Figure 31

Now we consider a more general case studied by A. N. Tihonov [267]. Suppose that

$$\left.\begin{array}{l} \dfrac{dx_i}{dt}=f_i(t,\ x_s,\ z_k), \\ \mu\,\dfrac{dz_j}{dt}=F_j(t,\ x_s,\ z_k), \\ x_s(t_0)=x_{s0},\ z_k(t_0)=z_{k0} \\ (i,\ s=1,\ 2,\ \ldots,\ m;\ j,\ k=1,\ 2,\ \ldots,\ p), \end{array}\right\} \tag{9}$$

where for brevity we have written

$$f_i(t,\ x_s,\ z_k)=f_i(t,\ x_1,\ x_2,\ \ldots,\ x_m,\ z_1,\ z_2,\ \ldots,\ z_p),$$
$$F_j(t,\ x_s,\ z_k)=F_j(t,\ x_1,\ x_2,\ \ldots,\ x_m,\ z_1,\ z_2,\ \ldots,\ z_p),$$

and assume that μ is a small positive parameter, that all the functions f_i and F_j have bounded derivatives in all the arguments in the region in question and that $(D(F_1, F_2, \cdots, F_p)/D(z_1, z_2, \cdots, z_p)) \neq 0$ in the same region.

A degenerate system is a system

$$\left.\begin{array}{r} \dfrac{dx_i}{dt}=f_i(t,\ x_s,\ z_k), \\ F_j(t,\ x_s,\ z_k)=0, \\ x_s(t_0)=x_{s0}\quad (i,\ s=1,\ 2,\ \ldots,\ m;\ j,\ k=1,\ 2,\ \ldots,\ p). \end{array}\right\} \tag{10}$$

Suppose that on solving the system $F_j = 0$ $(j = 1, 2, \cdots, p)$ with respect to

z_k we get one or several real solutions:

$$z_j = \varphi_j(t, x_1, x_2, \ldots, x_m) \qquad (j = 1, 2, \ldots, p).$$

We shall determine first under what conditions the integral curve

$$x_i = x_i(t, \mu), \quad z_j = z_j(t, \mu)$$
$$(i = 1, 2, \ldots, m;\ j = 1, 2, \ldots, p)$$

of the original system (9) approximates and then remains close to the surface

$$z_j = \varphi_j(t, x_1, x_2, \ldots, x_m) \qquad (j = 1, 2, \ldots, p).$$

For simplicity we shall suppose that the system $F_j = 0$ $(j = 1, 2, \cdots, p)$ has a unique solution $z_j = \varphi_j(t, x_1, x_2, \cdots, x_m)$, the functions φ_j having bounded partial derivatives in all the arguments. Consider the function

$$\rho = \sum_{j=1}^{p} [z_j(t, \mu) - \varphi_j(t, x_1(t, \mu), x_2(t, \mu), \ldots, x_m(t, \mu))]^2$$

representing the square of the distance of a point of the integral curve of the complete system (9) from the point of the surface $z_j = \varphi_j(t, x_1, x_2, \cdots, x_m)$ $(j = 1, 2, \cdots, p)$ with the same coordinates $t, x_1, x_2, \cdots, x_m$. The derivative

$$\frac{d\rho}{dt} = 2\sum_{j=1}^{p}(z_j - \varphi_j)\left(\frac{dz_j}{dt} - \frac{d\varphi_j}{dt}\right) = 2\sum_{j=1}^{p}(z_j - \varphi_j)\left(\frac{F_j}{\mu} - \frac{d\varphi_j}{dt}\right).$$

If the point of the integral curve of the system (9) does not lie close to the surface $F_j(t, x_1, x_2, \cdots, x_m) = 0$ $(j = 1, 2, \cdots, p)$, then for sufficiently small μ the sign of the derivative $d\rho/dt$ coincides with the sign of the sum $\sum_{j=1}^{p}(z_j - \varphi_j)F_j$, while for $\mu \to 0$ the modulus of the derivative $|d\rho/dt| \to \infty$. Accordingly, if $\sum_{j=1}^{p}(z_j - \varphi_j)F_j < 0$, then $d\rho/dt < 0$ and for small μ the function ρ rapidly decreases until the integral curve of the system (9) approaches the surface $z_j = \varphi_j(t, x_1, x_2, \cdots, x_m)$ $(j = 1, 2, \cdots, p)$, after which, if $\sum_{j=1}^{p}(z_j - \varphi_j)F_j \leqq 0$, the integral curve cannot leave the neighborhood of this surface, since away from the neighborhood in question, for sufficiently small μ, the derivative $d\rho/dt < 0$. It is natural to call this case stable.

Thus, the condition of stability for the root $z_j = \varphi_j(t, x_1, x_2, \cdots, x_m)$ $(j = 1, 2, \cdots, p)$ is $\sum_{j=1}^{p}(z_j - \varphi_j)F_j \leqq 0$ in the neighborhood of that root. For $p = 1$ and $m = 0$ we obtain the condition for stability: $(z_1 - \varphi_1)F_1 \leqq 0$, equivalent to the one given above on page 93.

If the integral curve entered the neighborhood of the surface

$$z_j = \varphi_j(t, x_1, x_2, \ldots, x_m) \qquad (j = 1, 2, \ldots, p),$$

then $z_j(t, \mu)$ may be represented in the form

$$z_j(t, \mu) = \varphi_j(t, x_1(t, \mu), x_2(t, \mu), \ldots, x_m(t, \mu)) + \varepsilon_j(t, \mu),$$

where the function $\epsilon_j(t,\mu) \to 0$ as $\mu \to 0$. It remains to be determined whether for small μ the function $x_i(t,\mu)$, defined by the complete system (9), is close for $t > t_0$ to the functions $x_i(t)$ defined by the degenerate system (10). The functions $x_i(t)$ are solutions of the system of equations

$$\frac{dx_i}{dt} = f_i(t, x_i, \varphi_k(t, x_1, x_2, \ldots, x_m)); \quad x_i(t_0) = x_{i0} \tag{11}$$
$$(i = 1, 2, \ldots, m).$$

The functions $x_i(t,\mu)$ satisfy the system

$$\frac{dx_i}{dt} = f_i(t, x_i, \varphi_k(t, x_1, x_2, \ldots, x_m) + \varepsilon_k(t, \mu)),$$
$$x_i(t_0 + \delta) = x_{i0} + \eta_i \quad (i = 1, 2, \ldots, m), \tag{12}$$

where δ and η_i may be made arbitrarily small for sufficiently small μ, since for small μ outside the neighborhood of the surfaces $F_j = 0$ ($j = 1, \cdots, p$) the modulus of the derivative $d\rho/dt$ is very large, so that in the stable case the integral curve of the system (9), for values of the coordinates $t, x_1, x_2, \cdots, x_m$ which are almost unchanged from $t_0, x_{10}, x_{20}, \cdots, x_{m0}$ approaches the surface

$$z_j = \varphi_j(t, x_1, x_2, \ldots, x_m) \qquad (j = 1, 2, \ldots, p).$$

Comparing systems (11) and (12) and the initial values of the solutions $x_i(t)$ and $x_i(t,\mu)$, we arrive, on the basis of the continuous dependence of the solutions on the parameters and on the initial values, at the conclusion that $x_i(t,\mu) \to x_i(t)$ as $\mu \to 0$ for $t_0 \leqq t \leqq T$.

In [268], [269] A. N. Tihonov considers a still more general case, where there are a number of small coefficients on the derivatives:

$$\left.\begin{aligned} &\frac{dx_i}{dt} = f_i(t, x_j, z_{kj_k}),\\ &\mu_s \frac{dz_{sj_s}}{dt} = F_{sj_s}(t, x_j, z_{kj_k}),\\ &x_j(t_0) = x_j^0, \quad z_{kj_k}(t_0) = \overset{0}{z}_{kj_k}\\ &(i = 1, 2, \ldots, m; \ s, k = 1, 2, \ldots, p;\\ &\qquad j_k = 1, 2, \ldots, p_k). \end{aligned}\right\} \tag{13}$$

Here on the right side we have employed the notation already encountered above, indicating that the right sides of the equations are functions of t and of all the x_j and z_{kj_k}.

The small parameters μ_s tend simultaneously to zero (for example, all the μ_s are functions of one parameter μ and $\lim_{\mu\to\mu_0}\mu_s(\mu)=0$), while $\lim_{\mu\to\mu_0}(\mu_s/\mu_{s+1})=0$. The functions f_i and F_{sj_s} are continuous and satisfy conditions guaranteeing the uniqueness of the solution of the problem. The degenerate system corresponding to the system (13) has the form

$$\left.\begin{aligned} \frac{dx_i}{dt}&=f_i(t,\ x_j,\ z_{kj_k}),\\ F_{sj_s}(t,\ x_j,\ z_{kj_k})&=0,\\ x_j(t_0)&=x_j^0. \end{aligned}\right\} \tag{14}$$

Here and in what follows all the indices will take on the same values as in the system (13). For simplicity we shall suppose that the system $F_{sj_s}=0$ has a unique real solution $z_{sj_s}=\varphi_{sj_s}(t,x_j)$, where all the partial derivatives of the functions φ_{sj_s} are bounded.

We shall prove that the solution $x_i(t,\mu_s)$, $z_{kj_k}(t,\mu_s)$ of (13) tends to a solution of the degenerate system for $t>t_0$ if

$$\sum_{j_k=1}^{p_k}[z_{kj_k}-\varphi_{kj_k}]\,F_{kj_k}\leqslant 0 \qquad (k=1,\ 2,\ \ldots,\ p), \tag{15}$$

while the kth inequality is satisfied under the hypothesis that $F_{sj_s}=0$ for $s=1,2,\cdots,(k-1)$.

Consider the square of the distance from the points of the integral curve of the system (13) from the corresponding points (i.e., points with the same values of t and x_i) of the surface $z_{kj_k}=\varphi_{kj_k}(t,x_j)$:

$$\rho=\sum_{k=1}^{p}\ \sum_{j_k=1}^{p_k}[z_{kj_k}(t,\ \mu_s)-\varphi_{kj_k}(t,\ x_j(t,\ \mu_s))]^2.$$

Differentiating, we obtain

$$\begin{aligned} \frac{d\rho}{dt}&=2\sum_{k=1}^{p}\ \sum_{j_k=1}^{p_k}[z_{kj_k}-\varphi_{kj_k}]\left(\frac{dz_{kj_k}}{dt}-\frac{d\varphi_{kj_k}}{dt}\right)\\ &=2\sum_{k=1}^{p}\ \sum_{j_k=1}^{p_k}[z_{kj_k}-\varphi_{kj_k}]\left(\frac{F_{kj_k}}{\mu_k}-\frac{d\varphi_{kj_k}}{dt}\right). \end{aligned}$$

As long as the points of the integral curve do not approach the surface $z_{kj_k}=\varphi_{kj_k}(t,x_j)$, the sign of the derivative $d\rho/dt$ is determined by the sign of the term $2\sum_{k=1}^{p}\sum_{j_k=1}^{p_k}[z_{kj_k}-\varphi_{kj_k}](F_{kj_k}/\mu_k)$, which tends to $-\infty$ as $\mu_s\to 0$. Accordingly, for sufficiently small μ_s the integral curve of the system (13) for arguments t and x_i hardly varying from t^0 and x_i^0 enters an arbitrarily small ϵ-neighborhood of the surface $z_{kj_k}=\varphi_{kj_k}(t,x_j)$ and cannot leave this neighborhood. Accordingly, for $t>t_0$, $z_{kj_k}(t,\mu_s)=\varphi_{kj_k}(t,x_j(t,\mu_s))+\eta_{kj_k}(t,\mu_s)$, where the $|\eta_{kj_k}|$ can be taken arbitrarily small for sufficiently small μ_k. Therefore the first m functions $x_i(t,\mu_s)$ of the solution of (13) are also solutions of the following system:

$$\frac{dx_i}{dt}=f_i(t,\ x_j,\ \varphi_{kj_k}(t,\ x_j)+\eta_{kj_k}(t,\ \mu_s)),$$
$$x_i(t_0+\delta)=x_i^0+\eta_i \qquad (i=1,\ 2,\ \ldots,\ m), \tag{16}$$

where δ and η_i may be taken arbitrarily small, if the μ_s are small enough and $\delta>0$. The first m functions $x_i(t)$ of the solution of the degenerate system (14) form a solution of the system

$$\frac{dx_i}{dt}=f_i(t,\ x_j,\ \varphi_{kj_k}(t,\ x_j)),$$
$$x_i(t_0)=x_i^0 \qquad (i=1,\ 2,\ \ldots,\ m). \tag{17}$$

Comparing (16) and (17) and applying the theorem on the continuous dependence of the solution on a parameter and on the initial values, we conclude that the difference $x_i(t,\mu_s)-x_i(t)$ may be taken in modulus less than any positive number ϵ given in advance for $t_0+\delta\leqq t\leqq T$, if μ_s is sufficiently small.

Thus the solution of the complete equation (13) tends to a solution of the degenerate equation (14) for $\mu_s\to 0$ and $t_0+\delta\leqq t\leqq T$, if conditions (15) are satisfied. These conditions are thus called the stability conditions for the roots $z_{kjk}=\varphi_{kjk}(t,x_j)$ of the degenerate equation.

The theory of Tihonov presented in brief above supposes that $(D(F_1,F_2,\cdots,F_p)/D(z_1,z_2,\cdots,z_p))\neq 0$ (page 96). An analogous condition must be satisfied for the solvability of the system $F_{kj_k}=0$ (page 98). If these conditions are violated, entirely new possibilities arise, which are treated in [62-65; 126] and [127]. Consider for example the following equations, which appear quite frequently in the applications:

$$\mu\ddot{x}=f(t,x), \tag{18}$$

where μ is a small positive parameter.

This equation can appear in the investigation of motions of a body with a small inertia under the action of a force depending on the time and on the position of the body.

Putting $\dot{x} = z$, we replace this equation by the system

$$\dot{x} = z; \quad \mu\dot{z} = f(t, x),$$

in which $\partial f/\partial z \equiv 0$, so that Tihonov's theory is inapplicable.

In order to describe the possibilities appearing here, we consider first the following two equations:

$$\mu\ddot{x} = -a^2x \qquad \text{and} \qquad \mu\ddot{x} = a^2x \qquad \text{for constant } a. \tag{19}$$

The solutions of these equations satisfying $x(t_0) = 0$ and $\dot{x}(t_0) = \dot{x}_0$ have respectively the following forms:

$$x(t, \mu) = x_0 \cos\frac{a}{\sqrt{\mu}}(t-t_0) + \sqrt{\mu}\frac{\dot{x}_0}{a}\sin\frac{a}{\sqrt{\mu}}(t-t_0)$$

and

$$x(t, \mu) = x_0 \operatorname{ch}\frac{a}{\sqrt{\mu}}(t-t_0) + \sqrt{\mu}\frac{\dot{x}_0}{a}\operatorname{sh}\frac{a}{\sqrt{\mu}}(t-t_0). \tag{20}$$

For $\mu \to 0$ the solution of the first equation does not tend to any limit. As μ decreases it oscillates with unboundedly increasing frequency and with an amplitude tending to x_0 around the solution $x \equiv 0$ of the degenerate equation. If x_0 is small, then the modulus $|x(t, \mu)|$ is small and in this sense the solution of the original equation for small μ is close to the solution $x \equiv 0$ of the degenerate equation. The solution of the second of the equations (19) for μ tending to zero increases unboundedly in modulus; thus the integral curve for small μ rapidly recedes from the root of the degenerate equation $x \equiv 0$.

In neither case does the solution $x(t, \mu)$ tend to a solution of the degenerate equation as $\mu \to 0$. However, the first case may be called stable in the sense that if the initial point (t_0, x_0) is close to a root of the degenerate equation, then the solution $x(t, \mu)$ remains close in the sense of closeness of the zeroth order to a solution of the degenerate equation.

The investigations of V. M. Volosov showed that the two examples just considered are typical for the equation of the form $\mu\ddot{x} = f(t, x)$, and also for equations of order n with a small coefficient of the highest derivative in which there is no derivative of order $(n - 1)$, and for systems of such equations.

Volosov's results in [62; 63] are quite complicated, so that we must restrict ourselves to a comparatively elementary and incomplete analysis of this case.

We suppose for simplicity that the degenerate equation $f(t, x) = 0$ has a unique solution $x = \varphi(t)$. Replacing the equation $\mu\ddot{x} = f(t, x)$ with initial conditions $x(t_0) = x_0$, $\dot{x}(t_0) = \dot{x}_0$ by the equivalent equation

$$\dot{x} = \dot{x}_0 + \frac{1}{\mu}\int_{t_0}^{t} f(t, x)\,dt, \quad x(t_0) = x_0 \tag{21}$$

or

$$x = x_0 + \dot{x}_0(t - t_0) + \frac{1}{\mu}\int_{t_0}^{t}\int_{t_0}^{t} f(t, x)\,dt^2,$$

we note that outside the neighborhood of the root $x = \varphi(t)$ of the degenerate equation $|\dot{x}(t, \mu)|$ increases unboundedly as $\mu \to 0$, while for sufficiently small μ for $t \geqq t_0 + \delta$, $\delta > 0$ the sign of the derivative $\dot{x}(t, \mu)$ coincides with the sign of the function f. Thus, if $(x - \varphi)f \geqq 0$, then for given x_0 and $\dot{x}_0$, where $f(t_0, x_0) \neq 0$, and for sufficiently small μ the integral curve for $t \geqq t_0 + \delta$ rapidly recedes from the graph of the root $x = \varphi(t)$ (Figure 32).

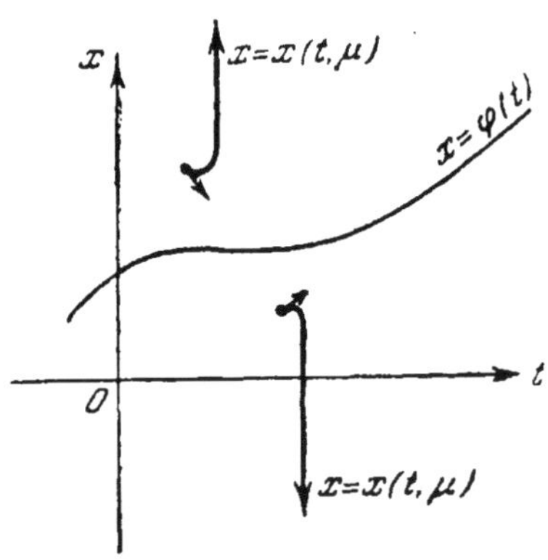

Figure 32

In this case the root $x = \varphi(t)$ of the degenerate equation is said to be unstable. But if $(x - \varphi)f \leqq 0$, then it follows from (21) that for sufficiently small μ and for an arbitrarily small interval of time the integral curve in the neighborhood of the initial point changes its initial direction into a direction almost parallel to the Ox axis, tends to the graph of the root of the degenerate equation and passes through it, after which the sign of the function f changes, and after some interval of time the sign of the derivative $\dot{x}(t, \mu)$ also changes, as is clear from (21). Accordingly the integral curve again tends to the graph of the root, again passes through it and continues to oscillate around the graph of the root $x = \varphi(t)$, generally speaking with varying amplitude (Figure 33).

For completeness of the qualitative description in this case we need to clear up further the question of the dependence of the amplitudes of these oscillations on the initial values and on the parameter μ. First we must discover in which cases the amplitude of the oscillations remains bounded as μ approaches zero. A complete but rather cumbersome solution of this question was given by Volosov in the papers [62; 63] mentioned above.

If the degenerate equation $f(t,x) = 0$ has several stable roots $x = \varphi_i(t)$, then the question arises as to their regions of influence. However, this question has not yet been studied.

Now we consider the equation

$$\mu x^{(n)} = f(t, x), \quad n \geqslant 3. \tag{22}$$

The argument and conclusions obtained above for $n = 2$ are fully applicable also to equation (22).

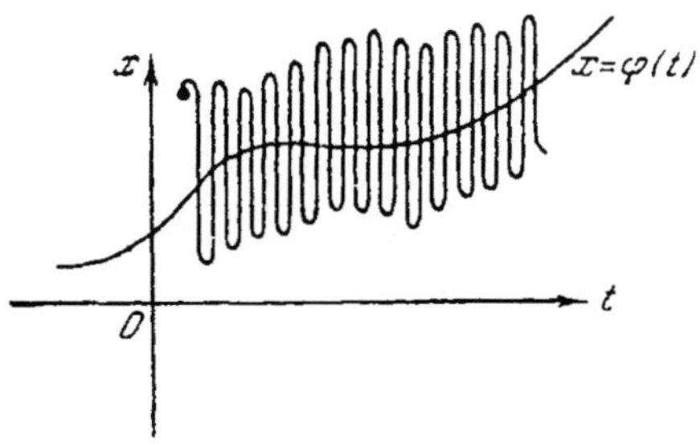

Figure 33

Indeed, if $f(t_0, x_0) \neq 0$ at some point, then in the neighborhood of this point $x^{(n)}(t,\mu) \to \infty$ as $\mu \to 0$, so that in the same neighborhood all the $x^{(k)}(t,\mu) \to \infty$ for $\mu \to 0$ and for $t > t_0 + \delta$, where $\delta > 0$, and any initial conditions we have

$$x^{(k)}(t_0) = x_0^{(k)} \qquad (k = 0, 1, 2, \ldots, (n-1)),$$

since

$$x^{(k)}(t) = x_0^{(k)} + (t - t_0)\, x_0^{(k-1)} + \ldots + \frac{(t-t_0)^{n-k-1}}{(n-k-1)!}\, x_0^{(n-1)} + \frac{1}{\mu}\int_{t_0}^{t}\int_{t_0}^{t} \cdots \int_{t_0}^{t} f(x, t)\, dt^{n-k}.$$

But if $f(t_0, x_0) < 0$, then all the $x^{(k)}(t) \to -\infty$ $(k = 0, 1, \cdots, n)$ for $\mu \to 0$ and for $t > t_1 > t_0$. Accordingly, if $f(t, \varphi(t) + \epsilon) > 0$, where $x = \varphi(t)$ is an isolated root of the degenerate equation, and $f(t, \varphi(t) - \epsilon) < 0$ for sufficiently small $\epsilon > 0$, then the root is unstable, since the integral curve for sufficiently small $\mu > 0$ recedes from the root $x = \varphi(t)$ of the degenerate equation. But if $f(t, \varphi(t) + \epsilon) < 0$, and $f(t, \varphi(t) - \epsilon) > 0$, $\epsilon > 0$, then the solutions oscillate around the root, since for sufficiently small μ the solution has an extremum around the initial point (t_0, x_0). Then the integral curve

$x = x(t, \mu)$ approaches the graph of the root $x = \varphi(t)$ and intersects it, whereupon $x^{(n)}(t, \mu)$ changes sign, and then successively all the first-order derivatives change sign and $x(t)$ takes on a minimum, while the integral curve again begins to approach the graph of the root $x = \varphi(t)$ and so forth.

Here, however, we must note that the possibility of increasing amplitude of the oscillations, and even its approach to infinity as $\mu \to 0$, is by no means excluded.

We present two typical examples:

(1) $\mu \dddot{x} = x$, since $\partial f / \partial x = 1 > 0$ the root $x \equiv 0$ of the degenerate equation is not stable.

(2) $\mu \dddot{x} = -x$, $\partial f / \partial x = -1 < 0$. The solution $x = x(t, \mu)$ has an oscillatory character around the root $x = 0$ of the degenerate equation.

The general solution of this equation has the form

$$x = c_1 e^{-\frac{t}{\sqrt[3]{\mu}}} + e^{\frac{t}{2\sqrt[3]{\mu}}}\left(c_2 \cos \frac{\sqrt{3}t}{2\sqrt[3]{\mu}} + c_3 \sin \frac{\sqrt{3}t}{2\sqrt[3]{\mu}}\right),$$

and for $t > 0$ and $\mu \to 0$ the amplitude and frequency of the oscillations increase unboundedly.

For linear equations with constant coefficients of type (22) for $n > 2$ the amplitude of the oscillations for $\mu \to 0$ always increases unboundedly. This follows directly from the fact that in this case the sum of the roots of the characteristic equation and the sum of their conjugate products is equal to zero, so that there exist roots with positive real parts, while for the other types of equations of type (22) the amplitude of the oscillations for $\mu \to 0$ can remain bounded.

The boundary problem with a small coefficient on the highest derivative. Suppose that $x(t, \mu)$ is the solution of the equation

$$\mu \ddot{x} + q(t) x = 0, \quad \mu > 0, \tag{23}$$

satisfying the conditions $x(t_0) = x_0$, $x(t_1) = x_1$. The usual question of this section arises, as to whether the solution $x(t, \mu)$ converges to the solution $x \equiv 0$ of the degenerate equation as μ tends to zero.

Suppose that the continuous function $q(t)$ is positive for $t_0 \leqq t \leqq t_1$, so that, as is well known, the solution of the equation $\ddot{x} + (q(t)/\mu)x = 0$ for small μ will rapidly oscillate, with the frequency of the oscillations increasing unboundedly as $\mu \to 0$.

The amplitude of the oscillations depends on μ and on the initial values, and if at least one of the numbers x_0, x_1 is not equal to zero, then the amplitude of the oscillations does not tend to zero and the solution $x(t, \mu)$

does not tend, as μ approaches zero, to a solution of the degenerate equation. But if $q(t) < 0$ for $t_0 \leqq t \leqq t_1$, then all the solutions of the equation do not oscillate. Suppose for definiteness that both $x_0 > 0$ and $x_1 > 0$.

In the upper half plane $x > 0$ the second derivative

$$\ddot{x}(t, \mu) > 0 \text{ and } \lim_{\mu \to 0} \ddot{x}(t, \mu) = -\lim_{\mu \to 0} \frac{q(t)\, x}{\mu} = \infty,$$

from which it follows that the integral curve, which is convex in the upper half plane and for sufficiently small μ almost coincides with a straight line parallel to the Ox axis, approaches the solution of the degenerate equation $x \equiv 0$, but cannot go through this solution, since in the region $x < 0$ the integral curves are concave and so cannot satisfy the condition $x(t_1) = x_1 > 0$. Thus, for a value $\bar{t_0}$ arbitrarily close to t_0, and for sufficiently small μ the integral curve enters a given one-sided ϵ-neighborhood U_ϵ of the solution of the degenerate equation $0 \leqq x \leqq \epsilon$, which it can leave only close to $t = t_1$, since outside this neighborhood the modulus $|\dot{x}(t, \mu)|$ for sufficiently small μ may be made arbitrarily large, so that if the integral curve left the limits of U_ϵ away from the point $t = t_1$, then for sufficiently small μ the integral curve, which almost coincides with a curve parallel to the Ox axis, will recede from the line $x \equiv 0$ and the condition $x(t_1) = x_1$ cannot be satisfied. Thus, in the case $q(t) < 0$ the solution $x(t, \mu)$ of (23) tends as $\mu \to 0$ to a discontinuous function $x(t_0) = 0$, $x \equiv 0$ for $t_0 < t < t_1$ and $x(t_1) = x_1$, while on the segment $t_0 + \delta \leqq t \leqq t_1 - \delta$, where $\delta > 0$, the solution $x(t, \mu)$ uniformly tends to a solution of the degenerate equation $x \equiv 0$.

If the numbers x_0 and x_1 are negative or have opposite signs, then the argument is almost the same, except that in the last case the integral curve $x = x(t, \mu)$ passes through the graph of the degenerate equation.

This argument also applies almost without change to the nonlinear case. Indeed, suppose that for every sufficiently small μ there exists a unique continuous solution $x(t, \mu)$ of the boundary problem

$$\mu \ddot{x} = f(t, x); \quad x(t_0) = x_0, \quad x(t_1) = x_1, \quad \mu > 0, \tag{23a}$$

where $f(t, x)$ is a continuous function, $f(t_0, x_0) \neq 0$ (or $f(t_1, x_1) \neq 0$). Suppose that the degenerate equation $f(t, x) = 0$ has a unique continuous solution $x = \varphi(t)$.

If on passing through the graph of this solution with increasing x for fixed t the function f changes from plus to minus ($(x - \varphi)f \leqq 0$), then the solution of equation (23a) $x(t, \mu)$ can oscillate around $x = \varphi(t)$, while as $\mu \to 0$ the frequency of the oscillation increases unboundedly. The amplitude is determined by the initial conditions and does not converge to zero. Accordingly, in this case the solution of the boundary problem $x(t, \mu)$ does not tend to a solution of the degenerate equation.

But if on the passage through a root with increasing x with fixed t the sign of the function f changes from minus to plus ($(x - \varphi)f \geqq 0$), then the integral curve, almost coinciding with a straight line parallel to the Ox axis, approaches the graph of the root $x = \varphi(t)$ and can leave its ϵ-neighborhood only close to $t = t_1$, since departure from the limits of the ϵ-neighborhood for sufficiently small μ means that the integral curve, which almost coincides with a straight line parallel to the Ox axis, recedes from the root. Hence if that departure takes place away from $t = t_1$, the conditions $x(t_1) = x_1$ cannot be satisfied.

These and substantially more general results (for equations of higher order, for systems of equations, for partial differential equations), may also be obtained from analysis of the corresponding variational problems.

In the case of the equation $\mu\ddot{x} = f(t, x)$, $x(t_0) = x_0$, $x(t_1) = x_1$, we need to consider the problem of extremalizing the functional

$$v_\mu[x(t)] = \int_{t_0}^{t_1} [\mu\dot{x}^2 + F(t, x)]\, dt$$

with the same boundary conditions, where $F'_x(t, x) = f(t, x)$. The functional $v_0[x(t)]$ takes on a minimum on the discontinuous curve

$$x(t_0) = x_0, \quad x = \varphi(t) \quad \text{for} \quad t_0 < t < t_1, \quad x(t_1) = x_1,$$

if $\partial f/\partial x > 0$ on the curve $x = \varphi(t)$, and takes on a minimum on the same curve if $\partial f/\partial x < 0$ there. For $\mu > 0$ the functional $v_\mu[x(t)]$ takes on a minimum on the solution $x(t, \mu)$ of the equation $\mu\ddot{x} = f(t, x)$, $x(t_0) = x_0$, $x(t_1) = x_1$, if $\partial f/\partial x > 0$, since in this case the Jacobi-Weierstrass condition is satisfied. But if $\partial f/\partial x < 0$, there is no extremum on the curve $x = x(t, \mu)$. In the first case the sequence of solutions $x(t, \mu_n)$ for $\mu_n \to 0$ will be minimizing for the functional $v_0[v(t)]$, so that it easily follows that $x(t, \mu_n)$ tends to $\varphi(t)$ in the mean.

For proof of the uniform convergence on the segment $t_0 + \delta \leqq t \leqq t_1 - \delta$ ($\delta > 0$) we need further to take account of the fact that the minimum of the functional, not only on the whole segment $[t_0, t_1]$ but also on any portion of this segment, must be assumed on the curve $x = x(t, \mu_n)$.

In conclusion, we observe that in all the problems of this section there is considerable interest, quite apart from the qualitative investigation, in the quantitative estimates of the error which arises in the stable case by replacing the complete system of equations with the degenerate system. Readers interested in these questions, which do not lie within the scope of qualitative analysis, are referred to the papers of A. B. Vasil'eva [**41-45**].

3. **Some asymptotic properties of the solutions of dynamical systems.** Many properties of the dynamical systems

$$\frac{dx_i}{dt} = f_i(x_1, x_2, \ldots, x_n) \qquad (i = 1, 2, \ldots, n), \tag{24}$$

where the functions f_i satisfy conditions guaranteeing the existence and uniqueness of the solution of the Cauchy problem are consequences of the following three fundamental properties of such systems:

(1) the initial point $(t_0, x_{10}, \cdots, x_{n0})$ defines a unique trajectory;

(2) the solution $x_i(t)$ is a continuous function of t and of the initial values $t_0, x_{10}, x_{20}, \cdots, x_{n0}$;

(3) the motions along the trajectory form a group.

Therefore it is appropriate to axiomatize these properties and to consider the so-called abstractly defined dynamical systems, of which the system (24) is a special case.

An abstractly defined dynamical system is a one-parameter group $f(p, t)$ $(-\infty < t < \infty, p \subset R)$ of transformations of a metric space R into itself, with the following properties:

(1) $f(p, 0) = p$;

(2) $f(p, t)$ is continuous in p and t;

(3) $f[f(p, t_1), t_2] = f(p, t_1 + t_2)$ (the group property).

The parameter t will be called the time. The set of points $f(p, t)$ of the space R for fixed p and all possible values of t, $-\infty < t < \infty$, will be called a *trajectory*.

The portion of the trajectory corresponding to the values $0 \leqq t < \infty$ of time (or $(-\infty < t \leqq 0)$) will be called the *positive* (*negative*) *semitrajectory*.

The point $f(p, t)$ is called the *image* of the point p at the moment of time t.

The images under the mapping $f(p, t)$ of all the points of the set A at the moment t form a set denoted by $f(A, t)$, or A_t for short.

If for arbitrary t the set $f(A, t) = A$, i.e., if the set A contains, together with any point p, the entire trajectory passing through p, then A is called an *invariant set*.

Consider any increasing sequence of values of t:

$$t_1 < t_2 < \cdots < t_n < \cdots,$$

with $\lim_{n\to\infty} t_n = \infty$.

If the sequence $f(p, t_n)$ has a limit point q, then this point is called the *ω-limit* of the trajectory $f(p, t)$. The collection of all ω-limit points of a trajectory $f(p, t)$ is called the *ω-limit* set of that trajectory and is denoted by Ω_p.

Analogously, considering the limit points of the negative semitrajectory, passing through the point p, i.e., the limit points of all possible sequences $f(p, t_n)$, where $t_n \to -\infty$ as $n \to \infty$, we obtain a set of α-limit points A_p, or the α-limit set of the trajectory $f(p, t)$.

For example, the α- and ω-limit sets for trajectories of periodic motions or for stationary points coincide with the points of the same trajectory. The ω-limit for a spiral, coiling inward as t increases to some limit cycle

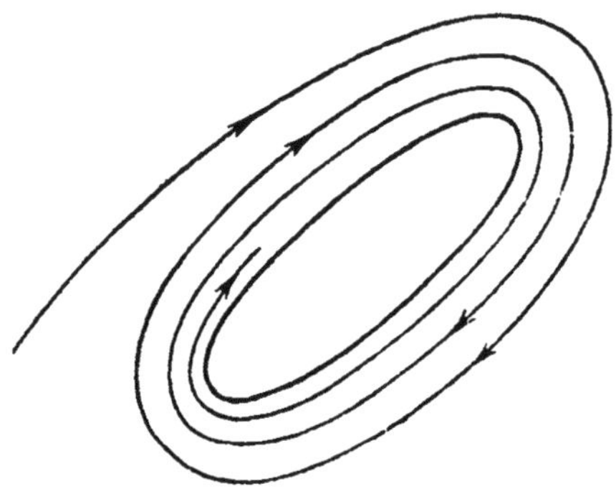

Figure 34

(Figure 34), is the set of points of this limit cycle. The ω-limit set for all trajectories having points in a sufficiently small neighborhood of a stable node q (Figure 35) is a single point q.

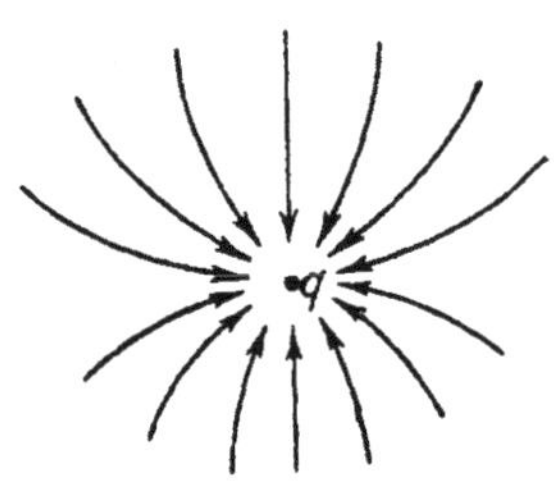

Figure 35

THEOREM I. *The α- and ω-limit sets A_p and Ω_p of some trajectory $f(p, t)$ are invariant.*

PROOF. Suppose that the point p lies in the set Ω_p (or A_p). We need to prove that the entire trajectory $f(q,t)$ lies in Ω_p. Suppose that p_n is a sequence of points of the trajectory $f(p,t)$ converging to q. Then the sequence of points $f(p_n,\tau)$ of the trajectory $f(p,t)$ converges to some point $f(q,\tau)$ of the trajectory $f(q,t)$ (Figure 36), since from property (2) of the

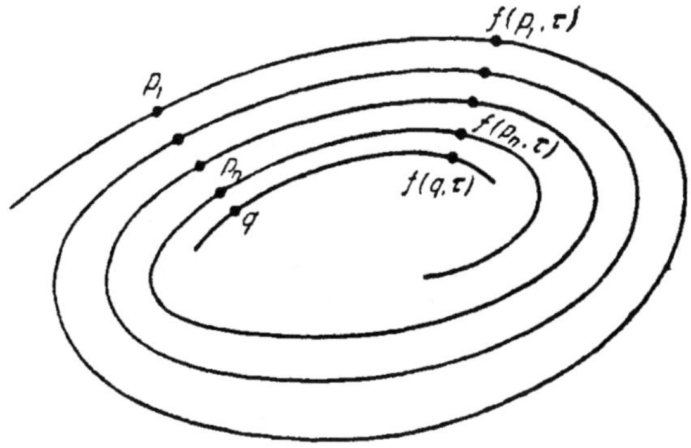

Figure 36

function f we have $\lim_{n\to\infty} f(p_n,\tau) = f(q,\tau)$. Accordingly the point $f(q,\tau)$ for any τ is an ω-limit of the trajectory $f(p,t)$ and thus belongs to Ω_p.

The trajectory $f(p,t)$ is called *positively stable* in the sense of Lagrange if the closure of the positive semitrajectory is *compact*. In the case of n-dimensional Euclidean space this requirement reduces to the requirement that the positive semitrajectory must lie in a bounded portion of that space. Analogously one defines negative stability in the sense of Lagrange by taking the negative semitrajectory in place of the positive semitrajectory.

Trajectories positively or negatively stable in the sense of Lagrange are called *stable in the sense of Lagrange*.

Evidently the α- and ω-limit sets of Lagrange-stable trajectories are not empty and are compact.

Moreover, it is not difficult to prove that in this case each of the limit sets is *connected*.[3]

Knowledge of the limit sets of a trajectory, or at least of some properties of these limit sets, already enables us to make inferences concerning the asymptotic properties of trajectories, since in the Lagrange-stable case the trajectories lie close to the limit sets for t large in absolute value.

For the classification of the possible types of asymptotic behavior of trajectories it would be desirable as a first step to give a complete classi-

[3] Trajectories not Lagrange-stable may have nonconnected α- or ω-limiting sets.

fication of the possible types of α- and ω-limiting sets. However, in such a general form this problem is still very far from its solution.

Only for two-dimensional Euclidean space and the two-dimensional sphere can this problem be considered as solved.

Both of these two-dimensional surfaces have no one-dimensional cycles which are not homologous to zero. The following theorem is a two-dimensional consequence of this.

THEOREM II. *If on the sphere or on the Euclidean plane an α- or ω-limiting set contains no stationary points, then that set consists of one periodic trajectory.*

Thus this theorem asserts that on those surfaces there are only the following types of compact limiting sets:

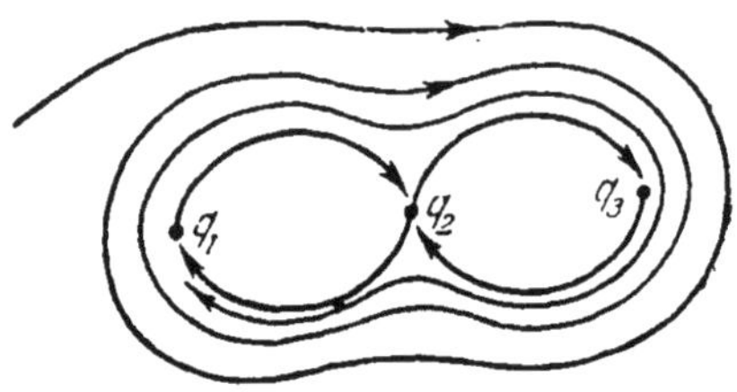

Figure 37

(1) periodic trajectories (see Figure 34);

(2) stationary points (see Figure 35);

(3) more complicated structures consisting of several trajectories, to which the stationary points necessarily belong (Figure 37).

In Figure 37 we show a trajectory and its α- and ω-limiting sets, consisting of semitrajectories, among which there are the three stationary points q_1, q_2, q_3.

An outline of the proof of this theorem is given on page 114.

We note that Theorem II gives the justification for the frequently employed method of finding periodic trajectories of a dynamical system defined on the Euclidean plane or on the sphere

$$\frac{dx_1}{dt} = f_1(x_1, x_2), \quad \frac{dx_2}{dt} = f_2(x_1, x_2), \tag{25}$$

where f_1 and f_2 satisfy the conditions of the theorems of existence and uniqueness for the solution of the Cauchy problem. This method consists in trying to find a ring-shaped region D, i.e., a region bounded by two closed nonintersecting and non-selfintersecting curves, at each point of the

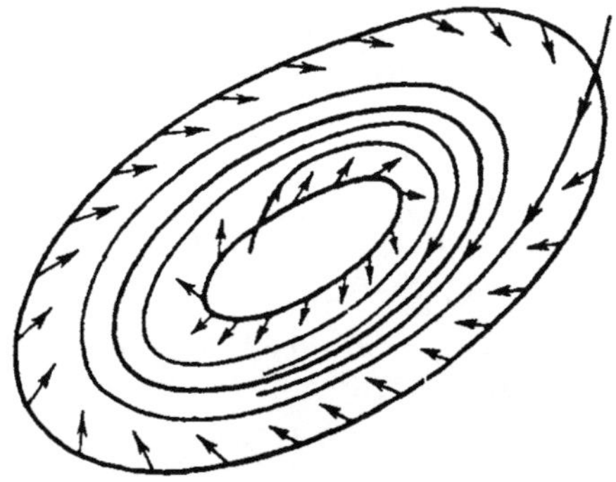

Figure 38

the boundary of which the vector tangent to the trajectories of the system (25) will be directed into the region D (Figure 38).

If there is no stationary point inside the ring D, then inside D there is at least one periodic trajectory.

Indeed, any of the trajectories entering into D must remain with increasing t inside that ring. Thus the trajectory is positively stable in the sense of Lagrange and has a nonempty ω-limit set, which by Theorem II must therefore be a periodic trajectory.

If we now shrink the ring-shaped region D but retain all its above properties guaranteeing the existence of a periodic trajectory, we thereby determine more and more precisely the position of the desired periodic trajectory.

It is easy to see that Theorem II ceases to be valid not only in spaces of three and higher dimensions, but even in two-dimensional spaces which

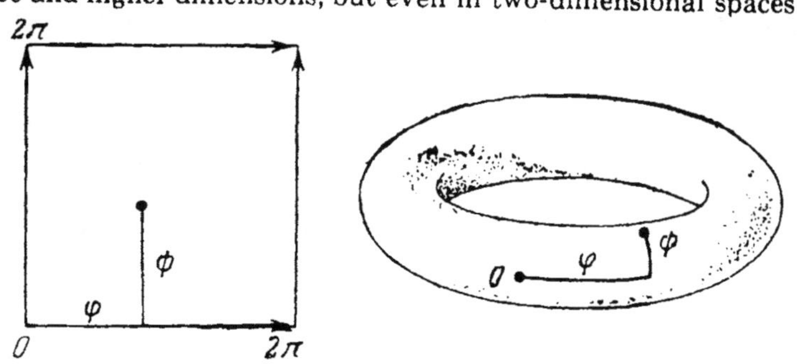

Figure 39

are more complicated from the topological point of view, for example on the torus.

Indeed, let us define on the surface of the torus T, on which two cyclical coordinates φ and ψ have been introduced, varying from 0 to 2π (Figure 39) (the points with coordinates (φ, ψ) and $(\varphi + 2n\pi, \psi + 2m\pi)$, where n and m are integers, are identified) the dynamical system $d\varphi/dt = 1$, $d\psi/dt = \alpha$, where α is a constant quantity. The equation of the trajectories may be obtained by eliminating t and integrating the equation $d\psi/d\varphi = \alpha$, so that $\psi = \psi_0 + \alpha(\varphi - \varphi_0)$. Accordingly, with a rational α the trajectories are spirals which close after one or several turns around the torus (Figure 40). However if α is irrational the trajectories are spirals which turn around

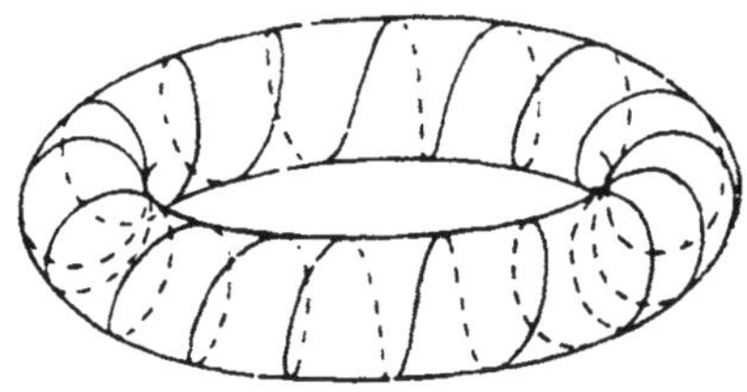

Figure 40

the torus an infinite number of times, neither closing nor intersecting.

This case of the distribution of trajectories will be called the irrationally wound torus.

We consider the intersection points of one of the trajectories of the irrationally wound torus, for example $\psi = \alpha\varphi$ with the meridian $\varphi = 0$ (the remaining trajectories are obtained from this by simple shifts along the φ and ψ axes). These points are $(0, 2n\pi\alpha)$ $(n = 0, \pm 1, \pm 2, \cdots)$. For an irrational α, as one easily verifies, they are everywhere dense on the meridian $\varphi = 0$, even if n takes on only positive values or only negative integer values. Thus every point of the meridian $\varphi = 0$ belongs both to the α- and ω-limiting sets of the trajectory.

Analogously, taking the intersection point of the trajectory with any other meridian $\varphi = \varphi_0$, we find that all the points of these meridians, and thus all the points of the torus, belong to the α- and ω-limiting sets of the trajectory in question.

In the case of an irrationally wound torus the trajectory is not closed. The motion is not periodic. However, its properties strongly suggest a periodic motion, since after a sufficiently large number of turns around the torus the trajectory goes into any arbitrarily small neighborhood of the initial point, i.e., it is almost closed, so that the trajectory almost repeats itself. More precisely this can be formulated as follows:

Given any ϵ, there is a segment of the trajectory of sufficiently great length $l(\epsilon)$ which approximates the entire trajectory to within ϵ.

Trajectories having this property are generalized closed trajectories. They are called recurrent trajectories.

DEFINITION. *The trajectory $f(p,t)$ is said to be recurrent if for any $\epsilon > 0$ one can find an $l(\epsilon)$ such that for any pair of values t_1 and $\bar{t}_1$ there is a number τ, $|\tau| \leqq l(\epsilon)$, such that the distance $\rho[f(p,t_1), f(p,t_1+\tau)] < \epsilon$.*

This definition, which convenient though rather complicated in form reduces to the statement that a segment of the trajectory of time length $2l$ ($|\tau| \leqq l$) approximates the entire trajectory to within ϵ. Trivial examples of recurrent trajectories are stationary points and trajectories of periodic motions. A nontrivial example is given by the trajectory of an irrationally wound spiral.

Now we shall prove a theorem containing Theorem II as a special case and showing the exceptionally great importance of recurrent trajectories in the study of asymptotic properties of trajectories of dynamical systems.

THEOREM III. *The α- and ω-limiting sets of a Lagrange-stable trajectory $f(p,t)$ always contain recurrent trajectories (see* [**33**]).

PROOF. The α- and ω-limiting sets A_p and Ω_p of a stable trajectory in the sense of Lagrange are invariant and compact. Both of these properties are inductive. Therefore, from Brouwer's theorem[4] there exist in A_p and Ω_p minimal sets having the same two properties, i.e., subsets F of the set A_p or of Ω_p which are invariant and compact but do not have proper subsets having these two properties.

We shall show that any trajectory of the minimal invariant and compact set F is recurrent. Suppose that the trajectory $f(q,t)$ passing through the point q of the set F is not recurrent. Then for some $\epsilon > 0$ there will for any l_i be a pair of points of the trajectory, $f(q,t_i)$ and $f(q,\bar{t}_i)$, such that for any $|\tau| < l_i$ the following inequality will hold:

$$\rho\,[f(q,t_i), f(q,\bar{t}_i+\tau)] \geqslant \epsilon. \tag{26}$$

Consider the increasing sequence of numbers

$$l_i (l = 1, 2, \ldots), \quad \lim_{i\to\infty} l_i = \infty$$

[4] On the concept of inductive properties and on Brouwer's theorem, see for example Hausdorff, *Theory of sets*, English transl., Chelsea, N. Y. 1957.

and the corresponding sequence of points

$$f(q, t_i) \text{ and } f(q, \bar{t}_i) \qquad (i = 1, 2, \ldots),$$

for which inequality (26) holds. We select from the sequence $f(q, t_i)$ a converging subsequence $f(q, t_{\alpha_i})$ $(i = 1, 2, \cdots)$ and write $\lim_{i\to\infty} f(q, t_{\alpha_i}) = \bar{q}$. Because of the invariance and compactness of F we have $\bar{q} \subset F$. Passing to the limit in the inequality

$$\rho[f(q, t_{\alpha_i}), \; f(q, \bar{t}_{\alpha_i} + \tau)] \geqslant \varepsilon, \quad |\tau| < l_{\alpha_i}$$

as $i \to \infty$, we obtain $\rho[\bar{q}, f(q, t)] \geqq \epsilon$ for arbitrary values of t, i.e., the entire trajectory $f(q, t)$ is distant not less than ϵ from the point $\bar{q}$. Accordingly, the closure of the trajectory $f(q, t)$ will not contain $\bar{q}$. At the same time it is an invariant compact subset of the set F, which contradicts the minimality of that set F.

Theorem II follows from Theorem III, if we observe that on the Euclidean plane and on the sphere there are no nontrivial (i.e., not reducing to a stationary point or to a periodic trajectory) recurrent trajectories $f(q, t)$, since on such a trajectory one would be able to select a segment C_1 of sufficiently great length with arbitrarily close boundary points A and B

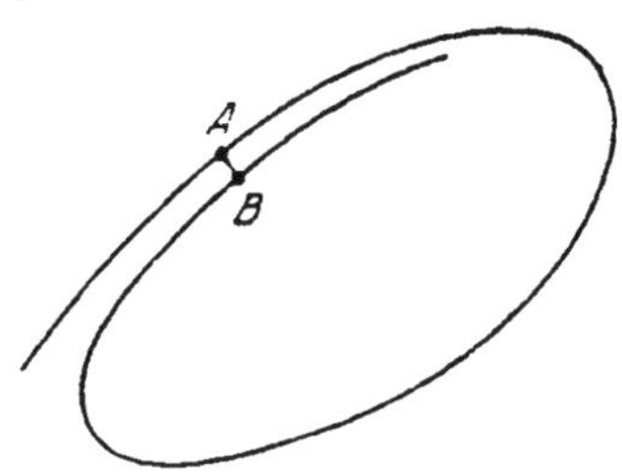

Figure 41

(see Figure 41), and, joining the points A and B by the shortest geodesic arc C_2, we would obtain a closed curve $C = C_1 + C_2$ dividing the Euclidean plane (or sphere) into two components, from one of which it is impossible to get into the other without intersecting the curve C, since on the Euclidean plane or sphere every closed curve is the boundary of some portion of the plane. Accordingly, that portion of the trajectory $f(q, t)$ which lies inside C can when extended approximate to any given degree of accuracy the portion of the trajectory lying outside C only after intersection with the curve C. But the part of the trajectory lying inside C cannot on its prolongation intersect the curve C either on the arc C_1, since the trajectories cannot intersect or selfintersect, nor on the arc C_2, since this would contradict the continuous dependence of the motion $f(q,t)$ on the first argument, C_2 having been taken arbitrarily short.

In their asymptotic behavior as $t \to \infty$ the trajectories may roughly be broken into the following three types:

(1) **The set Ω_p is empty.** In this case the trajectory is said to be *divergent in the positive direction.*

(2) **The set Ω_p is not empty, but the intersection of Ω_p with the trajectory in question is empty.** In this case the trajectory is said to be *positively asymptotic.*

(3) **The intersection of Ω_p with the trajectory is not empty.** In this last case the trajectory is said to be *positively stable in the sense of Poisson.*

It is not difficult to verify that any neighborhood of the point p of a Poisson-stable trajectory intersects with the points of that trajectory for arbitrarily large values of t, i.e., the trajectory returns for arbitrarily large values of t into arbitrarily small neighborhoods of its initial position.

Analogously one classifies the trajectories as $t \to -\infty$. A nontrivial example of trajectories positively and negatively stable in the sense of Poisson (in this case the trajectory is said to be stable in the sense of Poisson) is given by the irrationally wound torus.

4. **Dynamical systems with integral invariants.** Suppose that the dynamical system $f(p,t)$ describes the motion of an incompressible fluid. Then the volume of the set $f(A,t)$, denoted for short by A_t, remains invariant for any t. In this case one says that the dynamical system $f(p,t)$ has an *invariant volume.*

If the dynamical system $f(p,t)$ describes the motion of a compressible fluid or of a gas, while there are no sources or sinks, then not the volume, but the mass of the set $f(A,t)$ remains invariant under the motion for every t.

The presence of an invariant volume or mass or of other invariant quantities with similar properties puts a considerable restriction on the dynamical system, which necessarily has a number of particular properties.

Consequently it is worthwhile to select for special investigation those dynamical systems which either have an invariant volume or else a generalization of it called an invariant measure or an integral invariant. These dynamical systems are highly specialized from the mathematical point of view, but in view of the numerous applications to problems of hydrodynamics, gas dynamics, and to other questions, they are of great importance in applied mathematics.

An *integral invariant* for the dynamical system $dx_i/dt = f_i(x_1, \cdots, x_n)$ $(i = 1, 2, \cdots, n)$ is a set function $V(A_t)$ of the following form:

$$V(A_t) = \int\int_{A_t} \cdots \int \rho(x_1, x_2, \ldots, x_n)\, dx_1\, dx_2 \ldots dx_n, \tag{27}$$

where $\rho > 0$ and the integral is taken over the set A_t and does not depend on t.

If $V(A_t)$ does not depend on t for $\rho \equiv 1$, then in Euclidean space with rectangular coordinates $x_1, x_2, \cdots, x_n$ we have an invariant volume. If $\rho(x_1, x_2, \cdots, x_n)$ in the same coordinates is a density, then the system will have an invariant mass $V(A_t)$.

For an abstractly given dynamical system, along with the concept of integral invariant one introduces the concept of invariant measure. The *measure of the set* A (Carathéodory measure) is a set function $\mu(A)$ which for any set A of the metric space R for which $\mu(A)$ is defined, satisfies the following conditions:

(1) $\mu(A) \geqq 0$, while there exist sets of positive finite measure. The measure of the empty set is equal to zero.

(2) $\mu(A) \leqq \mu(B)$ if $A \subset B$.

(3) $\mu[\sum_{i=1}^{\infty} A_i] \leqq \sum_{i=1}^{\infty} \mu(A_i)$.

(4) $\mu(A + B) = \mu(A) + \mu(B)$, if $\rho(A, B) > 0$, where ρ is the distance in the space R.

Evidently the volume and mass and any other function V of the form (27) have all the properties of measure.

The measure is said to be *invariant* if $\mu(A_t)$ does not depend on t. The set A is called *μ-measurable* if for any set B with finite measure $\mu(B) = \mu(AB) + \mu(B - AB)$. It follows directly from this definition that if the set A is μ-measurable and the intersection AB is empty, then $\mu(A + B) = \mu(A) + \mu(B)$. For this it is sufficient that as the set B in the definition of measurability we take the set $A + B$. If A and B are measurable and $\mu(AB) = 0$, then, representing $A + B$ in the form $(A - AB) + (B - AB) + AB$, we easily see that $\mu(A + B) = \mu(A) + \mu(B)$.

It can be proved that measurable sets form a very wide class of sets. All open sets, all closed sets, all sets measurable in the sense of Borel are μ-measurable.

THEOREM ON RECURRENT SETS. *If the dynamical system $f(p, t)$, defined in the metric space R, has an invariant measure μ and the measure μ of the entire space R is finite, then for any μ-measurable set A with $\mu(A) > 0$ there are* *arbitrarily large values of t for which the intersection of A with A_t is nonempty and even has a positive measure.*

PROOF. Consider a sequence of sets $A = A_0, A_N, A_{2N}, \cdots, A_{nN}, \cdots$, where N is a positive number which we take to be arbitrarily large.

Some of the sets of this sequence necessarily intersect one another in sets of positive measure, since in the contrary case the measure of the sum $\mu(\sum_{j=0}^{n-1} A_{jN}) = n\mu(A)$ would increase unboundedly as n increases, which would contradict the finiteness of the measure of the entire space.

Thus $\mu(A_{iN} \cdot A_{jN}) \neq 0$ for some i and j, while $i \neq j$. Suppose for definiteness $j > i$. Then, shifting each point of these sets along its trajectory by the time distance $-iN$, we convert the set A_{iN} into $A_0 = A$, and the set A_{jN} into the set $A_{(j-i)N}$, while the common part of the sets A_{iN} and A_{jN} goes into the common part of the sets A_0 and $A_{(j-i)N}$, so that, from the invariance of the measure, the intersection $A_0 \cdot A_{(j-i)N}$ not only is not empty, but even $\mu(A_0 \cdot A_{(j-i)N}) > 0$.

An analogous property holds as $t \to -\infty$, for which it is sufficient to take $N < 0$.

The theorem on the recurrence of sets may be considerably strengthened. It is possible to prove the following theorem on the recurrence of points.

If the metric space R, on which there is defined the dynamical system $f(p, t)$ with the invariant measure μ, has a countable basis and $\mu(R)$ is finite, then almost all the points (i.e., points not lying in a set of μ-measure zero) lie on trajectories stable in the sense of Poisson.

In the theory of dynamical systems with invariant measure there has been a great development of the so-called probability or ergodic theorems, which find their application in statistical physics or mechanics.

Let us consider the simplest concepts and theorems concerned with this range of ideas.

Consider the trajectory $f(p,t)$ for $t \geqq 0$. Given a set A, it is natural to consider the expression $1/T\sum_i T_i$, where the sum is extended over all the intervals of time during which the trajectory, for $0 \leqq t \leqq T$, lies in the set A (Figure 42); this expression gives the fraction of the time interval $(0, T)$ during which $f(p,t) \in A$. The *probability* $P_A(p)$ that the point $f(p,t)$ lies in the set A is then defined as the limiting value of this fraction as $T \to \infty$.

In comparatively simple cases, however, this definition does not have any meaning (recall, for example, the rather complicated intersection of the trajectories of an irrational winding of the torus with any meridian or neighborhood of a meridian). Thus it is desirable to define the probability $P_A(p)$ in a somewhat different way, which at the same time is a natural generalization of the definition given above.

We introduce the characteristic function $\varphi_A(p)$ of the set A, defined by

$$\varphi_A(p) = \begin{cases} 1, & \text{if } p \subset A, \\ 0, & \text{if } p \not\subset A. \end{cases}$$

Then by definition

$$P_A(p) = \lim_{T \to \infty} \frac{1}{T} \int_0^T \varphi[f(p, t)]\, dt,$$

where the integral is understood in the sense of Lebesgue (Radon).

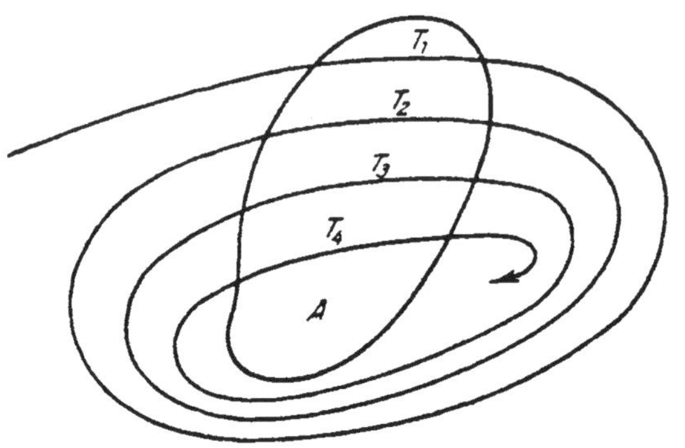

Figure 42

It is possible to prove that $P_A(p)$ is defined for almost all points of the space R in the sense of the measure μ. With this definition the value of $P_A(p)$ depends at least formally on the trajectory $f(p,t)$, on the choice of the initial point on the trajectory, and on the set A.

We shall show that in fact $P_A(p)$ does not depend on the choice of the initial point on the trajectory.

Indeed, if we take another point of the same trajectory $f(p,t)$, corresponding for example to the value $t = t_0$, as the initial point, then the probability $\overline{P}_A(p)$ that a point of the trajectory lies in the set A has the form:

$$\overline{P}_A(p) = \lim_{T\to\infty} \frac{1}{T} \int_{t_0}^{t_0+T} \varphi[f(p,t)]\,dt$$

$$= \lim_{T\to\infty} \frac{T+t_0}{T} \cdot \frac{1}{T+t_0} \int_0^{t_0+T} \varphi[f(p,t)]\,dt$$

$$- \lim_{T\to\infty} \frac{1}{T} \int_0^{t_0} \varphi[f(p,t)]\,dt$$

$$= \lim_{T\to\infty} \frac{1}{t_0+T} \int_0^{t_0+T} \varphi[f(p,t)]\,dt$$

$$= \lim_{T\to\infty} \frac{1}{T} \int_0^{T} \varphi[f(p,t)]\,dt = P_A(p).$$

For the applications there is a considerable interest in the case when the probability for almost all points does not depend on the choice of the trajectory, but depends only on the choice of the set A. This case is exceptional from the mathematical point of view, but arises frequently in the applications.

We shall call the dynamical system $f(p,t)$ defined in the space R *indecomposable* (or transitive) with respect to the measure μ if the space R cannot be decomposed into the sum of two invariant sets of positive measure without common points. An example of an indecomposable dynamical system is a system describing a motion with irrational windings about the torus: $d\varphi/dt = 1$, $d\psi/dt = \alpha$, where α is an irrational number (see page 112).

THEOREM. *For an indecomposable dynamical system with fixed A the probability $P_A(p)$ is the same for all points of the space R, with the exclusion of points of a set of measure zero* [**290**].

PROOF. Fix the set A. Suppose that M is the supremum of the values $P_A(p)$ with the exception of a set of measure zero (i.e., the set of points at which $P_A(p) > M$ has measure zero, but the set of points at which $P_A(p) > M - \epsilon$ has positive measure for any $\epsilon > 0$), and m the infimum of the values of $P_A(p)$ with the exception of a set of measure zero. It is necessary to prove that $M = m$. Suppose that $M \neq m$. Then the space R admits a decomposition into two invariant sets of positive measure

$$(P_A(p) \leqslant \alpha)^{5} \quad \text{and} \quad (P_A(p) > \alpha),$$

where $m < \alpha < M$, which contradicts the hypothesis that the system is indecomposable.

Readers wishing a more detailed acquaintance with the theory of dynamical systems, of integral invariants, or with the ergodic theorems, are referred to the following books and articles: V. V. Nemyckiĭ and V. V. Stepanov [**210**], A. Ja. Hinčin [**290-294**], and A. N. Kolmogorov [**110**].

5. **Stability of the solutions of differential equations.**

1. In order to make a mathematical description of any phenomenon it is necessary to simplify, to idealize the phenomenon being studied, selecting and taking into account only the most essential of the factors influencing it and rejecting other less essential factors.

In this connection the question as to how well the simplifying assumptions were chosen is inavoidable. It is possible that the simplification was done poorly, that the rejected factors strongly influence the phenomenon in question, significantly changing the quantitative or even the qualitative characteristics of that phenomenon.

Usually this question is resolved from the practical point of view by studying the relation of the solutions to the experimental data. Nevertheless, one should mention an important, though of course insufficient, criterion which must almost always be satisfied by any well chosen simplification of description of phenomena. As applied to phenomena described by any system of equations, this criterion comes down to the requirement that a small change (the meaning of smallness being determined by the nature of the problem in hand) in the equations can produce only a small change in the solutions.

If this condition for correct posing of problem is not satisfied, then small changes in the system of equations may give rise to considerable changes in the solutions, and since the system of equations is not set up exactly but only approximately, the solutions may turn out to be quite unsuitable even for an approximate description of the phenomena.

For the solution of the system of differential equations

$$\frac{dx_i}{dt} = f_i(t, x_1, x_2, \ldots, x_n) \qquad (i = 1, 2, \ldots, n) \tag{28}$$

[5] Here we retain the notations of Chapters I and II, thus $(P_A(p) \leqq \alpha)$ denotes the set of points $p \subset R$ where $P_A(p) \leqq \alpha$.

with the initial conditions $x_i(t_0) = x_{i0}$ $(i = 1, 2, \cdots, n)$ the question of the correctness of the posing of the problem comes down to a description of the conditions under which:

(a) a small change in the initial values gives rise to a small change in the solution on a finite or infinite interval of variation of the argument;

(b) a small change in the right sides of equations (28) corresponds to a small change in the solution on a finite or infinite interval of variation of t.

The answer to these questions with t varying on a finite interval $t_0 < t < T$ is given by well-known theorems about the continuous dependence of the solutions of the system (28) on the initial values and on the parameters (see, for example, [**226**]).

If now t varies on the infinite interval $t_0 \leqq t < \infty$, then another question arises, which is the subject of the theory of stability.

DEFINITION. *The solution* $\bar{x}_i(t)$ $(i = 1, 2, \cdots, n)$ *of the system* (28) *is said to be stable in the sense of A. M. Ljapunov if for every* $\epsilon > 0$ *there exists a* $\delta(\epsilon)$ *such that for arbitrary* x_{i0} *the inequality*

$$\sum_{i=1}^{n} [x_{i0} - \bar{x}_i(t_0)]^2 < \delta^2(\varepsilon)$$

implies the inequality

$$\sum_{i=1}^{n} [x_i(t) - \bar{x}_i(t)]^2 < \varepsilon^2 \quad \textit{for} \quad t \geqslant t_0,$$

where $x_i(t)$ $(i = 1, 2, \cdots, n)$ *is a solution of the system* (28) *satisfying the initial conditions* $x_i(t_0) = x_{i0}$ $(i = 1, 2, \cdots, n)$. In place of $t \geqq t_0$ we may write $t \geqq T$ under the conditions of the theorem about the continuous dependence of the solution on the initial conditions, and in place of

$$\sum_{i=1}^{n} [x_{i0} - \bar{x}_i(t_0)]^2 < \delta^2(\varepsilon) \text{ and } \sum_{i=1}^{n} [x_i(t) - \bar{x}_i(t)]^2 < \varepsilon^2$$

we may write

$$|x_{i0} - \bar{x}_i(t_0)| < \delta_1(\varepsilon)$$

and

$$|x_i(t) - x_i(t)| < \varepsilon \qquad (i = 1, 2, \ldots, n).$$

DEFINITION. *A solution* $\bar{x}_i(t)$ $(i = 1, 2, \cdots, n)$ *of the system* (28) *is said to be asymptotically stable if it is stable and if, moreover,*

$$\lim_{t \to \infty} \sum_{i=1}^{n} [x_i(t) - \bar{x}_i(t)]^2 = 0,$$

when the initial values, as in the definition of stability, satisfy the condition

$$\sum_{i=1}^{n} [x_{i0}(t) - \bar{x}_i(t_0)]^2 < \delta^2.$$

We note that the condition

$$\lim_{t\to\infty} \sum_{i=1}^{n} [x_i(t) - \bar{x}_i(t)]^2 = 0$$

for $\sum_{i=1}^{n} [x_{i0} - \bar{x}_i(t_0)]^2 < \delta^2$ can be satisfied even by unstable solutions of the system (28), for example by a stationary point of the type of a nodal region of instability (see Figure 20, page 84), although the condition

$$\lim_{t\to\infty} \sum_{i=1}^{n} [x_i(t) - \bar{x}_i(t)]^2 = 0$$

is satisfied by all solutions defined by initial points near the stationary point.

The question of stability of the solutions arises in almost every practical case that is defined by a system of differential equations. Only in rare cases is it possible to integrate the system in finite form, and accordingly the question of stability of the solution of a system of equations usually cannot be resolved by integrating the system, a fact which greatly complicates the problem. By a change of variables we can reduce the question of stability of a solution $y_i = \varphi_i(t)$ $(i = 1, 2, \cdots, n)$ of the system of equations

$$\frac{dy_i}{dt} = \Phi_i(t, y_1, y_2, \ldots, y_n) \qquad (i = 1, 2, \ldots, n)$$

to the question of the stability of the trivial solution, namely the case of a stationary point at the origin of coordinates.

In fact, putting $x_i = y_i - \varphi_i(t)$ $(i = 1, 2, \cdots, n)$ we obtain the system

$$\frac{dx_i}{dt} = \frac{d\varphi_i}{dt} + \Phi_i(t, x_1 + \varphi_1(t), x_2 + \varphi_2(t), \ldots, x_n + \varphi_n(t))$$
$$(i = 1, 2, \ldots, n),$$

whose solution $x_i \equiv 0$ $(i = 1, 2, \cdots, n)$ corresponds to the solution $y_i = \varphi_i(t)$ $(i = 1, 2, \cdots, n)$ of the original equations. Let us consider some examples:

(1) The stationary points of the dynamical system

$$\frac{dx_i}{dt} = f_i(x_1, x_2) \qquad (i = 1, 2),$$

considered in §1 of this chapter have the following properties with respect to stability: the stable node and the stable focus are asymptotically stable in the sense of A. M. Ljapunov: the center is stable, but not asymptotically stable; the unstable node, the unstable focus, the saddle point, the nodal region, the multiple saddle points, and the multiple nodes are not stable in the sense of A. M. Ljapunov.

(2) A stationary point of the dynamical system

$$\frac{dx_i}{dt} = f_i(x_1, x_2, \ldots, x_n) \qquad (i = 1, 2, \ldots, n).$$

of the type of a k-saddle (see page 82) is unstable for $k < n$ and asymptotically stable for $k = n$.

2. *The second method of Ljapunov.* To study the stability of the stationary point $x_i \equiv 0$ $(i = 1, 2, \cdots, n)$ of the system of differential equations

$$\frac{dx_i}{dt} = f_i(t, x_1, x_2, \ldots, x_n) \qquad (i = 1, 2, \ldots, n).$$

A. M. Ljapunov [**152**], in 1892 applied a very general method based on the following theorem and its generalizations:

If there exists a differentiable function $V(x_1, x_2, \cdots, x_n)$ (called the Ljapunov function) attaining a strict minimum at the origin of coordinates and having in the neighborhood of the origin a nonpositive derivative with respect to t along every integral curve of the system of differential equations in question, then the stationary point $x_i \equiv 0$ $(i = 1, 2, \cdots, n)$ is stable.

By adding a constant to the function V, if necessary, we may suppose without loss of generality that $V(0, 0, \cdots, 0) = 0$. We shall confine ourselves to a purely geometric interpretation of the proof, which can be translated without difficulty into the language of analysis. In the neighborhood of the minimum point $x_i = 0$ $(i = 1, 2, \cdots, n)$ the level surfaces $V = c$ form a family of closed surfaces containing the origin of coordinates in their interiors.

In virtue of the continuity of the function V we may choose a $\bar{c} > 0$ so small that the level surface $V = \bar{c}$, or at least one of its components, lies wholly in an ϵ-neighborhood of the origin of coordinates.

We choose a δ-neighborhood of the origin such that in it the function V satisfies the inequality $V < \bar{c}$. Then any arbitrary solution $x_i = x_i(t)$ $(i = 1, 2, \cdots, n)$ defined by an initial value $x_i(t_0) = x_{i0}$ lying in the δ-neighborhood of the origin for $t \geqq t_0$, cannot go outside the boundary of the ϵ-neighborhood of the origin nor even outside the level surface $V = \bar{c}$. For, if the trajectory went outside the surface $V = \bar{c}$, then in some part of the range of variation of t the function $V(x_1, x_2, \cdots, x_n)$ would have to increase along the trajectory, which contradicts the assumption that $dV/dt \leqq 0$ along the integral curves.

EXAMPLE 1. Investigate the stability of the stationary point $x \equiv y \equiv 0$ of the system of equations

$$\frac{dx}{dt} = y - x^3,$$
$$\frac{dy}{dt} = -x - y^3.$$

The function $V = x^2 + y^2$ satisfies the conditions of Ljapunov's theorem: $V \geqq 0$, $dV/dt = 2(x\,dx/dt + y\,dy/dt)$ and, along the integral curves of the system we have, by the defining equations,

$$\frac{dV}{dt} = 2\,[x(y - x^5) + y(-x - y^3)] = -(x^6 + y^4) \leqslant 0.$$

Accordingly, the stationary point $x \equiv 0$, $y \equiv 0$ is stable.

EXAMPLE 2. Investigate the stability of the stationary point $x_i \equiv 0$ $(i = 1, 2, \cdots, n)$ of the system of equations

$$\frac{dx_i}{dt} = \frac{\partial U}{\partial x_i} \qquad (i = 1,\ 2,\ \ldots,\ n),$$

where the function $U(x_1, x_2, \cdots, x_n)$ has a strict maximum at the origin of coordinates.

The Ljapunov function may be taken as $V = -U(x_1, x_2, \cdots, x_n)$. For, the function V has a strict minimum at the origin of coordinates and

$$\frac{dV}{dt} = -\sum_{i=1}^{n} \frac{\partial U}{\partial x_i} \frac{dx_i}{dt} = -\sum_{i=1}^{n} \left(\frac{\partial U}{\partial x_i}\right)^2 \leqslant 0.$$

Accordingly, the stationary point is stable.

EXAMPLE 3. Investigate the stability of the stationary point $x_i \equiv 0$ $(i = 1, 2, \cdots, n)$ of the system of equations

$$\frac{dx_i}{dt} = \sum_{j=1}^{n} a_{ij}(t)\, x_j \qquad (i = 1,\ 2,\ \ldots,\ n),$$

where the continuous functions $a_{ij}(t)$ satisfy the condition

$$a_{ij}(t) = -a_{ji}(t) \ \text{ for } i \neq j \text{ and } a_{ii}(t) \leqslant 0$$
$$(i,\ j = 1,\ 2,\ \ldots,\ n).$$

As the Ljapunov function we may take $V = \sum_{i=1}^{n} x_i^2$, which has a minimum at the origin of coordinates, and satisfies

$$\frac{dV}{dt} = 2\sum_{i=1}^{n} x_i \frac{dx_i}{dt} = 2\sum_{i=1}^{n} a_{ii}(t)\, x_i^2 \leqslant 0.$$

Accordingly, the stationary point $x_i \equiv 0$ $(i = 1, 2, \cdots, n)$ is stable.

Let us now prove the fundamental theorem of A. M. Ljapunov's second method on the stability or instability of stationary points.

THEOREM OF A. M. LJAPUNOV ON STABILITY. *A stationary point $x_i \equiv 0$ of a system of differential equations $dx_i/dt = f_i(t, x_1, x_2, \cdots, x_n)$ $(i = 1, 2, \cdots, n)$ is stable if there exists a differentiable function $V(t, x_1, x_2, \cdots, x_n)$ satisfying the following conditions:*

(1) *$V(t, x_1, x_2, \cdots, x_n) \geqq W(x_1, x_2, \cdots, x_n)$ in the neighborhood of the stationary point $x_i \equiv 0$ $(i = 1, 2, \cdots, n)$ for arbitrary $t \geqq T$, where W is a continuous function, not depending on t, which satisfies $W \geqq 0$ in the same neighborhood and has a strict minimum at the origin of coordinates, $W(0, 0, \cdots, 0) = 0$.*

(2) *Along the integral curves the derivative of V satisfies*

$$\frac{dV}{dt} \leqslant 0 \quad \left(\frac{dV}{dt} = \frac{\partial V}{\partial t} + \sum_{i=1}^{n} \frac{\partial V}{\partial x_i} f_i\right)$$

in the neighborhood of the origin of coordinates for arbitrary $t \geqq T$.

PROOF. The inequality $W \leqq V$ implies that the closed level surface $W = C$ contains within it (in the space of the $x_1, x_2, \cdots, x_n$) the level surface $V(t, x_1, x_2, \cdots, x_n) = C$, for arbitrary values of $t \geqq T$. By the continuity of the function W we may choose $C_0 > 0$ so small that the level surface $W = C_0$ or one of its components is contained in an arbitrarily small ϵ-neighborhood of the origin of coordinates. Since $C_0 > 0$, there exists a sufficiently small δ-neighborhood of the origin of coordinates, $\delta > 0$, lying entirely inside the closed level surface $W = C_0$, and such that at all its points we have $V(T_0, x_1, x_2, \cdots, x_n) < C_0$, where T_0 is some fixed value of t, greater than or equal to T (Figure 43). The trajectory defined by the initial point P_0, corresponding to the value $t = T_0 \geqq T$, lying in the δ-neighborhood of the origin of coordinates cannot, for $t \geqq T_0$, exit from the ϵ-neighborhood of the origin of coordinates, nor even from the level surface $W = C_0$, since $dV/dt \leqq 0$, and therefore the Ljapunov function V does not increase along any integral curve and in the neighborhood of the origin the whole level surface $V = C$, with $C < C_0$, lies within the surface $W = C_0$.

THEOREM OF A. M. LJAPUNOV ON ASYMPTOTIC STABILITY. *If the conditions of the preceding theorem are satisfied, and if, in the neighborhood of the origin of coordinates and for $t \geqq t_0$, the Ljapunov function $V(t, x_1, x_2, \cdots, x_n)$ satisfies, the following conditions:*

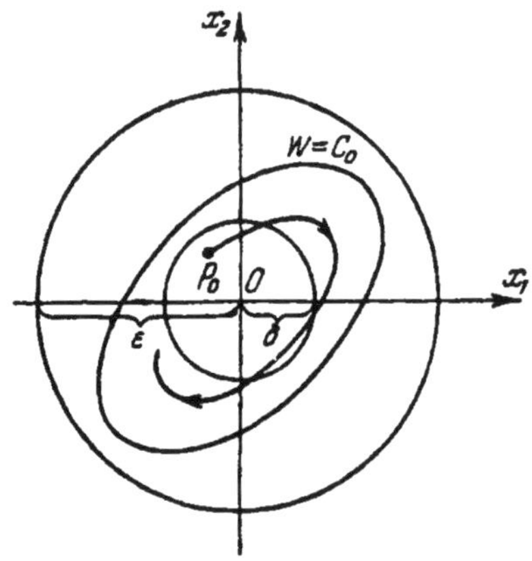

Figure 43

(a) *the derivative along the integral curves satisfies*

$$\frac{dV}{dt} = -W_1(x_1, x_2, \ldots, x_n) \leqslant 0,$$

where $W(x_1, x_2, \cdots, x_n)$ is a continuous function vanishing only at the origin;

(b) *the function V admits an infinitesimal upper limit, that is, V tends, uniformly in t, to the limit zero as $\sum_{i=1}^{n} x_i^2 \to 0$;*
then the stationary point $x_i \equiv 0$ $(i = 1, 2, \cdots, n)$ of the system (28) *is asymptotically stable.*

PROOF. Since the conditions of the preceding theorem are satisfied, every solution $x_i(t)$ $(i = 1, 2, \cdots, n)$ for which the initial value lies in a $\delta(\epsilon)$-neighborhood of the origin of coordinates never leaves the ϵ-neighborhood of the stationary point, and therefore along the integral curves the inequality (a) is satisfied for $t \geqq t_0$; accordingly, the limit $\lim_{t\to\infty} V(t, x_1(t), x_2(t), \cdots, x_n(t)) = \alpha$ exists and

$$V(t, x_1(t), x_2(t), \ldots, x_n(t)) \geqslant \alpha \quad \text{for} \quad t \geqslant t_0.$$

We shall show that $\alpha = 0$, whence in view of property (1) of the preceding theorem it follows that $\lim_{t\to\infty} x_i(t) = 0$ $(i = 1, 2, \cdots, n)$, that is, that the stationary point $x_i \equiv 0$ $(i = 1, 2, \cdots, n)$ is asymptotically stable.

Let us suppose that $\alpha > 0$; then by (b) every trajectory beginning at some t_0 lies within some neighborhood of the stationary point and thus by inequality (a) we have, for all $t \geqq t_0$, along the trajectory,

$$\frac{dV}{dt} \leqslant -l_1 < 0.$$

Integrating this inequality along the trajectory, we obtain

$$V(t, x_1(t), x_2(t), \ldots, x_n(t)) - V(\bar{t}_0, x_1(\bar{t}_0), x_2(\bar{t}_0), \ldots, x_n(\bar{t}_0)) \leqslant -l_1(t-\bar{t}_0), \quad (29)$$

which leads to a contradiction, since from the inequality (29) there follows, for sufficiently large t, the inequality $V(t, x_1(t), \cdots, x_n(t)) < 0$, contrary to our hypothesis.

THEOREM OF N. G. ČETAEV ON INSTABILITY. *If there exists a differentiable function $V(t, x_1, x_2, \cdots, x_n)$ defined in the neighborhood of the origin of coordinates and satisfying the following conditions:*

(1) *In an arbitrarily small neighborhood of the origin of coordinates there exists a region $(V > 0)$ independent of t in which $V > 0$ for $t \geqq t_0$;*

(2) *in the region $(V > 0)$ the function V is bounded;*

(3) *in the region $(V > 0)$ for $t \geqq t_0$ the derivative along the integral curves satisfies $dV/dt > 0$, and moreover in the region $V \geqq \alpha > 0$ we have $dV/dt \geqq \sigma > 0$, where α and σ are constants;*

then the stationary point $x_i \equiv 0$ $(i = 1, 2, \cdots, n)$ of the system (28) *is unstable.*

PROOF. We choose an initial point $(x_{10}, x_{20}, \cdots, x_{n0})$ in an arbitrarily small neighborhood of the origin of coordinates within the region $(V > 0)$ for $t \geqq t_0$. Suppose that $V(t_0, x_{10}, x_{20}, \cdots, x_{n0}) = \alpha > 0$. By condition (2) the function V increases along the trajectory with increasing t and therefore the trajectory remains in the region $(V \geqq \alpha)$, where $dV/dt \geqq \sigma > 0$. Integrating this inequality along the trajectory we obtain: $V(t, x_1(t), x_2(t), \cdots, x_n(t)) - V(t_0, x_1(t_0), \cdots, x_n(t_0)) \geqq \sigma(t - t_0)$ or $V \geqq V_0 + \sigma(t - t_0)$, whence it is obvious that V increases without limit along the trajectory with increasing t and accordingly leaves the neighborhood of the origin of coordinates, where by hypothesis the function V is bounded.

3. *Converse of Ljapunov's Theorem.* The fundamental theorems on stability and asymptotic stability admit converses under certain restrictions. Therefore, under these restrictions, the second method of Ljapunov may be thought of as a general method for investigating stability.

K. P. Persidskiĭ [221] proved the possibility of inverting the Ljapunov theorem on stability. He showed that if the system (28) has a stable stationary point at the origin of coordinates, there exists a function V satisfying the conditions of the Ljapunov stability theorem.

A more complete converse of the Ljapunov theorem on asymptotic stability (page 126) was recently given by I. G. Malkin [171]. The possibility of inverting the theorems of A. M. Ljapunov and N. G. Četaev on instability of motion in the stationary case has been investigated in a paper by N. N. Krasovskiĭ [118].

4. *Generalized Ljapunov functions.* One frequently encounters the problem of estimating the error that arises when the system of equations

$$\frac{dx_i}{dt} = f_i(t, x_1, x_2, \ldots, x_n), \; x_i(t_0) = x_{i0} \qquad (i = 1, 2, \ldots, n) \quad (30)$$

is replaced by the system

$$\frac{d\bar{x}_i}{dt} = F_i(t, \bar{x}_1, \bar{x}_2, \ldots, \bar{x}_n), \; \bar{x}_i(t_0) = \bar{x}_{i0} \quad (i = 1, 2, \ldots, n), \quad (31)$$

in which the functions f_i are in some sense close to the functions F_i.

For this purpose it will often be useful to consider a function $V(t, x_1, x_2, \cdots, x_n, \bar{x}_1, \bar{x}_2, \cdots, \bar{x}_n)$ of $(2n+1)$ variables (for example, the square of the distance ρ^2 between the points $(x_1, x_2, \cdots, x_n)$ and $(\bar{x}_1, \bar{x}_2, \cdots, \bar{x}_n)$

$$\rho^2(x_1, x_2, \ldots, x_n, \bar{x}_1, \bar{x}_2, \ldots, \bar{x}_n) = \sum_{i=1}^{n} (x_i - \bar{x}_i)^2),$$

and if the decrease of the function V implies in some sense the coming together of the points $(x_1, x_2, \cdots, x_n)$ and $(\bar{x}_1, \bar{x}_2, \cdots, \bar{x}_n)$, it is worthwhile to calculate the derivative dV/dt along the integral curves; that is, to calculate the derivative under the assumption that the arguments x_i $(i = 1, 2, \cdots, n)$ of the function V have been replaced by the solution $x_i(t)$ $(i = 1, 2, \cdots, n)$ of the system (30) and the arguments $\bar{x}_i$ $(i = 1, 2, \cdots, n)$ have been replaced by the solution $\bar{x}_i(t)$ $(i = 1, 2, \cdots, n)$ of the system (31):

$$\frac{dV}{dt} = \frac{\partial V}{\partial t} + \sum_{i=1}^{n} \frac{\partial V}{\partial x_i} f_i + \sum_{i=1}^{n} \frac{\partial V}{\partial \bar{x}_i} F_i,$$

and if this derivative turns out to be negative, the function V decreases along the integral curves with increasing t and therefore the integral curves of the systems (30) and (31) are close in the above sense.

We have already used a similar method in §2, Chapter IV to study the behavior of solutions of systems of equations with small coefficients of the derivative.

5. *Study of first-order stability.* The method of studying the stability of the stationary point $x_i \equiv 0$ $(i = 1, 2, \cdots, n)$ of the system of differential equations (28) to a first approximation consists in supposing the functions f_i to be differentiable in $x_1, x_2, \cdots, x_n$ in the neighborhood of the origin of coordinates $x_i = 0$ $(i = 1, 2, \cdots, n)$, and thus representing the system (28) in the form

$$\frac{dx_i}{dt} = \sum_{j=1}^{n} a_{ij}(t)\, x_j + R_i(t, x_1, x_2, \ldots, x_n) \tag{32}$$

$$(i = 1, 2, \ldots, n)$$

and then eliminating the terms R_i, which are of order higher than the first in $\sqrt{\sum_{i=1}^{n} x_i^2}$. We study the stability of the stationary point $x_i \equiv 0$ $(i = 1, 2, \cdots, n)$ of the linear system

$$\frac{dx_i}{dt} = \sum_{j=1}^{n} a_{ij}(t)\, x_j \qquad (i = 1, 2, \ldots, n), \tag{33}$$

which we call the first-order approximation to the system (32).

The conditions of applicability of this method, which was used for a long time without any foundation, were investigated in detail by A. M. Ljapunov, and have been considered recently by O. Perron [218], I. G. Malkin [159; 160; 165; 169], K. P. Persidskiĭ [224], N. G. Četaev [313], and others.

6. *Systems stationary to the first order.* If the system (33) is stationary, that is, if all the coefficients a_{ij} are constants, the system (32) is said to be stationary to the first order.

The following theorem states the conditions for application of the method of first approximation for the study of stability in this case, which is extremely important for applied work:

(1) *If all roots of the characteristic equation*

$$\begin{vmatrix} a_{11}-k & a_{12} & \cdots & a_{1n} \\ a_{21} & a_{22}-k & \cdots & a_{2n} \\ \cdot & \cdot & \cdot & \cdot \\ \cdot & \cdot & \cdot & \cdot \\ a_{n1} & a_{n2} & \cdots & a_{nn}-k \end{vmatrix} = 0$$

of the system of first-order approximations have their real part negative, the stationary point $x_i = 0$ $(i = 1, 2, \cdots, n)$ *of the systems* (33) *and* (32) *is asymptotically stable*;

(2) *if at least one root of the characteristic equation has a positive real part, the stationary point* $x_i = 0$ $(i = 1, 2, \cdots, n)$ *of the systems* (33) *and* (32) *is unstable*.

It is assumed here that in a sufficiently small neighborhood of the origin of coordinates and for $t \geqq T$ we have $|R_i| \leqq N(\sum_{i=1}^n x_i^2)^{(1/2)+\alpha}$, where $\alpha > 0$, and N is a constant.

These two conditions of applicability of the method of first-order approximation do not cover the so-called critical case, in which the real parts of all the roots of the characteristic equation are nonpositive and the real part of at least one root is zero.

In the critical case the stability of the trivial solution of the system (32) is in general influenced by the nonlinear terms R_i, and therefore the method of first-order approximation is in general not applicable.

In most cases the applicability of the method of first-order approximations for the study of stability is most easily determined by the use of Ljapunov's second method, since in the simplest cases the Ljapunov function can be taken as some quadratic form or other.

Suppose, for example, that all roots of the characteristic equation of the first-order approximation (33) to the system (32) are negative and distinct. We write the stationary system (33) in vector form:

$$\frac{dX}{dt} = AX,$$

and the system (32) in the form

$$\frac{dX}{dt} = AX + R,$$

where

$$X = \begin{Vmatrix} x_1 \\ x_2 \\ \cdot \\ \cdot \\ x_n \end{Vmatrix}, \quad A = \begin{Vmatrix} a_{11} & a_{12} & \cdots & a_{1n} \\ a_{21} & a_{22} & \cdots & a_{2n} \\ \cdot & \cdot & \cdots & \cdot \\ \cdot & \cdot & \cdots & \cdot \\ a_{n-1} & a_{n2} & \cdots & a_{nn} \end{Vmatrix}, \quad R = \begin{Vmatrix} R_1 \\ R_2 \\ \cdot \\ \cdot \\ R_n \end{Vmatrix}.$$

By means of a nondegenerate linear transformation with constant coefficients

$$X = BY, \text{where } B = \begin{Vmatrix} b_{11} & b_{12} & \cdots & b_{1n} \\ b_{21} & b_{22} & \cdots & b_{2n} \\ \cdot & \cdot & \cdot & \cdot \\ \cdot & \cdot & \cdot & \cdot \\ b_{n1} & b_{n2} & \cdots & b_{nn} \end{Vmatrix}, \quad Y = \begin{Vmatrix} y_1 \\ y_2 \\ \cdot \\ \cdot \\ y_n \end{Vmatrix},$$

the system (33) is transformed into $dY/dt = B^{-1}YB$; we choose the matrix B in such a way that the matrix $B^{-1}AB$ is diagonalized

$$B^{-1}AB = \begin{Vmatrix} k_1 & 0 & \cdots & 0 \\ 0 & k_2 & \cdots & 0 \\ \cdot & \cdot & \cdot & \cdot \\ \cdot & \cdot & \cdot & \cdot \\ 0 & 0 & \cdots & k_n \end{Vmatrix}.$$

Accordingly, the system (32) is converted into $dy_i/dt = k_i y_i$ $(i = 1, 2, \cdots, n)$, and the system (33) into $dy_i/dt = k_i y_i + \bar{R}_i(t, y_1, y_2, \cdots, y_n)$ $(i = 1, 2, \cdots, n)$, where $|\bar{R}_i| \leqq \bar{N}(\sum_{i=1}^n y_i^2)^{1/2+\alpha}$, $\bar{N}$ is a constant, and $\alpha > 0$, $t \geqq T$, $\sum_{i=1}^n y_i^2 < \delta_1^2$. The Ljapunov function for this system may be taken as $V = \sum_{i=1}^n y_i^2$. For along the integral curves we find that the derivative satisfies

$$\frac{dV}{dt} = 2\sum_{i=1}^n y_i \frac{dy_i}{dt} = 2\sum_{i=1}^n k_i y_i^2 + 2\sum_{i=1}^n k_i y_i \bar{R}_i \leqslant \sum_{i=1}^n k_i y_i^2 \leqslant 0$$

if the y_i are small enough in modulus, since all $k_i < 0$ and the sum $2\sum_{i=1}^n k_i y_i \bar{R}_i$, for sufficiently small y_i, is less in absolute value than the absolute value of the sum $\sum_{i=1}^n k_i y_i^2$; therefore the conditions of Ljapunov's theorem on asymptotic stability are fulfilled for the stationary point $y_i \equiv 0$ $(i = 1, 2, \cdots, n)$ (page 126).[6]

7. *The case of nonstationary first-order approximation.* The *characteristic number* $l(\varphi)$ of a function $\varphi(t)$ is defined as[7] $\overline{\lim}_{t\to\infty} (\ln|\varphi(t)|/t)$. For example, the characteristic number of the exponential function e^{pt} is equal to p, and the same value p serves as the characteristic number for the functions

---

[6] For the conditions of applicability of the method of first-order approximation in the case cited above (page 130), and for a detailed study of the fundamental critical cases, see the book by I. G. Malkin [170].

[7] Sometimes the characteristic number of a function $\varphi(t)$ is defined as $-l(\varphi)$.

$$e^{pt}\sin qt,\ e^{pt}\cos qt,\ t^n e^{pt},\ t^n e^{pt}\sin qt,\ t^n e^{pt}\cos qt.$$

The characteristic number of a solution $x_i(t)$ $(i = 1, 2, \cdots, n)$ of a system of equations is the largest of the characteristic numbers of the individual functions $x_i(t)$ $(i = 1, 2, \cdots, n)$.

For systems of linear homogeneous equations with variable coefficients the characteristic numbers of the solutions forming a fundamental system play the same role as the real parts of the roots of the characteristic equation for a system with constant coefficients, and in fact, one can prove the following theorem due to A. M. Ljapunov: *if the system of first-order approximations* (33) *is regular and if all its characteristic numbers are negative, then the trivial solution of the system* (33) *and* (32), *in which the functions* R_i *satisfy the conditions stated on page* 130, *is asymptotically stable.*

A few concepts are needed to clarify the formulation of this theorem.

A fundamental system of solutions of a linear homogeneous system of equations is said to be normal if the characteristic number of an arbitrary linear combination of the solutions belonging to the fundamental system coincides with the characteristic number of one of the solutions entering the linear combination in question.

The characteristic numbers of a linear homogeneous system of equations are defined as the characteristic numbers of a normal fundamental system of solutions.

A system is said to be regular, if the sum of its characteristic numbers is equal to the characteristic number of the function $\exp \int_{t_0}^{t}\sum_{i=1}^{n} a_{ii}(t)dt$ and if moreover the sum of its characteristic numbers of the two functions $\exp \int_{t_0}^{t}\sum_{i=1}^{n} a_{ii}(t)dt$ and $\exp\{-\int_{t_0}^{t}\sum_{i=1}^{n} a_{ii}(t)dt\}$ is equal to zero.

Regular systems form a very wide class among systems of linear homogeneous equations; in particular, all systems with periodic coefficients belong to this class.[8]

8. *Criteria for the real parts of all roots of a polynomial to be negative.* For the system of equations (32), stationary to the first order, a sufficient condition that the trivial solution be stable is, as shown above, that all roots of the characteristic polynomial of the first-order approximation system have negative real parts.

If the degree of the characteristic polynomial is high, the actual determination of the roots is a most complicated task; thus it is very important to find tests for negativeness of the real parts of all roots of a polynomial.

---

[8] A detailed exposition of the conditions for applicability of the method of first-order approximations for the study of stability in the case of nonstationary systems of equations of first-order approximation is given in the text by I. G. Malkin [170].

The best known test is due to Hurwitz: *in order that a polynomial* $f(z) = z^n + a_1 z^{n-1} + \cdots + a_{n-1}z + a_n$ *with real coefficients should have the real part of all its roots negative, it is necessary and sufficient that all principal diagonal minors* Δ_i *be positive in the Hurwitz matrix*

$$\begin{Vmatrix} a_1 & 1 & 0 & 0 & \ldots & 0 \\ a_3 & a_2 & a_1 & 1 & \ldots & \ldots \\ a_5 & a_4 & a_3 & a_2 & \ldots & \ldots \\ a_7 & a_6 & a_5 & a_4 & \ldots & \ldots \\ \cdot & \cdot & \cdot & \cdot & \cdot & \cdot \\ \cdot & \cdot & \cdot & \cdot & \cdot & \cdot \\ 0 & 0 & 0 & 0 & \ldots & a_n \end{Vmatrix}$$

$$\Delta_1 = a_1 > 0; \quad \Delta_2 = \begin{vmatrix} a_1 & 1 \\ a_3 & a_2 \end{vmatrix} > 0; \quad \Delta_3 = \begin{vmatrix} a_1 & 1 & 0 \\ a_3 & a_2 & a_1 \\ a_5 & a_4 & a_3 \end{vmatrix} > 0; \ldots$$

$$\ldots; \Delta_{n-1} > 0; \quad \Delta_n > 0,$$

where the last condition $\Delta_n > 0$ *can be replaced by the condition* $a_n > 0$, *since*

$$a_n > 0,$$
$$\Delta_n = \Delta_{n-1} \cdot a_n.$$

For example, a necessary and sufficient condition that the polynomial

$$z^3 + a_1 z^2 + a_2 z + a_3$$

have only roots with negative real part is that $a_1 > 0$, $a_1 \cdot a_2 - a_3 > 0$, and $a_3 > 0$.

Although the Hurwitz theorem also has a qualitative character, we shall not pause to prove it; the reader can find a proof in many texts on higher algebra.[9]

The Hurwitz test is excellent if the degree of the polynomial in question is not too high; when it is high, however, a more useful test is that of Čebotarev and Maĭman [178], based on the properties of a series which is analogous to the Sturm series, or else one of the various tests based on the principle of the argument (the amplitude-phase method, of Mihaĭlov [183; 200], and [178], etc.).

9. *Stability under a constantly operating perturbation*. If the system of equations

$$\frac{dx_i}{dt} = f_i(t, x_1, x_2, \ldots, x_n), \; x_i(t_0) = x_{i0} \tag{34}$$
$$(i = 1, 2, \ldots, n)$$

---

[9] See for example A. G. Kuroš, Kurs vysšeĭ algebry, [A course of higher algebra.] 4th ed. GITTL, Moscow, 1955. (Russian)

is subjected to a perturbation acting for a very short time (a so-called momentary perturbation), the system (34) must, during the short interval $\bar{t}_0 \leqq t \leqq \bar{\bar{t}}_0$ of variation of t, be replaced by the perturbed system

$$\frac{dx_i}{dt} = f_i(t, x_1, x_2, \ldots, x_n) + R_i(t, x_1, x_2, \ldots, x_n) \qquad (35)$$
$$(i = 1, 2, \ldots, n),$$

where all the R_i are small in absolute value. At the end of the perturbation, i.e., for $t \geqq \bar{\bar{t}}_0$ we return to the system (34); however, in view of the theorem on the continuous dependence of the solution on the parameters, the initial value on return, at the point $\bar{\bar{t}}_0$, is no longer $x_i(\bar{\bar{t}}_0)$, where $x_i(t)$ $(i = 1, 2, \cdots, n)$ is the solution of (34), but is instead $x_i(\bar{\bar{t}}_0) + \delta_i$ $(i = 1, 2, \cdots, n)$, where all the δ_i are small for small $|R_i|$ (see Figure 44).

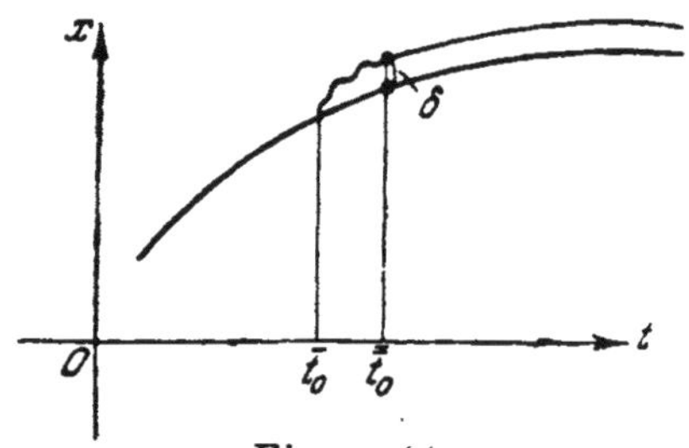

Figure 44

Thus, the effect of a momentary perturbation reduces, in the final analysis, to a perturbation of the initial value, and the question of stability under momentary perturbations reduces to the problem, considered above, of stability in the Ljapunov sense.

But if the perturbation is constantly at work, the system (34) must be replaced by (35) for all $t \geqq t_0$ and an entirely new question arises. The problem of stability under constantly operating perturbations has been studied by I. G. Malkin [**163**] and G. N. Dubošin [**93**]. Just as in the study of stability in the Ljapunov sense, a change of variables may be introduced to reduce the question of stability of a solution under constantly acting perturbation to the question of stability of the trivial solution $x_i \equiv 0$ $(i = 1, 2, \cdots, n)$ of the system (34).

DEFINITION. The trivial solution of the system (34) is said to be *stable under constantly acting perturbations* if for every $\epsilon > 0$ we can find numbers δ_1 and δ_2 such that the inequalities

$$\sum_{i=1}^{n} R_i^2 < \delta_1^2 \quad \text{for} \quad t \geqslant t_0 \text{ and } \sum_{i=1}^{n} x_{i0}^2 < \delta_2^2$$

imply that $\sum_{i=1}^{n} x_i^2(t) < \epsilon^2$ for $t \geqq t_0$, where $x_i(t)$ $(i = 1, 2, \cdots, n)$ is a solution of the system (35) satisfying the initial conditions $x_i(t_0) = x_{i0}$ $(i = 1, 2, \cdots, n)$.

THEOREM OF I. G. MALKIN. *If for the system of equations* (34) *there exists a differentiable Ljapunov function* $V(t, x_1, x_2, \cdots, x_n)$ *satisfying in the neighborhood of the origin of coordinates and for* $t \geqq t_0$ *the conditions:*

(1) $V(t, x_1, x_2, \cdots, x_n) \geqq W_1(x_1, x_2, \cdots, x_n) \geqq 0$; $V(t, 0, 0, \cdots, 0) = 0$;

(2) *the derivatives* $\partial V/\partial x_i$ *are bounded;*

(3)
$$\frac{dV}{dt} = \frac{\partial V}{\partial t} + \sum_{i=1}^{n} \frac{\partial V}{\partial x_i} f_i \leqslant - W_2(x_1, x_2, \ldots, x_n) \leqslant 0,$$

where the continuous functions W_1 *and* W_2 *have a strict minimum at the origin of coordinates, then the trivial solution of the system* (34) *is stable with respect to constantly operating perturbations.*

PROOF. We note that in view of the boundedness of the derivatives $\partial V/\partial x_i$ $(i = 1, 2, \cdots, n)$ the function V has an infinitesimal upper limit (that is, uniformly in t for $t \geqq t_0$, V tends to zero when $\sum_{i=1}^{n} x_i^2 \to 0$), since by the mean value theorem

$$V(t, x_1, x_2, \ldots, x_n) = \sum_{i=1}^{n} \left(\frac{\partial V}{\partial x_i}\right) x_i,$$

where the $(\partial V/\partial x_i)$ are evaluated at some point intermediate between 0 and x_i $(i = 1, 2, \cdots, n)$. We also remark that the conditions (2) and (3) imply that within some δ-neighborhood of the origin of coordinates the derivatives satisfy

$$\frac{dV}{dt} = \frac{\partial V}{\partial t} + \sum_{i=1}^{n} \frac{\partial V}{\partial x_i} f_i + \sum_{i=1}^{n} \frac{\partial V}{\partial x_i} R_i \leqslant -k < 0$$

when the R_i are sufficiently small in absolute value.

We prescribe $\epsilon > 0$ and choose some level surface $W_1 = l$, one component of which lies wholly within the ϵ-neighborhood of the origin of coordinates. The level surface $V = l$ moves with varying $t \geqq t_0$ so that it (or one of its components) lies wholly within the level surface $W_1 = l$, because of condition (1), and at the same time, because there exists an infinitesimal upper limit, it lies within some δ-neighborhood of the origin of coordinates. Accordingly, on the level surface $V(t, x_1, x_2, \cdots, x_n) = l$ and for arbitrary $t \geqq t_0$ the derivative satisfies

$$\frac{dV}{dt} = \frac{\partial V}{\partial t} + \sum_{i=1}^{n} \frac{\partial V}{\partial x_i}(f_i + R_i) \leqslant -k < 0 \quad \text{if} \quad \sum_{i=1}^{n} R_i^2 < \delta_1^2.$$

where δ_1 is sufficiently small. In view of the existence of an infinitesimal upper limit, we may choose a δ_2-neighborhood so small that within it $V < l$.

The trajectories of points lying within this δ_2-neighborhood at $t = t_0$ cannot leave the ϵ-neighborhood of the origin of coordinates for $t \geqq t_0$, since in view of the choice of δ_2 we have $V(t_0, x_{10}, x_{20}, \cdots, x_{n0}) < l$ and if for $t \geqq t_0$ the trajectory were to leave the ϵ-neighborhood, or even go outside the surface $W_1 = l$, it would have to intersect the level surface $V(t, x_1, x_2, \cdots, x_n) = l$ for some value of t and in the neighborhood of the point of intersection the function V would have to increase along the trajectory; but this contradicts the condition $dV/dt \leqq -k < 0$ along the trajectory at the points of the level surface $V(t, x_1, x_2, \cdots, x_n) = l$.

6. **Periodic solutions.**

1. The problem of finding and studying periodic solutions of systems of differential equations arises very often in applied mathematics; the difficulties standing in the way of a solution are so great, however, that notwithstanding the efforts of the greatest mathematicians, no methods that are both general and effective for the solution of this problem have up to now been found.

In the two-dimensional case a sufficiently general, though still incomplete, method exists for finding periodic solutions of the dynamical system

$$\frac{dx_1}{dt} = f_1(x_1, x_2); \quad \frac{dx_2}{dt} = f_2(x_1, x_2), \tag{36}$$

defined on the Euclidean plane or on the sphere; this is the method of annular regions, already mentioned in §3 of this chapter. The method comes down to the following: one tries to find an annular region (the topological product of a circle and a segment) such that in it and on its boundary there are no stationary points and such that on its boundary the vector field of directions defined by (36) is everywhere directed toward the interior of the annulus (or everywhere in the opposite direction) (see Figure 45). By Theorem II, §3, such an annular region contains at least one periodic trajectory, since the ω-limiting (or α-limiting) set of an arbitrary trajectory entering the annulus is a periodic trajectory by virtue of the theorem cited.

Having found such an annulus, one attempts to constrict it while preserving the properties described above and so to define more and more precisely the position of the periodic trajectory known to lie within the annulus. If one has no clue whatever as to the possible positions of the periodic trajectories, this method is clearly quite useless and it is also laborious in view of the difficulties in finding annular regions satisfying the conditions stated above.

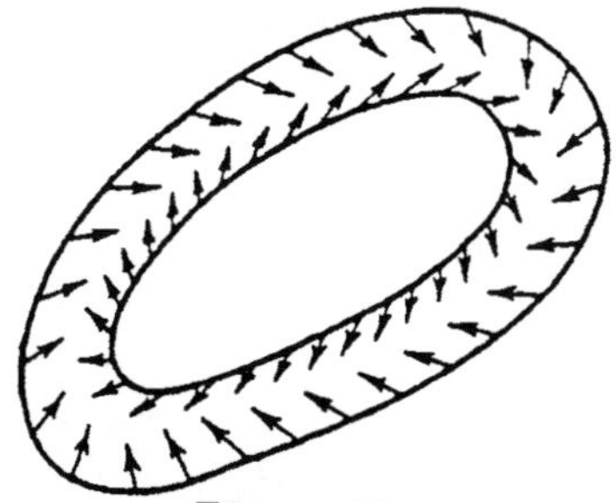

Figure 45

But if the physical, mechanical, or other considerations arising in a given practical case allow us to define a relatively small region in which the desired periodic trajectories may lie, the use of this method is greatly facilitated, and it becomes quite effective.[10]

Aside from these remarks, in the general case we may cite several tests for the presence or absence of periodic solutions of the system (36) in regions of different types:

(1) *If in a simply-connected region there is no stationary point, there are no periodic trajectories either*;

(2) *if in a simply-connected region all the stationary points have indices such that no combination of them adds up to unity (for example, if the indices are all negative or all even), then the region contains no periodic trajectories*;

(3) *if a simply-connected region contains only one stationary point, and if its index is unity (or only one combination of stationary points whose indices add up to unity), but on the other hand this stationary point (or one of the points of the combination whose indices add up to unity) is an α- or an ω-limit point for trajectories reaching the boundary of the region, then the region contains no periodic trajectories.*

These three tests for the absence of periodic trajectories are derived from the fact that the rotation of a vector field around a periodic trajectory is equal to 1, and therefore the sum of the indices of all stationary points lying within such a trajectory is equal to unity (see page 88). Therefore the periodic trajectories are to be found only among those

[10] We note that not every periodic trajectory lies in an annular region satisfying the stated conditions.

curves surrounding stationary points with index-sum equal to unity. It is still necessary, even if a region satisfies these tests, to choose an annular region within which one attempts to confine the periodic trajectories.

An n-dimensional analogue of Theorem II (page 110), on which one might hope to base a generalization of the method of annular regions, might be the following assertion (its validity, however, has not yet been proved):

If a field of directions, defined by the dynamical system of equations

$$\frac{dx_i}{dt} = f_i(x_1, x_2, \ldots, x_n) \qquad (i = 1, 2, \ldots, n), \tag{37}$$

satisfying the conditions of the theorems on existence and uniqueness, is everywhere directed toward the interior of the toroid region T (T is a topological product of the element E^{n-1} by the open set S^1), and if there are no stationary points in T or on its boundary, then T contains a periodic trajectory.

Despite all of these reservations, and even taking account of the existence of counterexamples [**283**] for $n > 3$, one meets many practical examples of systems of the type of (37) in which under precisely the stated conditions one can prove the existence of an element E^{n-1} which is the cross-section of a toroid body T and is such that trajectories originating in E^{n-1} do traverse the torus and return to E^{n-1}, so that these trajectories generate a continuous self-mapping of E^{n-1}. The fixed points of this mapping, which must exist by Brouwer's theorem, correspond to the periodic trajectories [**10**; **279**; **245**].

Another method, incomparably more effective and yet quite general, for finding the periodic solutions of the system

$$\frac{dx_i}{dt} = f_i(t, x_1, x_2, \ldots, x_n) \qquad (i = 1, 2, \ldots, n)$$

is the method of expansion in powers of a small parameter, which was outlined in its essentials by H. Poincaré [**236**; **237**] and by A. M. Ljapunov [**152**].[11] The procedure as applied to the system of equations

$$\frac{dx_i}{dt} = f_i(t, x_1, x_2, \ldots, x_n, \mu) \qquad (i = 1, 2, \ldots, n) \tag{38}$$

reduces to the following: assume that for $\mu = 0$ the system simplifies and can be integrated without great difficulty, and assuming that the func-

[11] This method is set forth in detail in I. G. Malkin's text [**164**].

tions $f_i(t, x_1, x_2, \cdots, x_n, \mu)$ are analytic in $x_1, x_2, \cdots, x_n$ throughout the region in which we shall later allow these variables to lie, for sufficiently small $|\mu|$ we expand a solution $x_i(t, \mu)$ $(i = 1, 2, \cdots, n)$ of the system (38) in the form of a power series in μ:

$$x_i(t, \mu) = x_{i0}(t) + \mu x_{i1}(t) + \ldots + \mu^n x_{in}(t) + \ldots \tag{39}$$

The existence of such an expansion is guaranteed for sufficiently small $|\mu|$ by the theorem on analytic dependence of the solution on its parameter.[12]

Substituting (39) into the system (38), in which the right-hand side is first arranged according to powers of $x_1, x_2, \cdots, x_n$ and μ, and comparing coefficients of equal powers of μ on the left and right sides of the equations (38), we obtain a system of equations defining the functions

$$x_{is}(t) \qquad (i = 1, 2, \ldots, n;\ s = 0, 1, 2, \ldots),$$

which in many cases will be much simpler than the system (38). In particular, we obtain as the definition of $x_{i0}(t)$ $(i = 1, 2, \cdots, n)$ the so-called generating system

$$\frac{dx_i}{dt} = f_i(t, x_1, x_2, \ldots, x_n, 0) \qquad (i = 1, 2, \ldots, n), \tag{40}$$

and this is, by hypothesis, relatively easy to integrate. This general procedure can in particular be applied to find periodic solutions of the system (38).

To be sure, we can find by this method only those periodic solutions that tend, as $\mu \to 0$, to periodic solutions of the generating system; therefore, to make use of this method we must show that to any periodic solution of the system there corresponds a periodic solution of the system (38) differing little from the desired solution when μ is sufficiently small. Such a correspondence, however, does not by any means occur in all cases. For example, the generating equation

$$\ddot{x}_0 + a^2 x_0 = 0 \tag{41}$$

for the equation

$$\ddot{x} + a^2 x = \mu f(x, \dot{x}) \tag{42}$$

has a two-parameter family of periodic solutions $x = c\cos(t - t_0)$, yet at the same time for arbitrarily small values of μ the equation (42) has at

[12] See, for example, A. N. Tihonov [267], where an especially simple proof is given for the theorem on analytic dependence of the solution on a parameter.

the origin of coordinates not a center, but a focus, which may be surrounded by a number of phase trajectories that are limiting cycles (Figure 46). These limiting cycles tend for $\mu \to 0$ to a completely defined

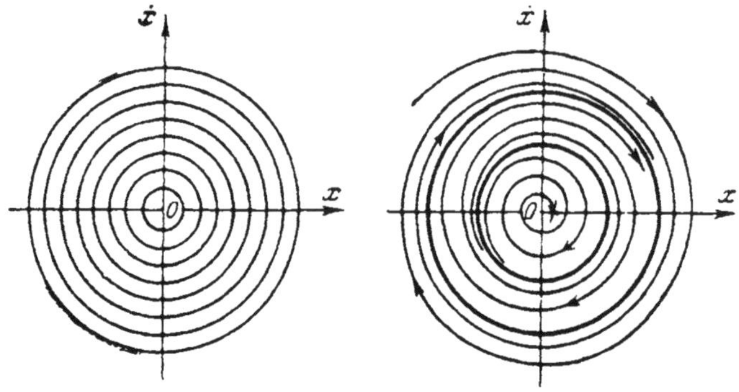

Figure 46

periodic trajectory of the generating equation (41), while the remaining periodic trajectories of the generating equation correspond to no periodic trajectory of the complete equation (42).

The fundamental theorem on the existence of periodic trajectories of the system (38) that tend for $\mu \to 0$ to periodic solutions of the generating system (40) was proved by H. Poincaré [**236**].

Aside from the above restrictions on the right-hand side of the system, we shall now suppose that the functions $f_i(t, x_1, x_2, \cdots, x_n)$ $(i = 1, 2, \cdots, n)$ are periodic in t. This condition is almost necessary for the existence of a periodic solution; more exactly, it is necessary on the periodic solution itself rather than for general $x_1, x_2, \cdots, x_n$.

In fact, suppose that $x_i(t)$ $(i = 1, 2, \cdots, n)$ is a periodic solution of the system (37) with period T; then by substituting this solution into (37) we obtain a system of identities; replacing t by $t + T$ in each of the identities we observe that all left-hand sides and the arguments $x_i(t)$ are unchanged, and therefore all the functions $f_i(t, x_1, x_2, \cdots, x_n)$ must be periodic in t with period T. By resorting if necessary to a linear transformation of the independent variables, we may without loss of generality suppose that the right-hand sides of the system and the desired periodic solutions have period 2π.

Suppose that $x_{i0}(t)$ $(i = 1, 2, \cdots, n)$ is a periodic solution, with period 2π, of the generating system

$$\frac{dx_i}{dt} = f_i(t, x_1, x_2, \ldots, x_n, 0) \qquad (i = 1, 2, \ldots, n). \tag{40}$$

We shall call $x_{i0}(t)$ $(i = 1, 2, \cdots, n)$ a generating solution.

We denote by $x_i(t, \mu, \beta_1, \beta_2, \cdots, \beta_n)$ $(i = 1, 2, \cdots, n)$ a solution of the system (38) defined by the initial values $x_i(0, \mu, \beta_1, \beta_2, \cdots, \beta_n) = x_{i0}(0) + \beta_i$ $(i = 1, 2, \cdots, n)$. Accordingly, the β_i are the deviations of the initial values of the desired functions x_i, which form the solution under study, from the initial values of the functions $x_{i0}(t)$ forming the generating solutions.

In order that for all sufficiently small μ there should exist a periodic solution $x_i(t, \mu, \beta_1, \beta_2, \cdots, \beta_n)$ $(i = 1, 2, \cdots, n)$ of the system (38) with period 2π which reduces to the generating solution as $\mu \to 0$, it is necessary that we be able to choose $\beta_i(\mu)$ $(i = 1, 2, \cdots, n)$ so that all $\beta_i(0) = 0$ and that the conditions of periodicity are satisfied:

$$x_i(2\pi, \mu, \beta_1, \beta_2, \ldots, \beta_n) - x_i(0, \mu, \beta_1, \beta_2, \ldots, \beta_n) = 0 \qquad (43)$$

$$(i = 1, 2, \ldots, n).$$

It is easy to see that these conditions, which we write in condensed form

$$\Phi_i(\beta_1, \beta_2, \ldots, \beta_n, \mu) = 0 \qquad (i = 1, 2, \ldots, n), \qquad (43')$$

are not only necessary, but also sufficient for the periodicity of the solution

$$x_i(t, \mu, \beta_1, \beta_2, \ldots, \beta_n) \qquad (i = 1, 2, \ldots, n).$$

In fact, in view of the periodicity of the right-hand side with respect to t, the initial values

$$x_i(t_0) = x_{i0} \qquad (i = 1, 2, \ldots, n) \qquad (44)$$

and

$$x_i(t_0 + 2k\pi) = x_{i0} \qquad (i = 1, 2, \ldots, n), \qquad (45)$$

$k = 1, 2, \cdots$ define in the respective intervals

$$0 \leqslant t \leqslant 2\pi, \quad 2\pi \leqslant t \leqslant 4\pi, \ \ldots$$

identical integral curves, or more exactly, curves that are periodic continuations of one another.

To ensure the existence of functions $\beta_i(\mu)$ satisfying the periodicity conditions and vanishing for $\mu = 0$, it is sufficient to require that the Jacobian satisfy

$$\frac{D(\Phi_1, \Phi_2, \ldots, \Phi_n)}{D(\beta_1, \beta_2, \ldots, \beta_n)} \neq 0 \qquad (46)$$

for $\mu = 0$ and $\beta_i = 0$ $(i = 1, 2, \cdots, n)$.

If for every periodic solution of the generating equation (40) the condition (46) is satisfied, then in the neighborhood of every periodic solution of equation (40), for arbitrary but sufficiently small μ there exists a unique periodic solution of the complete system (38) and in this sense the original problem for the system (38) differs little, qualitatively, from the corresponding problem for the generating equation.

From the viewpoint of qualitative theory a more interesting case is that in which the determinant (46) vanishes

$$\frac{D(\Phi_1, \Phi_2, \Phi_3, \ldots, \Phi_n)}{D(\beta_1, \beta_2, \ldots, \beta_n)} = 0 \text{ for } \mu = 0 \text{ and } \beta_i = 0 \qquad (47)$$
$$(i = 1, 2, \ldots, n),$$

since then, generally speaking, very few of the periodic solutions of the degenerate system (40) correspond for small μ to periodic solutions of the complete system; thus it is quite inadmissible to use the system (40) instead of (38) in order to find periodic solutions of the latter. The integral curves of the two systems are qualitatively distinct.

This case is also of great practical interest, since in applied mathematics, for example in the problems of oscillation, the above method is applied to nonlinear systems with small nonlinearities, and the generating system is usually chosen to be linear, yet the problem is often essentially nonlinear. This means that simplifying the system and treating it as though it were linear leads to a picture of the distribution of trajectories which is qualitatively quite different from the truth. As we have seen, this phenomenon is actually observed when the determinant (47) vanishes.

One frequently encounters the case where the Jacobian (47) vanishes because the generating system has a family of periodic solutions depending on one or more parameters c_i. It is evident that in this case $D(\Phi_1, \Phi_2, \cdots, \Phi_n)/D(\beta_1, \beta_2, \cdots, \beta_n) = 0$ for $\mu = \beta_1 = \beta_2 = \cdots = \beta_n = 0$, since in the neighborhood of these values the system (43) cannot be uniquely solved in terms of β_i $(i = 1, 2, \cdots, n)$.

For the sake of simplicity, then, we shall assume that the generating system (40) has a family of periodic solutions $x_i(t, c)$ depending on one arbitrary parameter c, and the Jacobian (47) has the rank $n - 1$. Suppose, for concreteness, that the minor

$$\frac{D(\Phi_1, \Phi_2, \cdots, \Phi_{n-1})}{D(\beta_1, \beta_2, \ldots, \beta_{n-1})} \neq 0 \text{ for } \mu = 0 \text{ and } \beta_i = 0 \quad (i = 1, 2, \ldots, n).$$

Using the first $n-1$ equations of the system (43), we express the β_i $(i = 1, 2, \cdots, (n-1))$ in terms of β_n and μ; then, substituting these expressions in the last equation of the system (43) we obtain an equation

$$F(\mu, \beta_n) = 0, \tag{48}$$

which reduces to an identity for $\mu = 0$.

If all the β_i are analytic functions of β_n and μ, then $F(\mu, \beta_n)$ is an analytic function of μ and β_n in the neighborhood of $\mu = 0$ and $\beta_n = 0$. Expanding the function F in a series according to powers of μ, and remembering that for $\mu = 0$ the equation (48) reduces to an identity, we conclude that the free term of this series is equal to zero, and therefore that equation (48) can be written in the form

$$\mu\Phi(\mu, \beta_n) = 0, \tag{48a}$$

where Φ is an analytic function of the arguments μ and β_n, not vanishing, in general, for $\mu = 0$ and $\beta_n = 0$. In order that the equation (48a) have a solution for arbitrary sufficiently small $\mu \neq 0$ it is necessary that β_n satisfy the equation $\Phi(\mu, \beta_n) = 0$, or, expanding in powers of μ and β_n, that

$$a_0(c) + a_{11}(c)\mu + a_{12}(c)\beta_n + \ldots = 0.$$

This equation can have a solution $\beta_n(\mu)$ reducing to zero for $\mu = 0$ only if

$$a_0(c) = 0. \tag{49}$$

In general, equation (49) determines values of the parameter c for which the neighborhood of a periodic solution of the generating equation contains periodic solutions of the complete equation (38) that tend, for $\mu \to 0$, to solutions of the generating equation. Accordingly, in this case the solutions of the generating system are qualitatively different from the solutions of the complete system, since the generating system has an infinite set of periodic solutions, depending on a parameter c, while the system (38) for small $|\mu|$ has a rule periodic solutions only near some of the periodic solutions of the generating system (40), i.e., those for which the value of the parameter c is defined by the equation $a_0(c) = 0$.

The case we have just looked at, in which the generating system admits a one-parameter family of periodic solutions is easily generalized to the case of a k-parameter family, depending on the parameters $c_1, c_2, \cdots, c_k$; supposing that $D(\Phi_1, \Phi_2, \cdots, \Phi_{n-k}) / D(\beta_1, \beta_2, \cdots, \beta_{n-k}) \neq 0$, and eliminating from the system (43) the variables $\beta_1, \beta_2, \cdots, \beta_{n-k}$, we obtain k equations of the form $F_j(\mu, \beta_{n-k+1}, \beta_{n-k+2}, \cdots, \beta_n) = 0$ $(j = 1, 2, \cdots, k)$ with analytic left-hand sides which, for $\mu = 0$, are satisfied by arbitrary values of

β_i $(i = n - k + 1, n - k + 2, \cdots, n)$ and can therefore be expressed in the form

$$\mu\bar{\Phi}_j(\mu, \beta_{n-k+1}, \beta_{n-k+2}, \ldots, \beta_n) = 0 \qquad (j = 1, 2, \ldots, k);$$

cancelling out the factor μ and expressing $\bar{\Phi}_j$ as a power series in $\mu, \beta_{n-k+1}, \cdots, \beta_n$ we again arrive at the conclusion that the free term of the series vanishes, i.e.,

$$a_{0j}(c_1, c_2, \ldots, c_k) = 0 \qquad (j = 1, 2, \ldots, k), \tag{50}$$

which, in general, defines the c_j $(j = 1, 2, \cdots, k)$.

If

$$\frac{D(a_{01}, a_{02}, \ldots, a_{0k})}{D(c_1, c_2, \ldots, c_k)} \neq 0 \tag{51}$$

for the values $c_1, c_2, \cdots, c_k$ which satisfy the equation (50), then, as may be easily seen, the system $\bar{\Phi}_j = 0$ $(j = 1, \cdots, k)$, has a unique solution for the β_i $(i = n - k + 1, n - k + 2, \cdots, n)$, which tends to zero as $\mu \to 0$, since the determinant $D(\Phi_1, \Phi_2, \cdots, \Phi_k)/D(\beta_{n-k+1}, \beta_{n-k+2}, \cdots, \beta_n)$ differs from the determinant (51) only by a nonvanishing factor when the c_i satisfy (50).

In addition to the solutions of periodic 2π, the system (38) may admit solutions of period $2\pi m$, where m is an integer $\geqq 2$, which do not coincide with the solutions of period 2π and which may, as $\mu \to 0$, tend toward solutions of the generating equation of period $2\pi m$ or toward solutions of smaller period, and in particular to solutions of period 2π.

For the solutions of period $2\pi m$ the periodicity conditions (43) have the form

$$x_s(2\pi m, \beta_1, \beta_2, \ldots, \beta_n, \mu) - x_s(0, \beta_1, \beta_2, \ldots, \beta_n, \mu) = 0 \tag{52}$$
$$(s = 1, 2, \ldots, n),$$

or, more briefly,

$$\varphi_s(\beta_1, \beta_2, \ldots, \beta_n, \mu) = 0 \qquad (s = 1, 2, \ldots, n). \tag{52'}$$

If the determinants (46) and

$$\frac{D(\varphi_1, \varphi_2, \ldots, \varphi_n)}{D(\beta_1, \beta_2, \ldots, \beta_n)} \tag{53}$$

do not vanish for $\mu = \beta_1 = \beta_2 = \cdots = \beta_n = 0$, there exists a unique solution of period 2π of the system (38) and a unique solution of this system of period $2\pi m$, both tending for $\mu \to 0$ to a periodic solution of the generating equation, and consequently these solutions coincide, since a solution of period 2π is in any case also a solution of period $2\pi m$.

On the other hand, if the determinant (53) vanishes, the system (38) may have solutions of period $2\pi m$ that do not coincide with any of its solutions of period 2π.

Let us apply the method of small parameters to the determination of periodic solutions of the so-called equations of quasi-linear oscillations. We limit ourselves to the case, often encountered in practice, of a single equation of second order, of the form

$$\ddot{x}+a^2x=f(t)+\mu F(t, x, \dot{x}, \mu), \tag{54}$$

where $a^2\neq 0$, and μ is a small parameter, with respect to which the function F is analytic for sufficiently small $|\mu|$. The function F, moreover, is analytic in the second and third arguments in those regions over which these arguments will range in our discussion. With respect to the first argument t, the function F is continuous, periodic with period 2π, and expandable in a Fourier series; and f has the same properties with respect to t. At least one of the functions F and f depends explicitly on t. The generating equation

$$\ddot{x}+a^2x=f(t) \tag{55}$$

is linear and its periodic solutions with period 2π may be expressed as a series

$$x_0(t)=\frac{\alpha_0}{2}+\sum_{n=1}^{\infty}(\alpha_n\cos nt+\beta_n\sin nt), \tag{56}$$

in which the coefficients are determined by equating the free terms, and the coefficients of $\cos nt$ and $\sin nt$, on the left-hand and right-hand sides of equation (55), in which the function $f(t)$ has been first expanded in the Fourier series

$$f(t)=\frac{A_0}{2}+\sum_{n=1}^{\infty}(A_n\cos nt+B_n\sin nt).$$

Here, if a is not an integer, we obtain a unique periodic solution of (56) of period 2π, with the Fourier coefficients

$$\alpha_0=\frac{A_0}{a^2},\quad \alpha_n=\frac{A_n}{a^2-n^2},\quad \beta_n=\frac{B_n}{a^2-n^2}. \tag{57}$$

If on the other hand a is an integer m, we encounter the so-called resonant case, in which periodic solutions exist only for $A_m=0$ and $B_m=0$, that is for

$$\int_0^{2\pi} f(t) \cos mt\, dt = 0, \quad \int_0^{2\pi} f(t) \sin mt\, dt = 0.$$

Then α_m and β_m are arbitrary, while the remaining α_n and β_n are determined by equation (57).

In this case the period 2π holds for all solutions of (55) for which $x = x_0(t) + c_1 \cos mt + c_2 \sin mt$, where c_1 and c_2 are arbitrary constants and $x_0(t)$ is a particular solution of (55) which is obtained from (56) for $\alpha_m = \beta_m = 0$.

2. *The nonresonant case.* We assume that in the equation

$$\ddot{x} + a^2 x = f(t) + \mu F(t, x, \dot{x}, \mu) \tag{54}$$

the constant a is not an integer.

We look for a periodic solution of this equation as the sum of a series

$$x(t, \mu) = x_0(t) + \mu x_1(t) + \ldots + \mu^n x_n(t) + \ldots; \tag{58}$$

substituting (58) in equation (54), where the function F has first been expanded in powers of μ and of the differences $x - x_0$ and $\dot{x} - \dot{x}_0$, and requiring that the equation should then reduce to an identity, we obtain for the determination of $x_0(t), x_1(t), x_2(t), \cdots$ the equations

$$\left.\begin{aligned} &\ddot{x}_0 + a^2 x_0 = f(t), \\ &\ddot{x}_1 + a^2 x_1 = F(t, x_0, \dot{x}_0, 0), \\ &\ddot{x}_2 + a^2 x_2 = \frac{\partial F}{\partial x}\Bigg|_{\substack{x = x_0 \\ \dot{x} = \dot{x}_0 \\ \mu = 0}} \cdot x + \frac{\partial F}{\partial \dot{x}}\Bigg|_{\substack{x = x_0 \\ \dot{x} = \dot{x}_0 \\ \mu = 0}} \cdot \dot{x}_1, \\ &\cdots\cdots\cdots\cdots\cdots\cdots\cdots\cdots \\ &\cdots\cdots\cdots\cdots\cdots\cdots\cdots\cdots \end{aligned}\right\} \tag{59}$$

In each of these equations the right-hand side is a known function of t, since the functions $x_k(t)$ entering into any one of the equations (59) have been defined in one of the preceding equations.

Therefore, a periodic solution of period 2π of each of the equations (59) can be found in just the same way as the periodic solutions of the generating equations (55). The solution is unique.

Thus we obtain a unique series

$$x_0(t) + \mu x_1(t) + \ldots + \mu^n x_n(t) + \ldots, \tag{58a}$$

in which the terms are periodic functions of period 2π and which formally satisfies equation (54).

If we now demonstrate the existence of a solution of (54) which is analytic in μ and which reduces to the generating solution (56) when $\mu \to 0$, the series (58a) necessarily converges and its sum must be the desired periodic solution.

To prove that equation (54) has a solution analytic in μ and of period 2π, it is sufficient to note that, as we have previously shown, the Jacobian does not vanish:

$$\frac{D(\Phi_1, \Phi_2, \ldots, \Phi_n)}{D(\beta_1, \beta_2, \ldots, \beta_n)} \neq 0 \text{ for } \mu = 0 \text{ and } \beta_i(0) = 0 \; (i = 1, 2, \ldots, n).$$

To evaluate this determinant we look for a trial solution of equation (54) which is an analytic function of $\beta_1, \beta_2, \cdots, \beta_n$ and μ, expressed as the sum of a power series in these variables:

$$x(t, \mu, \beta_1, \beta_2)$$
$$= x_0(t) + x_{11}(t)\beta_1 + x_{12}(t)\beta_2 + x_{13}(t)\mu + \ldots . \quad (60)$$

Substituting (60) in (54) and requiring that (54) reduce to an identity, we obtain a set of equations which in particular determine $x_{11}(t)$ and $x_{12}(t)$:

$$\ddot{x}_{11}(t) + a^2 x_{11}(t) = 0, \quad x_{11}(0) = 1, \quad \dot{x}_{11}(0) = 0,$$

$$\dot{x}_{12}(t) + a^2 x_{12}(t) = 0, \quad x_{12}(0) = 0, \quad \dot{x}_{12}(0) = 1,$$

whence $x_{11}(t) = \cos at$, $x_{12}(t) = 1/a \sin at$. The periodicity conditions (43) (page 140) appear as

$$(\cos 2a\pi - 1)\beta_1 + \frac{1}{a}\sin 2a\pi\beta_2 + \ldots = 0,$$

$$-a \sin 2a\pi\beta_1 + (\cos 2a\pi - 1)\beta_2 + \ldots = 0,$$

and therefore the determinant satisfies

$$\left.\frac{D(\Phi_1, \Phi_2)}{D(\beta_1, \beta_2)}\right|_{\beta_1=\beta_2=\mu=0} = (\cos 2a\pi - 1)^2 + \sin^2 2a\pi \neq 0.$$

In summary, there exists a solution of equation (54) which is analytic, periodic with period 2π, convergent to a periodic solution of the generating equation as $\mu \to 0$, and therefore identical with (58a).

3. *The resonant case* If a is an integer n or differs but little from an integer, then we have the so-called resonant case.

We shall suppose that the difference $n^2 - a^2$ is of order μ: $n^2 - a^2 = \mu\bar{a}$, where $\bar{a}$ is bounded. This case can be reduced to the case of exact resonance, for which $a^2 = n^2$. In essence, equation (54) may be rewritten in the form

$$\ddot{x} + n^2 x = f(t) + \mu F(t, x, \dot{x}, \mu) + (n^2 - a^2) x$$

or

$$\ddot{x} + n^2 x = f(t) + \mu [F(t, x, \dot{x}, \mu) + \bar{a} x],$$

or

$$\ddot{x} + n^2 x = f(t) + \mu F_1(t, x, \dot{x}, \mu). \tag{61}$$

The solution of (61) is now to be sought as a series

$$x(t, \mu) = x_0(t) + \mu x_1(t) + \ldots + \mu^n x_n(t) + \ldots. \tag{58}$$

In order to define the $x_k(t)$ $(k = 0, 1, 2, \cdots)$ we again generate the system (59), using the function F_1 instead of the function F.

The generating equation

$$\ddot{x}_0(t) + n^2 x_0(t) = f(t), \tag{62}$$

or

$$\ddot{x}_0(t) + n^2 x_0(t) = \frac{A_0}{2} + \sum_{k=1}^{\infty} (A_k \cos kt + B_k \sin kt),$$

as we have already noted, has a periodic solution only in case the resonant terms

$$A_n \cos nt + B_n \sin nt, \quad \text{that is, for} \quad A_n = 0 \text{ and } B_n = 0$$

or

$$\int_0^{2\pi} f(t) \cos nt \, dt = 0 \text{ and } \int_0^{2\pi} f(t) \sin nt \, dt = 0$$

are absent from the right-hand side. If the coefficients A_n and B_n are different from zero but have an order not less than μ, we associate them with the term μF_1 and express the function $f(t)$ as the sum of a series

$$\frac{A_0}{2} + \sum_{k=1}^{n-1} (A_k \cos kt + B_k \sin kt) + \sum_{k=n+1}^{\infty} (A_k \cos kt + B_k \sin kt).$$

If $A_n = B_n = 0$, all solutions of the generating equation are of period 2π:

$$x_0(t) = c_{01} \cos nt + c_{02} \sin nt + \varphi_0(t),$$

where $\varphi_0(t)$ is a particular solution of (62) and c_{01}, c_{02} are arbitrary constants determined by the condition that no resonant terms should be present in the right-hand sides of equations (59):

$$\ddot{x}_1(t)+n^2x_1(t)=F_1(t,\ x_0(t),\ \dot{x}_0(t),\ 0),\tag{63}$$

that is, these constants are determined by the equations

$$\left.\begin{aligned}&\int_0^{2\pi} F_1(t,\ x_0(t),\ \dot{x}_0(t),\ 0)\cos nt\,dt=0\\ \text{and}\qquad&\\ &\int_0^{2\pi} F_1(t,\ x_0(t),\ \dot{x}_0(t),\ 0)\sin nt\,dt=0,\end{aligned}\right\}\tag{64}$$

since equation (63) has a periodic solution only if these conditions are satisfied.

The family of periodic solutions of equation (63) again contains two arbitrary constants c_{11} and c_{12}

$$x_1(t)=c_{11}\cos nt+c_{12}\sin nt+\varphi_1(t),$$

which in turn are determined from the condition that there should be no resonant terms in the next equation occurring in the set (59):

$$\ddot{x}_2(t)+n^2x_2(t)=\frac{\partial F_1}{\partial x}\Bigg|_{\substack{x=x_0\\ \dot{x}=\dot{x}_0\\ \mu=0}}\cdot x_1+\frac{\partial F_1}{\partial \dot{x}}\Bigg|_{\substack{x=x_0\\ \dot{x}=\dot{x}_0\\ \mu=0}}\cdot\dot{x}_1$$

and so on.

In order to be able to assert that this method will lead to a periodic solution of equation (61), one must still prove that by using equations of the type of (64) one can determine the constants c_{i1} and c_{i2} (see page 144). We leave the verification to the reader; in case of difficulty, the reader is recommended to consult the above-mentioned monograph of I. G. Malkin.

4. *Resonance of genus n.* In systems defined by a nonlinear equation $\ddot{x}+a^2x=f(t)+\mu F(t,x,\dot{x},\mu)$, where the above notations and assumptions are retained, we sometimes encounter the phenomenon of resonance not only when a is close to an integer n, but also when a is close to the quantity $1/n$. This phenomenon has been given the name of *resonance of genus n*; it was first studied by L. I. Mandel′štam and N. D. Papaleksi [172].

We postulate that $a=1/n$; then equation (54) takes on the form

$$\ddot{x}+\frac{1}{n^2}x=f+\mu F\tag{65}$$

(if a is not equal to $1/n$, but differs little from it, then, adding a term $(a^2-1/n^2)x$ to the right-hand side, we are led to an equation of the type of (65)).

We must find the conditions under which equation (65) has a periodic solution $x(t, \mu)$ of period 2π, where $n \geqq 2$, not reducing to the periodic solution $\varphi(t)$ of period 2π, which latter, as we have indicated above, exists under certain restrictive conditions and can be found by the method of §2 (page 146).

We write the desired solution in the form

$$x(t, \mu) = x_0(t) + \mu x_1(t) + \mu^2 x_2(t) + \ldots + \mu^n x_n(t) + \ldots. \tag{58}$$

The generating equation $\ddot{x}_0 + (1/n^2)x_0 = f(t)$ has a two-parameter family of periodic solutions of period $2\pi n$:

$$x_0(t) = c_{01} \cos \frac{t}{n} + c_{02} \sin \frac{t}{n} + \varphi_0(t).$$

To define $x_1(t)$ we have the equation

$$\ddot{x}_1(t) + \frac{1}{n^2} x_1(t) = F(t, x_0(t), \dot{x}_0(t), 0), \tag{66}$$

in which the right-hand side is a known function of t of period $2\pi n$, and therefore, in order that there should exist a solution of (66) with period 2π it is necessary that in the Fourier expansion of the right-hand side there should be no resonant terms $\cos t/n$ or $\sin t/n$; in other words, it is necessary that

$$\left.\begin{aligned} \int_0^{2\pi n} F(t, x_0(t), \dot{x}_0(t), 0) \cos \frac{t}{n}\, dt = 0, \\ \int_0^{2\pi n} F(t, x_0(t), \dot{x}_0(t), 0) \sin \frac{t}{n}\, dt = 0; \end{aligned}\right\} \tag{67}$$

these equations, generally speaking, define c_{01} and c_{02}.

If the conditions (67) are satisfied, equation (66) will also have a family of solutions of period $2\pi n$, depending on two parameters c_{11} and c_{12}. These parameters will in general be determined by the condition that there shall be no resonant terms in the right-hand side of the equations determining $x_2(t)$

$$\ddot{x}_2(t) + \frac{1}{n^2} x_2(t) = \left.\frac{\partial F}{\partial x}\right|_{\substack{x = x_0 \\ \dot{x} = \dot{x}_0 \\ \mu = 0}} \cdot x_1 + \left.\frac{\partial F}{\partial \dot{x}}\right|_{\substack{x = x_0 \\ \dot{x} = \dot{x}_0 \\ \mu = 0}} \cdot \dot{x}_1$$

and so on.

Of course, it is necessary in this case, as it was in §3, to verify the existence of an analytic solution of period $2\pi n$ tending to the generating solution as $\mu \to 0$, and, in particular, to show that the systems of the type of (67) are solvable with respect to the constants c_{i1} and c_{i2} (see page 144).

5. *Quasi-linear oscillations of autonomous systems.* We assume that the right-hand side of equation (54) does not depend explicitly on t, and that therefore the equation has the form

$$\ddot{x}(t)+a^2x(t)=\mu F(x(t),\ \dot{x}(t),\ \mu), \tag{68}$$

where the function F satisfies the conditions stated earlier, (or $\ddot{x}+a^2x = A+\mu F(x,\dot{x},\mu)$, where A is a constant; this does not increase the generality, since the change of variables $\bar{x}=A/a^2+x$ frees the right-hand side of the term A).

It would seem at first glance that the study of the equation $\ddot{x}+a^2x = \mu F(x,x,\mu)$ should be simpler than the study of the equation

$$\ddot{x}+a^2x=f(t)+\mu F(t,\ x,\ \dot{x},\ \mu), \tag{54}$$

but in point of fact the absence of the argument t from the right-hand side complicates the problem instead of simplifying it. If the right-hand side depends explicitly on t, then, as we have already noted, a periodic solution can exist only when the right-hand side is periodic along a periodic solution with respect to the explicitly appearing argument t, whence the periods of the solution must be equal to or less than the period of the right-hand side.

Thus, if the right-hand side depends explicitly on t, we generally, although not always,[13] know the possible periods of the solution of the equation in question.

But if the argument t does not enter explicitly into the equation we are studying, the possibility of periodic solutions of arbitrary period cannot be excluded, and the period of the solution will in general be a function of the parameter μ.

For $\mu=0$ the equation $\ddot{x}+a^2x=\mu F(x,\dot{x},\mu)$ reduces to the equation $\ddot{x}+a^2x=0$, for which the origin of coordinates is a center (see Figure 39, page 111), while for arbitrarily small values of μ the center becomes in the general case a focus (see §1 Chapter IV) and the periodic trajectories encircling the center go over into spirals. Some of these may remain closed and go into limit cycles (see Figure 46, page 140).

One of the fundamental problems arising here consists in determining

[13] It may happen that along some solutions the right-hand side of (54) ceases to depend explicitly on the argument t; then there may be such solutions of arbitrary period.

those periodic trajectories of the generating equation that remain periodic when μ varies and that give rise to limit cycles.

For small μ these generating periodic trajectories will differ by very little from the desired limit cycles, and therefore, if we know the generating trajectories, we can locate approximately the position of the limit cycles, at least for small μ.

In what follows it will be useful to rewrite the equation $\ddot{x} + a^2x = \mu F(x, \dot{x}, \mu)$ by setting $t_1 = a \cdot t$ in the equation

$$\ddot{x} + x = \mu F_1(x, \dot{x}, \mu), \tag{69}$$

for which the solution of the generating equation has period 2π. Let the period of the solution of (69) be $2\pi + \alpha(\mu)$; it may be shown that $\alpha(\mu)$ (see [**164**] is an analytic function of μ for sufficiently small μ. Expanding $\alpha(\mu)$ in a power series in μ, we obtain:

$$2\pi + \alpha(\mu) = 2\pi(1 + h_1\mu + h_2\mu^2 + \dots + h_n\mu^n + \dots),$$

where h_j are some constant quantities. Since the period of the solution depends on μ, it is idle to look for a periodic solution of equation (69) in the form

$$x(t_1, \mu) = x_0(t_1) + \mu x_1(t_1) + \dots + \mu^n x_n(t_1) + \dots,$$

since the functions $x_k(t_1)$ are not in general periodic functions of t_1 and cannot be determined by the methods outlined on pages 145-146.

We therefore rewrite equation (69) using new variables such that a periodic solution of period $2\pi + \alpha(\mu)$ goes over into a solution with the constant period 2π.

This can be accomplished by the substitution of variables

$$t_1 = t_2(1 + h_1\mu + h_2\mu^2 + \dots + h_n\mu^n + \dots),$$

which transforms equation (69) into

$$\begin{aligned} \ddot{x} + x(1 + h_1\mu + \dots + h_n\mu^n + \dots)^2 \\ = \mu F(x, (1 + h_1\mu + \dots + h_n\mu^n + \dots)^{-1}\dot{x}, \mu) \\ \times (1 + h_1\mu + \dots + h_n\mu^n + \dots)^2. \end{aligned} \tag{70}$$

The periodic solution of this equation, if it exists, will also be an analytic function of μ and therefore may be found in the form

$$x(t_2, \mu) = x_0(t_2) + \mu x_1(t_2) + \dots + \mu^n x_n(t_2) + \dots, \tag{71}$$

where the $x_k(t_2)$ are periodic functions of period 2π, since the period of the solution $x(t_2, \mu)$ does not depend on μ and is equal to 2π. Since equation (70) is autonomous (i.e., does not contain the argument t explicitly), we may without loss of generality look for a periodic solution satisfying the condition $\dot{x}(0) = 0$. Substituting (71) in (70) and comparing coefficients of corresponding powers of μ on the left and right sides of the resulting identity, we obtain linear equations with known right-hand sides for the determination of the $x_k(t_2)$; $x_0(t)$ is determined by the equation $\ddot{x}_0 + x_0 = 0$, whence, taking account of the initial condition $\dot{x}_0(0) = 0$, we obtain $x_0 = c \cos t_2$.

We determine $x_1(t_2)$ from the equation

$$\ddot{x}_1 + x_1 = -2h_1 c \cos t_2 + F_1(c \cos t_2, \ -c \sin t_2, \ 0). \tag{72}$$

In order that this equation have a periodic solution, it is necessary and sufficient that the right-hand side contain no resonant terms (see page 146).

$$\left.\begin{aligned} &\int_0^{2\pi} F_1(c \cos t_2, \ -c \sin t_2, \ 0) \sin t_2 \, dt_2 = 0 \\ &\text{and} \\ &-2h_1 c + \frac{1}{\pi} \int_0^{2\pi} F_1(c \cos t_2, \ -c \sin t_2, \ 0) \cos t_2 \, dt_2 = 0. \end{aligned}\right\} \tag{73}$$

The first of these equations, which we abbreviate as $a(c) = 0$, defines the radius c of a circle with center at the origin of coordinates, in the interior of which there are limit cycles; the second equation defines h_1 and allows us to determine to a first approximation the period of the motion around a limit cycle:

$$2\pi + \alpha(\mu) \approx 2\pi(1 + h_1 \mu).$$

When condition (73) is satisfied there exists a two-parameter family of periodic solutions of equation (72). To determine the two parameters of this family we use the two conditions for existence of periodic solutions in the resonant case, applying them to the equation defining $x_2(t)$, and so on, as on page 149.

The question of the existence of periodic solutions of equation (70) that tend to some or other periodic solutions of the generating equation is solved by testing the condition $da(c)/dc \neq 0$ for values of c that are roots of the equation $a(c) = 0$.

In the theory of nonlinear oscillations especially significant contributions have been made by H. Poincaré [236; 237], A. M. Ljapunov [152],

A. A. Andronov [6; 11; 13], N. M. Krylov and N. N. Bogolyubov [122; 123], I. G. Malkin [162], L. I. Mandel'štam and N. D. Papaleksi [172].

A most complete exposition of the method of small parameters in the theory of nonlinear oscillations, with far-reaching generalizations, is given in the text by I. G. Malkin [164].

In this section we have by no means attempted to master all the fundamental methods and results connected with the theory of periodic solutions; in particular, we have not at all touched upon the methods for finding periodic solutions elaborated by N. M. Krylov and N. N. Bogolyubov [122; 123], nor the van der Pol method [40], nor the energy method [263], nor the many investigations of periodic solutions of equations of certain special types [27; 260; 276; 87; 95; 5], etc. Finally, we have not dealt with the question of the stability of periodic solutions.

Chapter V

Differential Equations with Deviating Arguments

Sundry differential equations with deviating arguments occur as long ago as in the works of L. Euler, but the systematic study of these equations was first undertaken in the twentieth century, to meet the demands of applied science, in particular of control theory.

Basically, the study has been devoted to differential equations with lagging arguments, which describe many processes with delayed effects. Almost all of the relevant papers were devoted to an investigation of linear differential equations with constant coefficients and with constant lags; in a few papers only were there studies of certain very special types of linear equations with variable lags, arising from particular and concrete geometrical or mechanical problems. Only very recently, beginning with the work of A. D. Myškis [192-198] (in 1949-1951), has there been any elaboration of a general theory of differential equations with lagging arguments.

Differential equations with deviating arguments are integrable in closed form only under very specialized circumstances, and therefore qualitative and approximate methods are of the utmost importance in studying them.

Since even the elements of the theory of differential equations with deviating coefficients cannot be assumed to be generally well known, we shall in the remainder of this exposition often have to go outside the bounds of qualitative theory.

1. Classification of differential equations with deviating arguments and establishment of the fundamental boundary-value problem. A differential equation with deviating arguments is a differential equation in which the unknown function enters with several different values of the argument. For example:

$$\dot{x}(t) = f\big(t,\ x(t),\ x(t-\tau(t))\big);$$

$$\ddot{x}(t) = f\big(t,\ x(t),\ x(t+\tau_1),\ \dot{x}(t),\ \dot{x}(t+\tau_2)\big);$$

$$\dot{x}(t) = f\left(t,\ x(t),\ x\left(\frac{t}{2}\right),\ x(t^2)\right).$$

If all deviations of the arguments τ_i in the equation

$$F(t,\ x(t+\tau_0),\ x(t+\tau_1),\ \ldots$$
$$\ldots,\ x(t+\tau_m),\ \dot{x}(t+\tau_0),\ \dot{x}(t+\tau_1),\ \ldots,\ \dot{x}(t+\tau_m),\ \ldots$$
$$\ldots,\ x^{(n)}(t+\tau_0),\ x^{(n)}(t+\tau_1),\ \ldots,\ x^{(n)}(t+\tau_m))=0,$$

or in a system of similar equations are constants, the equation is said to be a *differential-difference equation*. Sometimes this term is also applied to the case of variable deviation.

Differential equations with retarded argument are differential equations with deviating arguments in which the highest order derivatives of the unknown function enter with identical values of the argument and this value is not less than the arguments of the unknown function and its derivatives. For example:

$$\dot{x}(t)=f\big(t,\ x(t),\ x(t-\tau(t))\big),\quad \tau(t)\geqslant 0;$$

$$\ddot{x}(t+\tau)=f\big(t,\ x(t),\ x(t+\tau),\ \dot{x}(t),\ \dot{x}(t+\tau)\big),\quad \tau>0;$$

$$\dot{x}(t)=f\Big(t,\ x(t),\ x\Big(\frac{t}{2}\Big),\ x(t-e^{-t})\Big)\quad \text{for } t\geqslant 0.$$

Differential equations with advanced arguments are those with deviating arguments in which the highest order derivatives enter with identical values of the argument and the common value is not greater than the argument in the function or any of its derivatives that enter the equation. For example:

$$\dot{x}(t)=f\big(t,\ x(t),\ x(t+\tau(t))\big),\quad \tau(t)\geqslant 0;$$
$$\ddot{x}(t)=f\big(t,\ x(t),\ x(2t),\ \dot{x}(t),\ \dot{x}(2t)\big)\quad \text{for } t\geqslant 0;$$
$$\ddot{x}(t-\tau)=f\big(t,\ x(t),\ x(t-\tau),\ \dot{x}(t),\ \dot{x}(t-\tau)\big),\quad \tau>0.$$

It is possible that on some set of values of the independent variables a certain equation should appear as an equation with retarded argument while on another set it is an equation with advanced argument or perhaps belongs to neither of these types. For example $\dot{x}(t) = x(t) + x(t/2)$ is an equation with retarded argument for $t \geqq 0$ and an equation with advanced argument for $t \leqq 0$.

All remaining types of differential equations with deviating arguments in one unknown function are called *differential equations of neutral type*. For example:

$$\dot{x}(t) = f(t,\ x(t-\tau),\ x(t),\ x(t+\tau)),$$

where

$$\frac{\partial f}{\partial x(t-\tau)} \neq 0 \text{ and } \frac{\partial f}{\partial x(t+\tau)} \neq 0;$$

$$\ddot{x}(t) = f(t,\ x(t),\ \dot{x}(t),\ x(t-\tau),\ \dot{x}(t-\tau),\ \ddot{x}(t-\tau)),$$

where

$$\frac{\partial f}{\partial \ddot{x}(t-\tau)} \neq 0.$$

A similar classification by means of the arguments of the highest order derivatives of all the unknown functions is applicable to systems of differential equations with deviating arguments.[1] We remark that although from the point of view of applications the most important equations with deviating arguments are those with retarded argument, since such equations define processes with after effects, one must not think that other types of equation with deviating argument have no practical application.

Even in processes with after effects it is quite possible that the velocity at some instant of time t depends not only on the position of the point at some preceding instant of time $t - \tau_i(t)$, $\tau_i(t) \geqq 0$, but also on the velocity, or on both the velocity and the acceleration, at the time $t - \tau_i(t)$. In the simplest case we then have an equation of the form

$$\dot{x}(t) = f(t,\ x(t),\ x(t-\tau(t)),\ \dot{x}(t-\tau(t)))$$

or

$$\dot{x}(t) = f(t,\ x(t),\ x(t-\tau(t)),\ \dot{x}(t-\tau(t)),\ \ddot{x}(t-\tau(t))),$$

the first of these being an equation of neutral type, and the second an equation with advanced argument.

For the simplest equation with retarded argument,

$$\dot{x}(t) = f(t,\ x(t),\ x(t-\tau)), \tag{1}$$

where the retardation τ is assumed constant, the fundamental boundary-value problem consists in determining a continuous solution $x(t)$ of equation (1) for $t \geqq t_0$, under the condition that $x(t) = \varphi(t)$ for $t_0 - \tau \leqq t \leqq t_0$ where $\varphi(t)$ is a given continuous function called the initial function.

As we proved in §2, Chapter III and will prove again in the following

[1] G. A. Kamenskii has defined a somewhat different classification.

section, if we impose certain conditions on the functions f and φ, this fundamental boundary-value problem has a unique continuous solution.

For an nth order equation

$$x^{(n)}(t) = f(t, x(t), \dot{x}(t), \ldots$$
$$\ldots, x^{(n-1)}(t), x(t-\tau), \dot{x}(t-\tau), \ldots, x^{(n-1)}(t-\tau)),$$

where the previously stated restriction is imposed on τ, the initial conditions in the fundamental problem are the same, $x(t) = \varphi(t)$ for $t_0 - \tau \leqq t \leqq t_0$, where $\varphi(t)$ is continuously differentiable $(n-1)$ times if the solution has continuous derivatives up to the order $(n-1)$ inclusive.

In case the retardation $\tau = \tau(t)$ is variable, then for the first-order equation $\dot{x}(t) = f(t, x(t), x(t-\tau(t)))$ the initial function $x(t) = \varphi(t)$ in the fundamental problem must be defined on a so-called initial set E_{t_0} consisting of the point $t = t_0$ and of those values of the differences $t - \tau(t)$ for $t_0 \leqq t \leqq B$ which are less than t_0, if the solution is defined for values $t_0 \leqq t \leqq B$. For example, for the equation $\dot{x}(t) = f(t, x(t), x(t-\cos^2 t))$, where $0 \leqq t < \infty$, the initial set E_0 consists of all points of the interval $[-1, 0]$.

If there are several retardations, as for example in the equation

$$\dot{x}(t) = f(t, x(t), x(t-\tau_1(t)), \ldots, x(t-\tau_n(t)));$$
$$t_0 \leqslant t \leqslant B, \ \tau_i(t) \geqslant 0,$$

the initial function $x(t) = \varphi(t)$ must be defined on the initial set E_{t_0} consisting of the point $t = t_0$ and of all those differences $t - \tau_i(t)$ $(i = 1, 2, \cdots, n)$ which are less than t_0 for $t_0 \leqq t \leqq B$.

An nth order equation with variable retardations $\tau(t) \geqq 0$

$$x^{(n)}(t) = f(t, x(t), \dot{x}(t), \ldots, x^{(n-1)}(t), x(t-\tau(t)),$$
$$\dot{x}(t-\tau(t)), \ldots, x^{(n-1)}(t-\tau(t)))$$

normally determines an $(n-1)$-fold continuously differentiable solution $x(t)$, $t_0 \leqq t \leqq B$, and the initial conditions are as described above, $x(t) = \varphi(t)$ on E_{t_0} where the function $\varphi(t)$ is continuously differentiable $(n-1)$ times, except when the set E_{t_0} for $t_0 \leqq t \leqq B$ consists of the single point t_0 or when the point t_0 is isolated in the set E_{t_0}.

If the exceptional case occurs, then at the point t_0 it is necessary to define the values of the derivatives on the right-hand side up to the order $(n-1)$ inclusive.

We will also consider the analogous boundary problem for systems of equations and for differential equations with deviating arguments of other types.

2. **The method of successive integrations (method of steps).** Let us consider the fundamental boundary-value problem for the simplest differential equation with retarded argument

$$\dot{x}(t) = f(t,\ x(t),\ x(t-\tau)),$$

with constant retardation $\tau > 0$, and

$$x(t) = \varphi_0(t) \quad \text{for} \quad t_0 - \tau \leqslant t \leqslant t_0.$$

The most natural way to solve this problem is the so-called method of successive integrations (or method of steps), which consists of finding the continuous solution $x(t)$ of the equation in question as the solution of a number of equations without retardation:

$$\dot{x}(t) = f\big(t,\ x(t),\ \varphi_0(t-\tau)\big) \quad \text{for} \quad t_0 \leqslant t \leqslant t_0 + \tau,$$
$$x(t_0) = \varphi_0(t_0),$$
$$\dot{x}(t) = f\big(t,\ x(t),\ \varphi_1(t-\tau)\big) \quad \text{for} \quad t_0 + \tau \leqslant t \leqslant t_0 + 2\tau,$$
$$x(t_0+\tau) = \varphi_1(t_0+\tau),$$
$$\dots\dots\dots\dots\dots\dots\dots\dots\dots\dots$$
$$\dot{x}(t) = f\big(t,\ x(t),\ \varphi_n(t-\tau)\big) \quad \text{for} \quad t_0 + n\tau \leqslant t \leqslant t_0 + (n+1)\tau,$$
$$x(t_0+n\tau) = \varphi_n(t_0+n\tau),$$
$$\dots\dots\dots\dots\dots\dots\dots\dots\dots\dots$$

where $\varphi_j(t)$ are the solutions of the given problem in the interval

$$t_0 + (j-1)\tau \leqslant t \leqslant t_0 \leqslant t_0 + j\tau.$$

This method allows us to determine the solution $x(t)$ and at the same time proves the existence of a solution in the neighborhood of the point $t = t_0$, if the functions φ_0 and f are continuous, and also to prove its uniqueness if the function f satisfies one of the conditions guaranteeing the uniqueness of the solution of the equation $\dot{x}(t) = f(t, x(t), \varphi_0(t-\tau))$ without retardation: for example, if it satisfies Osgood's condition, or a Lipschitz condition, for the second argument. If the solution of the equation $\dot{x}(t) = f(t, x(t), \varphi_j(t-\tau))$ $(j = 0, 1, 2, \cdots)$ exists and is unique in the whole interval $t_0 + j\tau \leqq t \leqq t_0 + (j+1)\tau$, the method of steps allows us to determine the solution in an arbitrary finite interval. We note that even when the functions φ_0 and f possess continuous derivatives of arbitrarily high order, the solution of the given problem will in general have a discontinuity of the first kind in its kth order derivative at the point $t_0 + (k-1)\tau$, but the derivatives of lower order will be continuous at this point.

In fact, it is obvious that at the point t_0 the first derivative $\dot{x}(t)$ has in general a discontinuity of the first kind, since by integrating the equation

$$\dot{x}(t) = f(t, x(t), \varphi_0(t-\tau)), \quad t_0 \leqslant t \leqslant t_0 + \tau,$$

we can satisfy the condition $x(t_0) = \varphi_0(t_0)$ but we cannot in general also satisfy the condition $\dot{x}(t_0) = \dot{\varphi}_0(t_0 - 0)$. Only under special conditions on the initial function $\varphi_0(t)$ can we guarantee the continuity of the derivatives of the solution at the point t_0; for this, the function $\varphi_0(t)$ must satisfy the condition

$$\dot{\varphi}_0(t_0 - 0) = f(t_0, \varphi_0(t_0), \varphi_0(t_0 - \tau)).$$

At the point $t_0 + \tau$ the first derivative of the solution is already continuous. In fact, $\dot{x}(t) = f(t, x(t), x(t-\tau))$, and the right-hand side is a continuous function of t at the point $t_0 + \tau$, since $x(t)$ is continuous at the point t_0.

The second derivative, however,

$$\ddot{x}(t) = \frac{\partial f}{\partial t} + \frac{\partial f}{\partial x(t)} \dot{x}(t) + \frac{\partial f}{\partial x(t-\tau)} \dot{x}(t-\tau),$$

is in general discontinuous at the point $t_0 + \tau$, since the derivative $\dot{x}(t-\tau)$ is in general discontinuous at $t = t_0 + \tau$, as we have already shown; yet at the point $t = t_0 + 2\tau$ the second derivative $\ddot{x}(t)$ is continuous, since $\dot{x}(t-\tau)$ and $x(t-\tau)$ are continuous at the point $t = t_0 + 2\tau$. Continuing this line of reasoning, we note that at the point $t_0 + (k-1)\tau$ the derivative $x^{(k)}(t)$ is in general discontinuous, but the derivatives of lower order are continuous, under the assumption, of course, that the function f is differentiable a sufficient number of times.

At the inner points of the intervals $[t_0 + k\tau, t_0 + (k+1)\tau]$ the solutions tend to smooth out with increasing t.

In fact, if the function $\varphi_0(t)$ is continuous, then in the interval $t_0 < t < t_0 + \tau$ the solution of the equation $\dot{x}(t) = f(t, x(t), \varphi_0(t))$ has a continuous first derivative, and at the interior points of the interval $[t_0 + \tau, t_0 + 2\tau]$ the solution of the equation $\dot{x}(t) = f(t, x(t), \varphi_1(t))$ has already a continuous second derivative, since $\varphi_1(t)$ has a continuous first derivative and the function f is differentiable, by hypothesis.

Continuing the argument, we observe that in the interior of the interval $[t_0 + k\tau, t_0 + (k+1)\tau]$ the solution $x(t)$ will have continuous derivatives up to order $(k+1)$ inclusive, if the function f has continuous partial derivatives up to order k inclusive, and if the function $\varphi_0(t)$ is continuous.

Therefore, the solution $x(t)$ smooths out the increasing t. We note in passing that if the function f is infinitely differentiable it follows that a periodic solution of the equation under consideration is also infinitely differentiable.

Application of the method of successive integrations (method of steps) becomes difficult if the retardation τ is small compared to the range in which the solution is sought.

It is obvious that the method of successive integrations and its consequences can be applied also to differential equations with retarded argument having a variable deviation $\tau(t)$.

In applying the method to the solution of the equation

$$\dot{x}(t) = f(t, x(t), x(t - \tau(t))), \quad t \geqslant t_0,$$

with $x(t) = \varphi_0(t)$ on the initial set E_{t_0}, the solutions are determined by the unretarded equation $\dot{x}(t) = f(t, x(t), \varphi_0(t - \tau(t)))$, with $x(t_0) = \varphi_0(t_0)$, for t in the interval $F_{t_0}[t_0 \leqq t \leqq \gamma(t_0)]$, which is the maximal interval beginning at the point t_0 for which we have $t - \tau(t) \leqq t_0$.

We note that $\gamma(t)$ is the inverse of the function $t - \tau(t)$, if the inverse exists.

We define the solution in the interval $[\gamma(t_0), \gamma(\gamma(t_0))] = F_{\gamma(t_0)}$ by the equation

$$\dot{x}(t) = f(t, x(t), \varphi_1(t - \tau(t))), \quad x(\gamma(t_0)) = \varphi_1(\gamma(t_0)),$$

where $\varphi_1(t)$ is a solution of the original equation in the interval $[t_0, \gamma(t_0)]$; continuing this process, we reduce the problem to that of integrating an equation without retardation.

In what follows we shall write $\gamma(\gamma(t_0))$ as $\gamma_2(t_0)$, and $\gamma_n(t)$ will denote the n-fold iteration of the operation $\gamma(t)$: $\gamma_1(t) = \gamma(t)$, $\gamma_0(t) = t$.

This process of successive definition of the solution on the intervals $F_{\gamma_n}(t)$ cannot be carried further in the case when one of the intervals

$$F_{t_0}, \ F_{\gamma(t_0)}, \ F_{\gamma_2(t_0)}, \ \ldots, \ F_{\gamma_n(t_0)}, \ \ldots$$

reduces to a single point. We shall call this the special case. It can arise, obviously, only if the function $\tau(t)$ reduces to zero at some point. Proof of the existence and uniqueness of the solution in the special case will be given in the next section. Excluding the special case, we apply the method of successive integrations (method of steps) and prove the existence and uniqueness of the solution of (1) in the neighborhood of the point $t = t_0$, with the initial function $\varphi_0(t)$ given on the initial set E_{t_0}, under the assumption that the equation $\dot{x}(t) = f(t, x(t), \varphi_0(t - \tau(t)))$, where t ranges over the interval $\gamma_0(t) \leqq t \leqq \gamma_1(t)$, satisfies the conditions of the existence and uniqueness theorem. Under the same assumptions, but with the additional condition that $\tau'(t) < 1$, and arguing in the same way as on page 159 we observe that at the points t of the interval $\gamma_{p-1}(t_0) < t \leqq \gamma_p(t_0)$ the solution has continuous derivatives up to and including the pth order, if the function f is $p - 1$ times differentiable, while at the point $\gamma_{p-1}(t_0)$ the solution has in general continuous derivatives only up to order $p - 1$.

We remark that, by choosing sufficiently smooth functions f and φ_0, we can increase the smoothness of the solution at the inner points of the intervals $\gamma_{p-1}(t_0) \leqq t \leqq \gamma_p(t_0)$, but not at the boundary points of these intervals.

If the equation contains several retardations, e.g. if it has the form

$$\dot{x}(t) = f(t,\ x(t),\ x(t - \tau_1(t)),\ \ldots,\ x(t - \tau_n(t))),$$

then the interval $F_{t_0} = [t_0, \gamma(t_0)]$ becomes the largest interval of variation of t having its leftmost point at t_0 and such that all the inequalities $t - \tau_i(t) \leqq t_0$ are satisfied.

We use the same definition, word for word, in determining the intervals F_t on which we can carry out the method of successive integrations at each step of the process of solving an nth order equation, or a system of equations, with retarded arguments. Similarly, if $F_{\bar{t}}$ reduces to the single point $\bar{t}$, further application of the method of successive integrations becomes impossible. This case we shall call the special case.

If we apply the method of successive integrations, and if the unretarded equations which we integrate in the several steps of the solution of equation (1), or of a system of equations with retarded argument, all satisfy the conditions of the theorem on existence and uniqueness of the solution, and on the continuous dependence of the solution on its parameters, then the solution of the original retarded equation enjoys the same properties. In this case, the method of obtaining the solution immediately demonstrates that the solution in the space C_0 depends continuously on the choice of the initial function and on the retardation $\tau(t)$ if the right-hand side of the equation $\dot{x}(t) = f(t, x(t), x(t - \tau(t)))$ (or of an equation of more general type) is continuous and satisfies a Lipschitz condition in the second and third arguments.

3. **Method of successive approximations and the theorem on existence and uniqueness.** As we have already shown, in §2, Chapter III, a differential equation $\dot{x}(t) = f(t, x(t), x(t - \tau(t)))$, $t \geqq t_0$, $x(t) = \varphi(t)$ on E_{t_0} (or a system of similar equations with one or more retardations) can be replaced by

the equivalent integral equation

$$x(t) = \varphi(t_0) + \int_{t_0}^{t} f(t, x(t), x(t - \tau(t)))\, dt,$$

$$t \geqslant t_0, \quad x(t) = \varphi(t) \text{ on } E_{t_0}$$

(or a system of such equations), to which we may apply the method of successive approximations, if the functions f, τ, and φ are continuous and f satisfies a Lipschitz condition in the second and third arguments. In §2, Chapter III we applied the principle of contraction mappings to the operator $A(x(t)) = \varphi(t_0) + \int_{t_0}^{t} f(t, x(t), x(t - \tau(t)))dt$, and this allows us to assert that the solution of our equation, under the stated conditions, exists and is unique, and here the special case described on page 162 is not excluded. It is easy to verify that the solution depends continuously on the initial function. If we apply the same method of successive approximations to the equation $\dot{x}(t) = f(t, x(t), x(t - \tau(t)), \mu)$, in which the function f is continuous in all arguments and satisfies a Lipschitz condition in the second and third arguments for all $\mu, \mu_0 \leqq \mu \leqq \mu_1$, we obtain not only the theorem on the existence and uniqueness of the solution, but also the theorem on the continuous dependence of the solution on the parameter μ.

4. **Integrable types of differential equations with retarded argument.** One of the fundamental methods of integrating differential equations with retarded arguments, discussed above in §2, Chapter V, is the method of successive integrations (method of steps). The application of this method to an equation of the form

$$\dot{x}(t) = f(t, x(t), x(t - \tau(t))) \tag{1}$$

reduces to the integration of unretarded equations

$$\dot{x}(t) = f(t, x(t), \varphi_n(t - \tau(t))), \quad \gamma_{n-1}(t_0) \leqslant t \leqslant \gamma_n(t) \tag{2}$$

(see §2, Chapter V).

It is natural to say that equation (1) is integrable by quadrature if the method of successive integrations is applicable to it and if the equations (2) are integrable by quadrature, whatever the functions $\varphi_n(t)$ may be. It is useful to introduce a similar definition for the more general first-order equation

$$\dot{x}(t) = f(t, x(t), x(t - \tau_1(t)), \ldots, x(t - \tau_m(t))),$$

and also for equations of higher order.

As examples of first-order equations with retarded argument which are integrable by quadrature we may cite the following:

(1) $\dot{x}(t) = f(t, x(t-\tau_1(t)), x(t-\tau_2(t)), \cdots, x(t-\tau_m(t)))$.

This type is very simply integrated, since at every step the right-hand side appears as a known function of t.

(2) $M(x(t))dx(t) = N(t, x(t-\tau_1(t)), \cdots, x(t-\tau_m(t)))dt$ (or $\varphi_1(x(t))\psi_1(t, x(t-\tau_1(t)), x(t-\tau_2(t)), \cdots, x(t-\tau_m(t)))dx(t) = \varphi_2(x(t))\cdot \psi_2(t, x(t-\tau_1(t)), x(t-\tau_2(t)), \cdots, x(t-\tau_m(t)))dt$), namely, an equation with separable variables.

(3) $\dot{x}(t) = p(t, x(t-\tau_1(t)), \cdots, x(t-\tau_m(t)))\cdot x(t) + q(t, x(t-\tau_1(t)), \cdots, x(t-\tau_m(t)))$, namely a generalized linear equation.

(4) $\dot{x}(t) = p(t, x(t-\tau_1(t)), x(t-\tau_2(t)), \cdots, x(t-\tau_m(t)))\cdot x(t) + q(t, x(t-\tau_1(t)), x(t-\tau_2(t)), \cdots, x(t-\tau_m(t)))\cdot x^n(t)$ $n \neq 0$ and $n \neq 1$, namely a generalized Bernoulli equation, and so on.

In all these examples it is presupposed that the method of successive integrations is applicable (that is, that the set F_t does not reduce to a single point for any t) and that all $\tau_i(t) \geqq 0$, where $\tau_i(t)$ can vanish only at isolated points.

We remark that an equation may be integrable by quadrature, but that this fact usually does not simplify the investigation of the asymptotic behavior of its solution; often, even when the range of variation of the independent variable is finite, it is more useful to apply approximate or qualitative methods of solution than to make use of the method of steps.

5. **Approximate methods of integration of differential equations with retarded argument.** In this section we shall pause briefly to examine the qualitative basis of approximate methods for integrating differential equations with retarded argument.

In §2 of this chapter, and in §2, Chapter III, we have already laid down the conditions for the applicability of the method of successive approximations.

As in the unretarded case, the method of successive approximations in the pure form is rarely applied to practical numerical approximations, but it can be successfully applied in conjunction with certain extrapolation methods.

Euler's method, which replaces the desired integral curve by a broken line, the slopes of whose links coincide with the slopes of the tangents to the integral curves at the initial point of each link, can be applied successfully to equations with retarded arguments.

For the equation $\dot{x}(t) = f(t, x(t), x(t-\tau(t)))$, $x(t) = \varphi(t)$ on E_{t_0} (or $\dot{x}(t) = f(t, x(t), x(t-\tau_1(t)), \cdots, x(t-\tau_n(t)))$) the calculation is carried out by the iterative formula

$$x_{n+1} = x_n + \dot{x}_n h,$$

where x_n and x_{n+1} are the ordinates of the Euler polygon at the points t_n and t_{n+1} respectively, h is the length of the approximation step, $\dot{x}_n$ is the right derivative at the point $t = t_n$. The step h may be either constant or variable, $h = h_n = t_{n+1} - t_n$. The convergence of the Euler polygons, as $\max h_n \to 0$, to the integral curve of the original equation is proved in the same way as Peano's theorem for equations without retardation, and under similar assumptions.

We remark that when the retardation is variable the calculation is most suitably cai.. .. ..ie with a variable step, in order that, in calculating the value of $f(t, x(t), x(t - \tau(t)))$ at a point t_n we may already have available the value of $x(t_n - \tau(t_n))$, which would otherwise have to be determined by interpolation.

In applying Euler's method it is desirable to develop an iteration routine. We shall from now on use the term parabolic method to refer to approximation methods that are found on an approximation of the desired solution by a parabola of order $m \geqq 2$.

Among such methods we may include the approximation of the solution in an interval by its Taylor expansion

$$x(t) \approx x(t_0) + \dot{x}(t_0)(t - t_0) + \dots + x^{(m)}(t_0)\frac{(t-t_0)^m}{m!},$$

and the methods of Stürmer, Runge, Milne, or the many variants and elaborations of these.

Immediate application of these methods to the differential equation with retarded argument

$$\dot{x}(t) = f\big(t, x(t), x(t - \tau(t))\big), \quad x(t) = \varphi(t) \text{ on } E_{t_0}, \tag{1}$$

beginning at $t = t_0$ does not lead in general to a good approximation. It is easy to see why this should be so if we recall that, generally speaking the solutions of (1) have discontinuous derivatives of order $(m+1)$ at $t = \gamma_m(t_0)$, even though the functions f and φ are differentiable a sufficient number of times (see page 160).

The parabolic method yields a good approximation only for solutions that are differentiable a sufficient number of times, as may be seen at once from the formula for the approximation error, which for an mth order parabolic approximation generally contains the value of the mth or $(m+1)$-st order derivative of the approximating function at the intermediate point.

Therefore, the parabolic method can be applied, together with the usual formulas for the error of estimate, if the desired solutions are known to be sufficiently smooth.

In order that a solution of (1) be continuously differentiable $(m+1)$ times it is sufficient that for $\tau'(t) < 1$ the inequality $t > \gamma_m(t_0)$ be satisfied and that the function f be continuously differentiable m times.

Accordingly, we may recommend the following compound method of numerical integration: up to the point $t \geqq \gamma_m(t_0) + h$ use Euler's method with small step, either with or without iteration (either the method of successive integrations or the method of successive approximations), and from there on apply one of the parabolic methods based on an mth order parabola, e.g., Stürmer's method.

We remark that parabolic methods, beginning at $t = t_0$, may be successfully applied to equation (1) in the special case (when the set F_{t_0} reduces to a single point), if the function f is sufficiently smooth.

The method of decomposition in powers of the retardation may very often be applied to approximate integration and even to the investigation of the stability of the solution of differential equations with small retardation. When τ is a small constant, the method, as applied to the equation

$$\dot{x}(t) = f(t, x(t), x(t-\tau)), \quad x(t) = \varphi(t) \text{ on } E_{t_0}, \tag{3}$$

consists in replacing (3) by an unretarded equation

$$\dot{x}(t) = f\left(t, \ x(t), \ x(t) - \tau\dot{x}(t) + \ldots + \frac{(-1)^m \tau^m x^{(m)}(t)}{m!}\right). \tag{3a}$$

Since in general the solutions $x(t)$ of (3) have an angular point at $t = t_0$ we cannot choose $x^{(k)}(t_0) = \varphi^{(k)}(t_0 - 0)$ $(k = 0, 1, \cdots, (m-1))$ as initial points for $m > 1$. As initial values we must choose the derivatives from the right at $t = t_0$

$$\dot{x}(t_0 + 0) = f(t_0, \ \varphi(t_0), \ \varphi(t_0 - \tau)),$$
$$\ddot{x}(t_0 + 0) = \left[\frac{\partial f}{\partial t} + \frac{\partial f}{\partial x(t)} f + \frac{\partial f}{\partial x(t-\tau)} \dot{\varphi}(t-\tau)\right]_{t = t_0 + 0},$$
$$\cdots\cdots\cdots\cdots\cdots\cdots\cdots\cdots$$

Even in this case the method may not yield a satisfactory approximation to the solution of (3) for $t > t_0 + \tau$ for $m > 2$, because of the discontinuity of the second derivative of the solution at $t = t_0 + \tau$. One might expect that this method would give good results, like the parabolic methods, beginning with values of t for which sufficient smoothness of the solution is already guaranteed, especially in view of the fact that such smoothness is necessary if we are to expand the solution in a Taylor's series in τ.

Experimental investigations lead to a different answer, namely, that satisfactory approximations are obtained only for $m = 1$, regardless of the values of t for which it is applied. The reason for this paradox is as follows: if the solution is known to be sufficiently smooth, then

$$x(t-\tau) = x(t) - \tau \dot{x}(t) + \dots + \frac{(-1)^m \tau^m x^{(m)}(t)}{m!} + \frac{(-1)^{m+1} \tau^{m+1} x^{(m+1)}(t-\theta\tau)}{(m+1)!},$$
$$0 < \theta < 1,$$

and therefore equation (3) is equivalent to the equation

$$\dot{x}(t) = f\Big(t,\ x(t),\ x(t) - \tau\dot{x}(t) + \dots + \frac{(-1)^m \tau^m x^{(m)}(t)}{m!} + \frac{(-1)^{m+1}\tau^{m+1} x^{(m+1)}(t-\theta\tau)}{(m+1)!}\Big) \tag{4}$$

in the sense that any solution of (3) satisfying the condition $x(t) = \varphi(t)$ on E_{t_0} will satisfy (4) with the same initial function.

For $m > 1$, however, we cannot apply to equation (4) the theorem on continuous dependence of the solution on the parameters and yet neglect the term $(-1)^{m+1}\tau^{m+1}x^{(m+1)}(t-\theta\tau)/(m+1)!$, since this term is by no means small, as may be seen by reference to the theory of differential equations with small parameters in the high-order derivative, as given in §2, Chapter IV.

Let us assume, to simplify the exposition, that equation (3) can be written in the form

$$x(t-\tau) = F\big(t,\ x(t),\ \dot{x}(t)\big),$$

where F is continuously differentiable m times; then equation (4) has the form

$$x(t) - \tau\dot{x}(t) + \dots + \frac{(-1)^m \tau^m x^{(m)}(t)}{m!} + \frac{(-1)^{m+1}\tau^{m+1}x^{(m+1)}(t-\theta\tau)}{(m+1)!} = F\big(t,\ x(t),\ \dot{x}(t)\big),$$

or

$$\frac{\tau}{m} x^{(m)}(t) = x^{(m-1)}(t) + \dots + \frac{(-1)^{m-1}(m-1)!}{\tau^{m-1}} x(t) + \frac{(-1)^m (m-1)!}{\tau^{m-1}} F\big(t,\ x(t),\ \dot{x}(t)\big) + \frac{\tau^2}{m(m+1)} x^{(m+1)}(t-\theta\tau). \tag{5}$$

If we apply the method of successive integrations to this equation, we obtain an unretarded equation

$$x^{(m)}(t)=\frac{m}{\tau}x^{(m-1)}(t)+\ldots+\frac{(-1)^{m-1}m!}{\tau^{m}}x(t)$$
$$+\frac{(-1)^{m}m!}{\tau^{m}}F\left(t,\ x(t),\ \dot{x}(t)\right)+\frac{\tau}{m+1}\varphi_k^{(m+1)}(t-\theta\tau). \qquad \text{(5a)}$$

This equation is always of unstable type for $m > 2$, and is unstable for $m = 2$ when τ is sufficiently small,[2] and therefore, for $m > 1$, the neglected term $(\tau/m+1)\varphi_k^{(m+1)}(t-\theta\tau)$ is of order $1/\tau$ (since $x^{(m+1)}(t)=(m/\tau)x^{(m)}(t)+\cdots$, or $x^{(m+1)}(t)=(m/\tau^2)x^{(m-1)}(t)+\cdots$) and thus for small τ is never small itself. If, however, $m = 1$, equation (5a) has no small parameter in the derivatives, and, in view of the theorem on continuous dependence of the solution on the parameters, we may neglect the term $(\tau/2)\ddot{x}(t-\theta\tau)$ for sufficiently small τ (if $\ddot{x}(t-\theta\tau)$ is not too large in absolute value).

Thus the method of expansion in powers of τ can be applied only if $m = 1$.

Note that for $m = 1$ this method coincides with the method of substituting the approximation ratio $(x(t)-x(t-\tau))/\tau$ for the derivative $\dot{x}(t)$. Similar reasoning, applied to the kth order equation

$$x(t-\tau)=F\left(t,\ x(t),\ \dot{x}(t),\ \ldots,\ x^{(k)}(t)\right),\ \frac{\partial F}{\partial x^{(k)}}\neq 0$$

permits us to say that the method of expansion in powers of τ is applicable only for $m \leqq k$.

As noted earlier, the method of successive integrations is often applied in studying the stability of the solutions of differential equations with retarded argument. The most elementary examples prove, however, that the domains of stability of the respective solutions of (3) and (3a) may be totally distinct (see page 191).

6. **Dependence of the solution of a differential equation with retarded argument on a small coefficient of the leading derivative.** For the differential equation with retarded argument

$$\mu\dot{x}(t)=f\left(t,\ x(t),\ x\left(t-\tau(t)\right)\right),\ x(t)=\varphi(t)\ \text{on}\ E_{t_0}, \qquad (6)$$

---

[2] If we have the stable case for $m = 2$, the solution differs very little from the solution for $m = 1$.

where μ is a small positive parameter, and for equations or systems of equations of more general type, there arises the natural question: can we, for small μ, approximate the solution to (6) by the solution of the so-called degenerate equation

$$f\big(t,\ x(t),\ x(t-\tau(t))\big)=0. \tag{7}$$

The problem may be put more precisely as follows: To determine the conditions under which the solution $x(t,\mu)$ of equation (6) converges to a solution of (7) as $\mu\to 0$, and, if the latter has several solutions $\psi_i(t)$, to determine precisely which of them is the limit of $x(t,\mu)$ as $\mu\to 0$.

If the set F_{t_1} does not reduce to a single point for any $t_1 \geqq t_0$, the solution $x(t,\mu)$ of equation (6) can be found by the method of successive integrations (method of steps), that is, by solving unretarded equations of the form

$$\mu\dot{x}(t)=f\big(t,\ x(t),\ \varphi_k(t-\tau(t))\big)\ \text{ for }\ \gamma_k(t_0)\leqslant t\leqslant\gamma_{k+1}(t_0),$$
$$x\big(\gamma_k(t_0)\big)=\varphi_k\big(\gamma_k(t_0)\big)\quad (k=0,\ 1,\ 2,\ \ldots), \tag{8}$$

to which we may apply the results stated in §2, Chapter IV; that is, we may assert that the solution $x(t,\mu)$ of equation (8) tends, as $\mu\to 0$, to a stable root of the degenerate equation, if such a root exists, and to that root which contains in its domain of influence the initial values $x(\gamma_k(t_0)) = \varphi_k(\gamma_k(t_0))$.

We remark that the degenerate equation

$$f\big(t,\ x(t),\ x(t-\tau(t))\big)=0,\quad x(t)=\varphi_0(t)\ \text{on}\ E_{t_0},$$

in general has only discontinuous solutions, with points of discontinuity at $\gamma_k(t_0)$ $(k = 0,1,2,\cdots)$; these solutions can be found by iterating the solution of

$$f\big(t,\ x(t),\ \varphi_k(t-\tau(t))\big)=0\qquad (k=0,\ 1,\ 2)$$

(the method of steps); we also remark, in connection with the discontinuity of the solution of the degenerate equation, that the question which stable root (perhaps infinite) of the degenerate equation forms the limit as $\mu\to 0$ must be resolved afresh at each point of discontinuity $\gamma_k(t_0)$.

Example 1. $\mu\dot{x}(t) = -\frac{1}{2}x(t) + x(t-\tau)$, τ constant, $\tau > 0$, $\mu > 0$, $x(t) = a$ for $-\tau \leqq t \leqq 0$, where a is a constant.

The degenerate equation $-\frac{1}{2}x(t) + x(t-\tau) = 0$ has the stable roots

$$x(t) = 2a \quad \text{for} \quad 0 < t \leqslant \tau,$$
$$x(t) = 4a \quad \text{for} \quad \tau < t \leqslant 2\tau,$$
$$\cdots\cdots\cdots$$
$$x(t) = 2^n a \quad \text{for} \quad (n-1)\tau < t \leqslant n\tau,$$
$$\cdots\cdots\cdots$$

to which the solution of the original equation converges as $\mu \to 0$. (Figure 47).

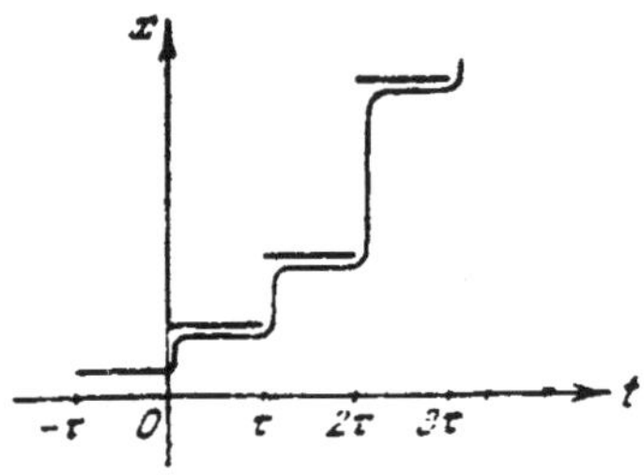

Figure 47

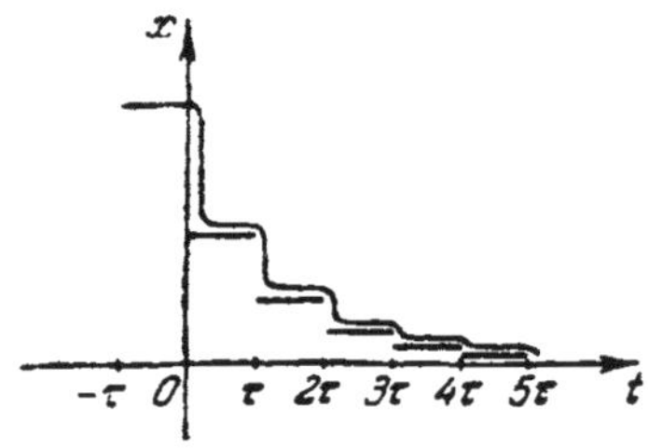

Figure 48

EXAMPLE 2. $\mu\dot{x}(t) = -2x(t) + x(t-\tau)$, τ constant, $\mu > 0$, $x(t) = a$ for $-\tau \leqq t \leqq 0$, where a is a constant.

The degenerate equation $-2x(t) + x(t-\tau) = 0$ has the stable roots

$$x(t) = 2^{-n}a \quad \text{for} \quad (n-1)\tau < t \leqslant n\tau \qquad (n = 1, 2, \ldots),$$

to which the solution of the original equation converges as $\mu \to 0$ (Figure 48).

A retarded equation with more than one retardation or a system of first order equations with retarded argument, which have small coefficients with respect to some or all of the derivatives, or higher order equations with retarded argument which have a small coefficient with respect to the leading derivative, all may be studied by reduction to an equation without retardation if the set F_{t_1} never reduces to a single point for $t_1 \geqq t_0$. Accordingly, the method of successive integrations (method of steps) is applicable. But an altogether new situation arises in the special case; the method of successive integrations is not applicable.

Suppose the set F_{t_0} in equation (6) consists of the single point $t = t_0$, the problem again being to study the relationship between the solution $x(t, \mu)$ of the equation

$$\mu\dot{x}(t) = f(t, x(t), x(t-\tau(t))), \quad x(t_0) = x_0$$

and the solutions of the degenerate equation

$$f(t,\ x(t),\ x(t-\tau(t)))=0$$

as $\mu\to 0$.

Only under highly specialized circumstances will the solution $x(t,\mu)$ tend to a solution of the degenerate equation as $\mu\to 0$. The most important cases are:

(1) The function f changes sign from $+$ to $-$, irrespective of the value of its third argument, as the representative point crosses the graph of the solution $x(t)=\varphi(t)$ of the degenerate equation in the direction of increasing x.

For example, $\mu x(t) = -x^2(t/2)x(t)$, $\mu>0$, $t\geqq 0$. The solution of the degenerate equation is $x(t)\equiv 0$ and this solution is the limit of all solutions $x(t,\mu)$ of the original equation as $\mu\to 0$.

(2) $f(t,x_0,x_0)\equiv 0$; then the original and the degenerate equation have the same solution $x\equiv x_0$.

For example, the equation $\mu\dot{x}(t) = x(t) - x(t/2)$, $x(0)=x_0$ and the degenerate equation $x(t)-x(t/2)=0$ have the solution $x(t)\equiv x_0$.

Now let us consider an equation[3] of the form

$$\mu\ddot{x}(t)=f(t,\ x(t),\ x(t-\tau(t)),\dot{x}(t-\tau(t))),$$
$$x(t)=\varphi(t)\ \text{on}\ E_{t_0} \tag{9}$$

$\mu>0$, which does not contain $\dot{x}(t)$.

If the sets F_{t_1} do not reduce to a single point for $t_1\geqq t_0$, the method of successive integrations (method of steps) reduces the study of equation (9) to the study of the unretarded equations

$$\mu\ddot{x}(t)=f(t,\ x(t),\ \varphi_k(t-\tau(t)),\ \dot{\varphi}_k(t-\tau(t))) \tag{10}$$
$$\text{for}\ \ \gamma_k(t_0)\leqslant t\leqslant\gamma_{k+1}(t_0),\ \ x(\gamma_k(t_0))=\varphi_k(\gamma_k(t_0)),$$
$$\dot{x}(\gamma_k(t_0))=\dot{\varphi}_k(\gamma_k(t_0)-0)$$

and, applying the results of Volosov, §2, Chapter IV (page 101) we may assert that the solution of equation (9) oscillates about that stable root (if it exists) of the degenerate equation which has the initial values in its domain of influence.

The solution of the degenerate equation is, generally speaking, discontinuous at the points $\gamma_k(t_0)$ $(k=0,1,2,\cdots)$, a fact which makes it necessary, at each point $\gamma_k(t_0)$ $(k=0,1,2,\cdots)$, to determine which stable root has the initial value in its domain of influence. Moreover, the amplitudes of oscillation of the solution $x(t,\mu)$ of equation (9) will in general undergo

[3] If the point t_0 is isolated in E_{t_0}, we also set $\dot{x}(t_0)=\dot{x}_0$.

a sharp change (Figure 49) upon passing through a discontinuity of the degenerate equation.

Thus, in the stable case, and for initial values close to the initial values of the solution of the degenerate equation, the unretarded equation $\mu\dot{x}(t) = f(t, x(t))$ has a solution $x(t, \mu)$ which for small μ remains near the solution of the degenerate equation, in the sense of zero-th order closeness; the analogous retarded equation (9) has this property only under highly exceptional circumstances, since as a rule the degenerate equation has a discontinuous solution and the amplitude of oscillation of $x(t, \mu)$ will in general change discontinuously as the trajectory goes through one of the discontinuities of the solution to the degenerate equation.

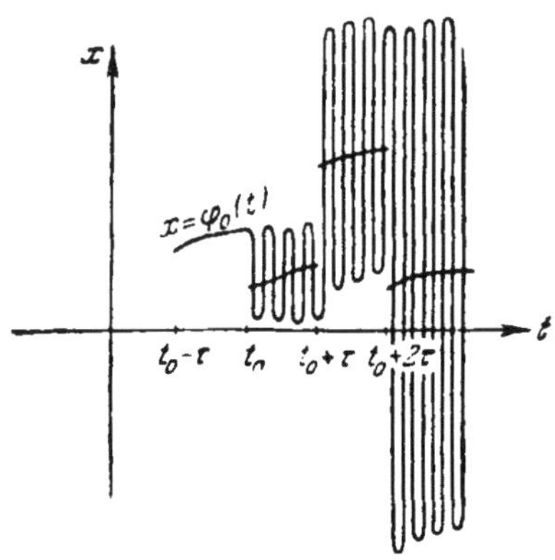

Figure 49

This special case for equation (9), (F_{t_0} consists of a single point), has been little studied, but with even a little thought on the problem it is not difficult to see that the stable case can occur only very rarely (for example, if the function f changes sign from plus to minus when the trajectory crosses the graph of the solution to the degenerate equation, in the direction of increasing values of the second argument, independently of the value of the third argument, and if f also satisfies conditions analogous to those of Volosov).

7. **Theorems on oscillations of the solution.** A. D. Myškis, in a series of papers [**193**; **198**], has studied the behavior of the solution of certain differential equations with retarded argument, of the form

$$\dot{x}(t) = M(t)\, x(t - \tau(t)) \tag{11}$$

and

$$\ddot{x}(t) = M(t)\, x\big(t - \tau(t)\big). \tag{12}$$

Recent work in this direction has been done by G. A. Kamenskiĭ [108], E. G. Cvang, and A. Ja. Lepin.

The basic approach consists in developing a comparison theorem among the solutions of the given class of equations with different $M(t)$ and $\tau(t)$; then the solutions of the equations with variable coefficients and variable retardations are compared to solutions of equations of the same form, but with constant coefficients and retardations. The particular solutions of the latter may be easily found by Euler's method.

We will not give the proofs for the fundamental theorems of A. D. Myškis, since they are set forth in great detail, and with extensive generalizations, in his book *Linear differential equations with retarded arguments*, but we shall nevertheless take time to formulate some of these theorems.

THEOREM I. *The solutions of the equation* $\dot{x}(t) = M(t)\,x(t - \tau(t))$, *where* $M(t)$ *and* $\tau(t)$ *are continuous non-negative functions,* $M(t) \geqq 0$, $\tau(t) \geqq 0$, *and are bounded in absolute value for* $t \geqq t_0$, *and the initial function* $\varphi(t)$ *is continuous, may be of one of the following types:*

(a) $x(t) \neq 0$ *for* $t \geqq t_0$, *and the rate of growth or decrease of the solution can be estimated;*

(b) $x(t)$ *changes sign on every interval* $[a, a + \tau_0)$, *where* $a \geqq t_0$, $\tau_0 = \sup \tau(t)$ *for* $t \geqq t_0$;

(c) $x(t) \equiv 0$;

(d) *mixed case: for* $t_0 \leqq a < b$ *we have case* (b) *and for* $t \geqq b + 2\tau_0$ *we have either case* (a) *or case* (c).

THEOREM II. *Under the same limitations, except that* $M(t) \leqq 0$, *the solutions* $x(t)$ *of equation* (11) *have the following properties:*

(1) *If* $\underline{\lim}_{t\to\infty} \tau(t) \cdot \underline{\lim}_{t\to\infty} M(t) > 1/e$, *the set of zeros of the solution* $x(t)$ *is unbounded from the right.*

This estimate is exact; that is, if $\underline{\lim}_{t\to\infty} \tau(t) \cdot \underline{\lim}_{t\to\infty} M(t) = 1/e$ there exist solutions that are known to have no zeros at all.

(2) *if* $\overline{\lim}_{t\to\infty} \tau(t) \cdot \overline{\lim}_{t\to\infty} M(t) < 3/2$, *all solutions tend to zero as* $t \to \infty$. This estimate is also exact.

(3) *The solutions may have one of the following forms:*

(a) $x(t) \neq 0$ *for* $t > t_0$;

(b) $x(t)$ *changes sign in every interval* $[a, a + \tau_0]$, *where* $a \geqq t_0$;

(c) $x(t) \equiv 0$;

(d) *mixed case: for* $t_0 \leqq a \leqq b$ *the solution* $x(t)$ *changes sign in every interval* $[a, a+\tau_0]$, *while for* $t \geqq b+\tau_0$ *the set of zeros of the solution is either connected or empty.*

THEOREM III. *The solutions* $x(t)$ *of the equation* $\ddot{x}(t) = M(t)x(t-\tau(t))$, *where* $M(t) \leqq 0$ *is a continuous function bounded in absolute value and the functions* $\tau(t)$ *and* $\varphi(t)$ *satisfy the conditions of Theorem* II, *have the following properties*:

(1) *If* $\operatorname{Inf} M(t) < 0$ *for* $t \geqq t_0$, *the set of zeros of each solution is unbounded on the right*;

(2) *if* $\tau_0\sqrt{\overline{\operatorname{Inf} M(t)}} < \pi/2 + \sqrt{2}$ *and the solution* $x(t)$ *is different from zero on some interval of length* τ_0, *then for all large values of the argument any two zeros of* $x(t)$ *are separated by an interval of length greater than* τ_0.

The properties of solutions of the equation $\ddot{x}(t) = M(t)x(t-\tau(t))$ when $M(t) \geqq 0$ have been studied by G. A. Kamenskiĭ, who obtained, in particular, the following theorem:

THEOREM IV. *The solutions of the equation* $\ddot{x}(t) = M(t)x(t-\tau(t))$, *where* $M(t)$ *and* $\tau(t)$ *are continuous non-negative functions and the initial function* $\varphi(t)$ *is continuous and* $\lim_{t\to\infty} \inf M(t) > 0$, *may have one of the following forms*:

(a) *as* $t \to \infty$ *the solution* $x(t)$ *tends to zero together with its derivative*;

(b) *as* $t \to \infty$ *the solution* $x(t)$ *and its derivative increase without bound in their absolute value*;

(c) *for arbitrary* t_1, *the solution* $x(t)$ *changes sign in the interval* (t_1, ∞).

There are examples to show that solutions of all three types (a), (b), and (c) exist.

Similar theorems exist for nonhomogeneous equations, of the form

$$\dot{x}(t) = M(t)\,x(t-\tau(t)) + P(t)$$

and

$$\ddot{x}(t) = M(t)\,x(t-\tau(t)) + P(t)$$

since the solution of a nonhomogeneous equation defined by the initial function $\varphi(t)$ on E_{t_0} can be considered to be the sum of the solution $x_1(t)$ of the corresponding homogeneous equation with the same initial function $\varphi(t)$ on E_{t_0} and the solution $x_2(t)$ of the nonhomogeneous equation with the initial function $\psi(t) \equiv 0$ on E_{t_0} (see page 177).

8. **Linear equations.**

1. *Linear homogeneous equations with retarded argument.* An nth order

linear homogeneous equation with retarded arguments can be written as

$$x^{(n)}(t)+\sum_{p=0}^{n-1}\sum_{j=1}^{m}a_{pj}(t)\,x^{(p)}(t-\tau_j(t))=0, \tag{13}$$

or, more briefly, as $L(x(t)) = 0$.

All the $\tau_j(t)$ are continuous and non-negative, the initial function $\varphi(t)$ is defined and continuous together with its derivatives up to and including the order $(n-1)$ on the initial set E_{t_0}.

For linear homogeneous equations with retarded arguments we have the following easily verified theorem:

(1) *If all the coefficients $a_{pj}(t)$ are continuous on the interval $t_0 \leqq t \leqq T$, there exists a unique solution $x(t)$ which, together with its derivatives up to and including the order $(n-1)$, satisfies* (13) *on the given interval and satisfies the initial conditions $x(t) = \varphi(t)$ on E_{t_0}. If the point t_0 is isolated in E_{t_0}, then at t_0 we must prescribe the values of the derivatives from the right, up to and including the order $(n-1)$.*

(2) *Linearity and homogeneity are conserved under a linear homogeneous transformation of the desired function.*

(3) *A linear combination* $\sum_{i=1}^{p} c_i x_i(t){:}\varphi_i$ of solutions of equation* (13), *defined by the initial functions $\varphi_i(t)$, is a solution of the same equation with the initial function $\sum_{i=1}^{p} c_i\varphi_i(t)$, and if the point t_0 is isolated in E_{t_0}, the values of the derivatives from the right corresponding to the solution $\sum_{i=1}^{p} c_i x_i(t){:}\varphi_i$ are given at this point by*

$$\sum_{i=1}^{p} c_i \underset{\varphi_i}{x_i^{(s)}}(t_0+0) \qquad (s=0,\ 1,\ \ldots,\ (n-1)).$$

Property (3) continues to hold even when $p \to \infty$ if the series so generated converges and admits an n-fold term-by-term differentiation.

(4) *If equation* (13), *having real coefficients, admits the complex solution $x(t) = u(t) + iv(t)$, then the real and imaginary parts $u(t)$ and $v(t)$ of the solution are themselves solutions of the equation.*

We consider the set $\{x(t){:}\varphi\}$ of solutions $x(t){:}\varphi$ $(t_0 \leqq t \leqq T)$ of equation (13), defined by some set $\{\varphi(t)\}$ of initial functions $\varphi(t)$ given on the initial set E_{t_0}.

We introduce into the sets $\{x(t){:}\varphi\}$ and $\{\varphi(t)\}$ a linear metric of some kind, and thereby convert these sets into linear spaces Ω_x and Ω_φ. We shall suppose that the metric satisfies the theorem on the continuous dependence

* *Translator's note*: In the following expression the notation $x_i(t){:}\varphi_i$ means that $x_i(t)$ satisfies the initial condition $x_i(t) = \varphi_i(t)$. The notation in the original is $\underset{\varphi_i}{x_i(t)}$, as in the displayed formulas in this translation.

of the solution on the initial function, that is, that for every $\epsilon > 0$ one may choose $\delta(\epsilon)$ so that $\rho_\varphi(\varphi_1, \varphi_2) < \delta(\epsilon)$ implies $\rho_x(x:\varphi_1, x:\varphi_2) < \epsilon$, where ρ_φ and ρ_x are distances in the metrics of Ω_φ and Ω_x respectively.

In particular, if the metrics of the spaces Ω_φ and Ω_x coincide with the metric of the space C_{n-1} (which is a frequently occurring case) then, as is well known, the theorem on the continuous dependence of the solution on the initial function is valid for arbitrary $T < \infty$.

Definition. The solutions $x_i(t):\varphi_i$ $(i = 1, 2, \cdots)$ of equation (13), defined by the initial functions $\varphi_i(t)$, form a *fundamental system of solutions* of equation (13) with respect to the space of initial functions Ω_φ if the φ_i $(i = 1, 2, \cdots)$ form a basis for the space Ω_φ.

Theorem. *Every solution $x(t):\varphi$ of equation* (13) *with continuous coefficients $a_{ij}(t)$, which has an initial function φ belonging to the space Ω_φ can be represented in the form $x(t):\varphi = \sum_{i=1}^{\infty} \alpha_i x_i(t):\varphi_i$, where $x_i(t):\varphi_i$ $(i = 1, 2, \cdots)$ is a fundamental system of solutions of equation* (13) *with respect to the space Ω_φ, and the convergence of the series $\sum_{i=1}^{\infty} \alpha_i x_i(t):\varphi_i$ with constant coefficients α_i is to be taken in the sense of the metric of the space Ω_x.*

Proof. Since the functions φ_i form a basis for the space Ω_φ, the function φ can be represented in the form $\varphi = \sum_{i=1}^{\infty} \alpha_i \varphi_i$.

In view of the continuous dependence of the solution on the initial function, the convergence of the series $\sum_{i=1}^{\infty} \alpha_i \varphi_i$ to the function φ implies the convergence of the series $\sum_{i=1}^{\infty} \alpha_i x_i(t):\varphi_i$ to the function $x(t):\varphi$, since the inequality

$$\rho_\varphi\Big(\sum_{i=1}^{n} \alpha_i \varphi_i(t), \varphi(t)\Big) < \delta(\varepsilon)$$

implies the inequality

$$\rho_x\Big(\sum_{i=1}^{n} \alpha_i \underset{\varphi_i}{x_i(t)}, \underset{\varphi}{x(t)}\Big) < \varepsilon.$$

2. *Linear nonhomogeneous equations of order n.* An nth order linear nonhomogeneous equation with retarded argument can be written in the form

$$x^{(n)}(t) + \sum_{p=0}^{n-1} \sum_{j=1}^{m} a_{pj}(t)\, x^{(p)}\big(t - \tau_j(t)\big) = f(t), \tag{14}$$

or in abbreviated notation

$$L(x(t)) = f(t), \tag{14'}$$

where the notation is the same as before, and $f(t)$ is a continuous function of t for $t_0 \leqq t \leqq T$.

The linear nonhomogeneous equation (14) has the following easily verified properties:

(1) If all coefficients $a_{pj}(t)$ and the function $f(t)$ are continuous in the interval $t_0 \leqq t \leqq T$, there exists a unique $(n-1)$-times differentiable solution of equation (14) in the same interval, satisfying the initial condition $x(t) = \varphi(t)$ on E_{t_0} (if the point t_0 is isolated in E_{t_0}, it is necessary to prescribe the derivatives from the right $x^{(p)}(t_0+0) = x_0^{(p)}$ $(p = 0, 1, \cdots, (n-1))$ at the point t_0).

(2) The linearity of equation (14) is conserved under a linear transformation of the desired function.

(3) The sum of a solution $x(t):\varphi$ corresponding to the homogeneous equation (13) and $\bar{x}(t):\psi$ corresponding to the nonhomogeneous equation (14) is a solution of the nonhomogeneous equation (14) with the initial function $\varphi + \psi$, if φ and ψ are the respective initial functions of the solutions $x(t):\varphi$ and $\bar{x}(t):\psi$.

(4) The superposition principle is valid: the solution $x(t):\varphi$ of the equation

$$L(x(t)) = f_1(t) + f_2(t)$$

is the sum of the solutions $x_1(t):\varphi_1$ and $x_2(t):\varphi_2$ of the equations $L(x(t)) = f_1(t)$ and $L(x(t)) = f_2(t)$ respectively, if the initial functions φ_1 and φ_2 satisfy the condition $\varphi_1 + \varphi_2 = \varphi$ on E_{t_0}.

(5) An arbitrary solution $x(t):\varphi$ of the nonhomogeneous equation (14) for $t_0 \leqq t \leqq T$ can be represented in the form

$$x(t):\varphi = x(t):\psi + \textstyle\sum_{i=1}^{\infty} \alpha_i x_i(t):\varphi_i,$$

if $\varphi - \psi$ belongs to the space Ω_φ, where $x(t):\psi$ is a particular solution of the nonhomogeneous equation (14) and the $x_i(t):\varphi_i$ are solutions of the corresponding homogeneous equation that form a fundamental system for the space of initial functions Ω_φ.

This theorem reduces to the corresponding theorem for homogeneous equations by the substitution $y(t) = x(t) - x(t):\psi$. Regarding the properties of the spaces Ω_φ, Ω_x and the meaning of convergence of the series $\sum_{i=1}^{\infty} \alpha_i x_i(t):\varphi_i$, see page 175.

3. *Linear equations with constant coefficients and constant retardations.* The simplest, and at the same time most important class of linear homogeneous equations with retarded argument consists of linear homogeneous equations with constant coefficients and constant retardations, which we obtain by supposing that in equation (13) all a_{pj} and τ_j are constants:

$$x^{(n)}(t)+\sum_{p=0}^{n-1}\sum_{j=1}^{m}a_{pj}x^{(p)}(t-\tau_j)=0. \tag{15}$$

Linear homogeneous equations with constant coefficients and constant retardations are integrable (see §4, Chapter V) and therefore difficulty in integrating or investigating equations of this type can arise only if the smallest nonzero retardation τ_j is small compared to the length of the interval (t_0, T) on which one attempts to determine the solution; it may also be difficult to determine the asymptotic properties of the solution, in particular with respect to questions of stability.

If we apply Euler's method to linear homogeneous equations with constant coefficients and constant retardations, we quite easily determine a countable set of linearly independent solutions which are analytic in the interval $t_0 \leqq t < \infty$.

In fact, we obtain the solution in the form e^{kt}. Substituting $x(t) = e^{kt}$ in (15), where all the a_{pj} and τ_j are constant, and cancelling out the e^{kt}, we obtain for the determination of k the following equation, which is called the characteristic equation:

$$k^n+\sum_{p=0}^{n-1}\sum_{j=1}^{m}a_{pj}k^p e^{-k\tau_j}=0. \tag{16}$$

Thus, k must be a root of the characteristic quasi-polynomial occurring on the left-hand side of equation (16).

If $k = k_i$ is a root of multiplicity α of this polynomial, then by methods analogous to those applied to linear homogeneous equations with constant coefficients and without deviating arguments, it is easy to show that the solutions of equation (15) will be $t^s e^{k_i t}$ $(s = 0, 1, \cdots, (\alpha - 1))$.

The characteristic quasi-polynomial has a countable set of roots. If all coefficients a_{pj} are real, the complex roots of the characteristic quasi-polynomial, $k_j = p_j + q_j i$, correspond to two real solutions, the real and imaginary parts of $e^{k_j t}$ (or, in the case of multiple roots the real and imaginary parts of $t^s e^{k_j t}$).

By the results of Leont'ev [**131**; **132**] and Hylb [**286**], the system so obtained forms a fundamental system for a fairly broad class of spaces of admissible initial functions, but, as is not hard to prove, *the system in question is not a fundamental system for the space C_{n-1}* of $(n-1)$-fold differentiable initial functions with the metric of C_{n-1}.

4. *Linear nonhomogeneous equations with constant coefficients and constant retardations*. Linear nonhomogeneous nth order equations with constant

coefficients and constant retardations have the form

$$x^{(n)}(t)+\sum_{p=0}^{n-1}\sum_{j=1}^{m} a_{pj}x(t-\tau_j)=f(t), \tag{17}$$

where all a_{pj} and τ_j are constant.

In view of property (5) of the solution of a nonhomogeneous equation (see page 177) the task of integrating equation (17) reduces to finding a particular solution of (17) and to integrating the corresponding homogeneous equation.

In many cases the particular solution of the inhomogeneous equation is comparatively easy to find or can be found by methods of the operational calculus.

All theorems of this section can be carried over practically without change to systems of linear equations with retarded argument.

9. **Classification of stationary points and estimates of their number.** By a *stationary point of the dynamical system* of equations with retarded argument

$$\dot{x}_i(t) = f_i(x_1(t-\tau_{11}(t)),\cdots,x_1(t-\tau_{1m}(t),\cdots,$$
$$\cdots,x_n(t-\tau_{n1}(t)),\cdots,x_n(t-\tau_{nm}(t))), \tag{18}$$

which we shall from now on write in the short form

$$\dot{x}_i(t) = f_i(x_j(t-\tau_{js}(t))) \tag{18'}$$

we shall mean a solution $x_i = x_{i0}$ $(i = 1, 2, \cdots, n)$ both for $t \geqq t_0$ and on the initial set E_{t_0} where the x_{i0} are constants.

For a linear homogeneous dynamical system with constant retardations τ_{js}

$$\dot{x}_i(t)=\sum_{s=1}^{m}\sum_{j=1}^{n} a_{ijs}x_j(t-\tau_{js}) \tag{A}$$

the stationary point $x_i \equiv 0$ $(i = 1, 2, \cdots, n)$ may be naturally classified according to the properties of the roots of the characteristic quasi-polynomial (see page 178). In fact, if the characteristic quasi-polynomial has p roots with negative real part (when we speak of the number of roots we mean the sum of their multiplicities), q roots with positive real part, and s roots for which the real part is zero, the singular point is said to be a stationary point of type (p, q, s) (some of the p, q, s may take on an infinite value). A stationary point of the type $(\infty, 0, 0)$ may naturally be called a generalized stable node (or focus); a point $(0, \infty, 0)$ a generalized unstable node (or focus); a point of the type $(p, q, 0)$, where $p \neq 0$ and

$q \neq 0$ may naturally be called a p-saddle. A full analogue of a center, that is, a point of the type $(0, 0, \infty)$, cannot exist (on the finite intervals of the imaginary axis there can be at most a finite number of zeroes of the characteristic quasi-polynomial of the system (A), and there cannot be purely imaginary roots iy_k which increase without limit in absolute value, since the absolute value of the leading term k^n of the characteristic quasi-polynomial becomes larger, for sufficiently large $|y_k|$, than the absolute value of the sum of all the other terms).

A more complete characterization of the stationary points may be made by listing the number p of negative roots, q of positive roots, r of complex roots with positive real part, s of complex roots with negative real part, t of purely imaginary roots, and u of roots that are zero.

For the system (18) with constant τ_{js}, if we suppose that the f_i are continuously differentiable twice, the classification of the stationary points in the noncritical case can be carried out by transferring the origin of coordinates to the stationary point and then, in the neighborhood of the point, studying the behavior of the system of first order approximations to the system (18); it must be observed, however, that if such a classification is to be correct there must be a topological equivalence between the set of integral curves of (18) in the neighborhood of the stationary point, on the one hand, and the set of integrals of the first order approximations on the other, under the assumption that we are in the noncritical case, i.e., the stationary point is of the type $(p, q, 0)$. There are only scattered, results, far from complete, along these lines (see §10, Chapter V).

The problem of estimating the minimal number of stationary points of the system (18), which we consider to be defined on some closed manifold M^n, so that the x_i $(i = 1, 2, \cdots, n)$ are local coordinates of the points of this manifold, can be solved as follows: corresponding to each equation of the system (18) we consider the equation without deviations

$$x_i(t) = f_i\big(x_1(t), \ldots, x_1(t), x_2(t), \ldots \\ \ldots, x_2(t), \ldots, x_n(t), \ldots, x_n(t)\big),$$

which we write briefly as

$$\dot{x}_i(t) = f_i\big(x_j(t)\big) \qquad (i = 1, 2, \ldots, n). \tag{19}$$

Clearly, the stationary points of the systems (18) and (19) coincide. It is also clear that to every system (19) there corresponds an infinite set of systems (18).

Therefore, an estimate of the number of stationary points of the system (18) can be replaced by an estimate of the number of stationary points of the system (19), and additional restrictions may be placed on the functions f_i and the retardations $\tau_{js}(t)$ so long as they permit us to find at least one system (18) corresponding to (19).

If we consider the index of a stationary point of (18) to be the index of the corresponding point of (19), then the theorems counting indices or sums of indices of stationary points continue to hold good on (18); in particular, Hopf's theorem stating that the sum of the indices of all stationary points is equal to the Euler characteristic of the manifold.

If we consider only such systems (18) as have corresponding systems (19) that are potential systems, i.e., $f_i(x_j(t)) = \partial u(x_1, x_2, \cdots, x_n)/\partial x_i$, or at least such systems (18) as have stationary points coinciding with the stationary points of some potential system (19), then the invariants of category type (homology class, homotopy class, unit length, number of elements, number of elements with regular contiguity, etc.) of the manifold M^n provide a lower bound to the number of geometrically distinct stationary points.

If we introduce in a potential system the concept of stationary point of type k, and define the type of a stationary point in the system (18) as equal to the type of the corresponding stationary point in the system (19), and similarly introduce the notion of multiple stationary point, then the number of analytically distinct stationary points of type k is not less than the k-dimensional Betti number, mod 2, of the manifold M^n. Under the same assumptions Morse's inequality holds, and we may estimate the number of stationary points of type k by Betti numbers relative to other moduli, with the aid of the Betti groups and the fundamental group.

The condition $\tau_{js}(t) \geqq 0$ has been used only in defining the initial set E_{t_0}, so that everything we have said may be easily generalized to the case of arbitrary continuous $\tau_{js}(t)$.

10. **Stability of solutions of differential equations with deviating arguments.**

1. Everything said in Chapter IV, §5, about the value of studying the stability of a solution of a differential equation in the Ljapunov sense, is equally true of solutions of differential equations with deviating arguments; it is therefore not surprising that the question of stability of the solutions is central to the theory of differential equations with deviating arguments.

In dealing with the simplest kind of first order equations

$$\dot{x}(t) = f\big(t, \ x(t), \ x(t-\tau(t))\big), \ \tau(t) \geqslant 0 \tag{20}$$

one generally starts with a definition of stability as follows: A solution $x(t):\varphi$ of (20), defined by the initial function $\varphi(t)$ on the initial set E_{t_0}, is said to be *stable* if to every $\epsilon > 0$ corresponds a $\delta > 0$ such that the inequality $\sup|\bar{\varphi}(t) - \varphi(t)| < \delta$ on E_{t_0} implies that $|x(t):\bar{\varphi} - x(t):\varphi| < \epsilon$ for $t \geqq t_0$, where $x(t):\bar{\varphi}$ is a solution of (20) determined by the initial function $\bar{\varphi}(t)$.

This definition is not altogether satisfactory, since it depends not only formally but also essentially on the choice of the point t_0. In actual systems, described by differential equations with or without deviating arguments, a perturbation may occur at any moment, not only on the initial set.

For differential equations without deviating arguments, the theorem about the continuous dependence of the solution on the initial value, combined with the possibility of extending the solution in the direction of decreasing time, implies that if the solution is stable against a perturbation at any one moment t_0 it is stable against a perturbation at any other moment of time $\bar{t}_0$.

The situation is altogether different when we come to differential equations with retarded arguments, since in general we cannot extend the solution in the direction of decreasing values of the time t.

Let us consider an example. We shall see that a solution may be stable against perturbations on one initial set and still unstable when perturbed on a different initial set.

Suppose $\dot{x}(t) = x(t) - x(te^{-t})$ for $t \geqq 0$ and that the initial set E_0 consists of the single point $t = 0$. The trivial solution $x \equiv 0$ of this equation is stable against perturbation on the set E_0, since the general solution of the equation is $x = C$, where C is an arbitrary constant, provided the initial point is taken as $t_0 = 0$. But if we take any other initial value $t_1 \geqq e^{-1}$, the trivial solution is no longer stable with respect to perturbations on E_{t_1}. In fact, if we put $\varphi(t) = \delta/2$ on E_{t_1} for $t \neq t_1$ and $\varphi(t_1) = \delta$, where δ is a constant, we obtain the solution of the equation in the form

$$\underset{\varphi}{x}(t) = \frac{\delta}{2}[e^{(t-t_1)} + 1].$$

The solution $x(t):\varphi$ increases in absolute value without limit as t increases, and the solution $x \equiv 0$ is therefore unstable.[4]

---

[4] We note that $\varphi(t)$ is continuous in E_{t_1}, since the point t_1 is isolated in E_{t_1}. Against a disturbance on the set E_{t_1} for $0 < t_1 < e^{-1}$ the trivial solution is also unstable, but the proof of its instability requires the choice of another initial function, since the point t_1 is then not an isolated point of E_{t_1}.

This example is not exceptional; the definition of stability that we have used should therefore be changed, at least with respect to a retarded equation of the form (20) or with respect to a system of equations

$$\begin{gathered}\dot{x}_i(t) = f_i\big(t, x_j(t - \tau_{kj}(t))\big)\\ (i, j = 1, 2, \ldots, n; \quad k = 1, 2, \ldots, m), \\ x_j(t) = \varphi_j(t) \text{ on } E_{t_0}.\end{gathered} \tag{21}$$

A solution $x_j(t):\varphi_j$ $(j = 1, 2, \cdots, n)$ of the system (21), defined by the initial functions $\varphi_j(t)$, is said to be stable if for every $\epsilon > 0$ there exists a $\delta(\bar{t}_0, \epsilon)$ such that if the inequality $|\psi_j(t) - x_j(t):\varphi_j| < \delta(\bar{t}_0, \epsilon)$ holds on the initial set $E_{\bar{t}_0}$, then

$$|\underset{\psi_j}{x_j(t)} - \underset{\varphi_j}{x_j(t)}| < \varepsilon \quad (j = 1, 2, \ldots, n) \quad \text{for} \quad t \geqslant \bar{t}_0$$

for all $\bar{t}_0 \geqq t_0$, where $x_j(t):\psi_j$ $(j = 1, 2, \cdots, n)$ is a solution of the system (21) defined by the initial functions $\psi_j(t)$ which are given on $E_{\bar{t}_0}$.

If δ can be chosen so that it depends on ϵ only, the stability is said to be uniform. If the solution $x_j(t):\varphi_j$ $(j = 1, 2, \cdots, n)$ is stable and if also $\lim_{t \to \infty} |x_j(t):\psi_j - x_j(t):\varphi_j| = 0$, then the solution $x_j(t):\varphi_j$ $(j = 1, 2, \cdots, n)$ is said to be asymptotically stable.

From now on, when speaking of stability, we shall have in mind this definition of stability and asymptotic stability. An equation of neutral type, of the form $\dot{x}(t) = f(t, x(t), x(t - \tau(t)), \dot{x}(t - \tau(t)))$, or a system of similar equations, will in general admit an extension of the solution in the direction of forward and backward time, although without additional conditions the extension to values of $t < t_0$ will not be unique; thus even for equations of this type it is desirable to adopt the above definition of stability.

We remark that for equations of neutral type and of first order it is much more natural, in connection with the definition of stability, to require that the initial functions be near to the first order rather than to the null order.

The above definition also holds for nth order differential-difference equations, but nearness is defined for initial functions and solutions in the sense of nearness of order $(n - 1)$; for equations of neutral type nearness is usually taken as nth order nearness. In many cases, of course, an nth order differential equation can be replaced by a system of first-order equations, for which the above definitions of stability apply directly.

When we make such a reduction to a system of equations, the class of initial functions becomes broader, since the new unknown functions may be defined arbitrarily on the initial set and need not be derivatives of order k ($k = 1, 2, \cdots, (n-1)$) of one initial function which determines the solution of the original nth order equation.

One might expect, therefore, that the requirement of stability imposed on the solution of an nth order equation is not equivalent to the requirement imposed on the solutions of the system of equations to which the original nth order equation is converted; it is not hard to see, however, that at least in the important cases, we encounter no such inequivalence, although it may be met when some one of the differences $t - \tau_i(t)$ takes on a value less than t_0 for sufficiently large values of t.

To sum up, if we apply the method of steps and $\tau'(t) < 1$,[5] an admissible set of initial functions $(\varphi_0, \varphi_1, \varphi_2, \cdots, \varphi_{n-1})$ on the initial set E_{t_0} induces a perturbation on the set $E_{\gamma(t)}$ which satisfies the system of equations and is small enough in absolute value; therefore, the set of solutions of the system of equations generated by a perturbation on E_{t_0} is included for $t \geqq \gamma(t_0)$ in the set $(x(t), \dot{x}(t), \cdots, x^{(n-1)}(t))$, where $x(t)$ is a solution of the nth order equation determined by a perturbation on the set $E_{\gamma(t_0)}$. Therefore the stability of the trivial solution of the nth order equation implies the stability of the trivial solution of the corresponding system of equations.

In some cases another definition of nearness is more natural for the initial functions. If the set E_t is bounded, it is sometimes useful to introduce the metric of the space L^2 into the set of initial functions.

2. *Stability of solutions of linear homogeneous equations with constant coefficients and constant retardations.* A linear homogeneous differential-difference equation of order n with constant coefficients can be written in the form

$$\sum_{\mu=0}^{m} \sum_{\nu=0}^{n} a_{\mu\nu} x^{(\nu)}(t+\tau_\mu) = 0, \tag{22}$$

where $a_{\mu\nu}$ and τ_μ are real constants, for which $0 \leqq \tau_0 < \tau_1 < \cdots < \tau_m$ and at least one of the $a_{\mu n} \neq 0$.

If $a_{mn} \neq 0$ and $a_{in} = 0$ for $i = 0, 1, 2, \cdots, (m-1)$, then equation (22) is an equation with retarded argument. If $a_{0n} \neq 0$ and $a_{in} = 0$ for $i = 1, 2, \cdots, m$, then we have an equation with advanced argument.

Particular solutions of such equations may be found by Euler's method (see page 178).

We find the solution in the form $x(t) = e^{kt}$; substituting this in equation

[5] This condition may be considerably weakened.

(22) and cancelling out e^{kt}, we obtain the so-called characteristic equation

$$\sum_{\mu=0}^{m}\sum_{\nu=0}^{n} a_{\mu\nu}k^{\nu}e^{k\tau_{\mu}} = 0, \tag{23}$$

whose roots k_j correspond to particular solutions of the form $e^{k_j t}$. Multiple roots of the characteristic equation, as we have already indicated earlier, correspond to solutions of the form $P_j(t)e^{k_j t}$, where $P_j(t)$ is a polynomial with arbitrary constant coefficients, of degree $\alpha_j - 1$, α_j being the multiplicity of the root k_j.

It is obvious that if one of the roots $k_j = p_j + iq_j$ of the characteristic quasipolynomial $\sum_{\mu=0}^{m}\sum_{\nu=0}^{n} a_{\mu\nu}k^{\nu}e^{k\tau_{\mu}}$ has a positive real part, the trivial solution of equation (22) is unstable.

In fact, in this case the solutions of equation (22) of the form $c_1 e^{p_j t}\cos q_j t$ and $c_2 e^{p_j t}\sin q_j t$ do not remain close to the solution $x \equiv 0$ as $t \to \infty$.

It may be shown that if the real parts $\mathrm{Re}(k_j)$ of all roots of the characteristic equation satisfy the condition $\mathrm{Re}(k_j) \leqq -\alpha < 0$, the solution $x \equiv 0$ of equation (22) is asymptotically stable. This assertion is far from obvious, even if only because not every solution of (22) can be represented in the form $\sum_{j=1}^{\infty} P_j(t)e^{k_j t}$. The proof of this criterion for stability of equations with retarded argument was given by Bellman [**28**; **29**], and for equations of neutral type by Wright [**239**; **240**]. For equations of the type of (22) with $a_{mn} = 0$, provided that some $a_{m\nu} \neq 0$, there is always at least one root of the characteristic equation with positive real part, and therefore in such equations the trivial solution $x \equiv 0$ is always unstable. These equations include in particular all equations with advanced arguments which are of type (22).

This instability criterion follows from some results of L. S. Pontrjagin [**230**], which were generalized in the text of N. G. Čebotarev and N. N. Meĭman [**178**] to the extent that a quasipolynomial without a principal term[6] has an infinite set of roots with real parts increasing unboundedly.

Bellman's proof, as applied to the simplest equation with retarded argument

$$x(t+\tau) = a_1 x(t+\tau) + a_2 x(t), \quad \tau > 0, \tag{24}$$

is based on the following idea: if in a purely formal way we apply a Laplace transformation, we obtain a function $x(t):\varphi$ which, if the appli-

[6] The principal term of the quasipolynomial $\sum_{\mu=0}^{m}\sum_{\nu=0}^{n} a_{\mu\nu}z^{\nu}e^{\lambda_{\mu}z}$ is the term containing simultaneously the largest values of ν and λ_{μ}.

cation of the Laplace transformation is in fact justifiable, will satisfy equation (24) for $t > \tau$ and will satisfy the initial conditions $x(t):\varphi = \varphi(t)$ for $0 \leqq t \leqq \tau$, where $\varphi(t)$ is a continuous function of bounded variation:

$$\underset{\varphi}{x}(t) = \frac{1}{2\pi i} \int_{\Gamma} \frac{\varphi(\tau) + (se^{\tau s} - a_1 e^{\tau s}) \int_0^{\tau} \varphi(u) e^{-su} du}{se^{\tau s} - a_1 e^{\tau s} - a_2} e^{ts} ds. \quad (25)$$

The contour of integration Γ has the equation $s = -c + iy$, where c is a constant satisfying the inequalities $0 < c < \alpha$, and y ranges over the infinite line, $-\infty < y < \infty$.

Let us verify that the function $x(t):\varphi$ does indeed satisfy equation (24) and the initial conditions.

We shall make use of the fact that all roots k_j of the characteristic equation satisfy the inequality

$$\operatorname{Re}(k_j) \leqslant -\alpha < 0.$$

It follows from (25) that for sufficiently large t, $|x(t):\varphi| < \epsilon$ and $\lim_{t\to\infty} x(t):\varphi = 0$, independently of the choice of $\varphi(t)$. In other words, the stationary point $x \equiv 0$ is not only asymptotically stable, but also the solutions of the equation are asymptotically stable in the large.

Using the same method, I. A. Fried proved the stability of the stationary point $x \equiv 0$ of equation (24) when the characteristic quasipolynomial not only has zeros satisfying the inequality $\operatorname{Re}(k_i) \leqq -\alpha < 0$ but may have also a finite set of simple zeros with real part equal to zero.

Wright **[241]** proved the asymptotic stability of the trivial solution of (22), using another method, under the assumption that all $\operatorname{Re}(k_i) \leqq -\alpha < 0$ and that in addition the initial function $\varphi(t)$ is close to $x \equiv 0$ in the sense of nth order nearness.

The results of Bellman, Wright, and Fried can be carried over without difficulty to systems of linear homogeneous differential-difference equations with constant coefficients.

3. *Study of stability in first-order approximation.* The papers of Bellman **[29]** and Wright **[241]** have provided the basis for a method frequently applied to the study of first-order stability of differential-difference equations which are stationary to the first order.

Up to the present time, no method nor basis for a method has been developed for studying the first-order stability of systems that have first order approximations with variable coefficients or variable retardations.

Bellman's results, as applied to a single equation of first order with one retarded argument can be formulated as follows: consider the equation

$$\dot{x}(t) = ax(t) + bx(t-\tau) + c(t)x(t) + d(t)x(t-\tau) + \sum_{i+j=2}^{\infty} a_{ij}(t)x^i(t)x^j(t+\tau),$$

where: a, b, and τ are constants; $\tau > 0$; $c(t)$ and $d(t)$ are continuous functions of t; $\lim_{t\to\infty} c(t) = \lim_{t\to\infty} d(t) = 0$; all the continuous functions $a_{ij}(t)$ are bounded in absolute value; the series

$$\sum_{i+j=2}^{\infty} a_{ij}(t)y^i z^j$$

converges for sufficiently small absolute values of y and z; i and j are positive integers. Under all these conditions our equation has an asymptotically stable trivial solution if the real parts of all the roots of the characteristic equation $k = a + be^{-k\tau}$ satisfy the inequality $\mathrm{Re}(k_i) \leqq -\alpha < 0$.

Bellman's theorem may be similarly formulated for systems of equations. Wright's result is a significant generalization of Bellman's theorem, and in particular embraces many equations of neutral type.

WRIGHT'S THEOREM. *The trivial solution of the equation*

$$\sum_{\mu=0}^{m}\sum_{\nu=0}^{n} a_{\mu\nu}x^{(\nu)}(t+\tau_\mu) + \Psi(t, x^{(\nu)}(t+\tau_\mu)) = 0 \tag{26}$$

is asymptotically stable if all roots k_i of the characteristic equation

$$\sum_{\mu=0}^{m}\sum_{\nu=0}^{n} a_{\mu\nu}k^\nu e^{\tau_\mu k} = 0 \tag{23}$$

satisfy the inequality $\mathrm{Re}(k_i) \leqq -\alpha < 0$.

In equation (26) all $a_{\mu\nu}$ and τ_μ are assumed constant; $a_{mn} \neq 0$; $0 \leqq \tau_0 < \tau_1 < \cdots < \tau_m$; the function

$$\Psi(t, x^{(\nu)}(t+\tau_\mu)) = \Psi(t, x(t+\tau_0), \dot{x}(t+\tau_0), \ldots, x^{(n-1)}(t+\tau_0), \ldots, x(t+\tau_m), \ldots, x^{(n-1)}(t+\tau_m))$$

is continuous in all its arguments in the domain D where $t \geqq 0$, and the remaining arguments are bounded in absolute value by some constant $\sigma > 0$. Moreover, the function Ψ is to satisfy the following condition:

$$\text{if} \quad \sum_{\mu=0}^{m}\sum_{\nu=0}^{n-1} |W_{\mu\nu}| \leqslant W, \tag{27}$$

$$\text{then } |\Psi(t, W_{\mu,i})| \leqslant W\chi(W), \tag{28}$$

where $\chi(W)$ is a bounded function of W such that $\int_0^a (\chi(w)/W)dW$ converges for $a > 0$.

The condition imposed on the function Ψ is significantly broader than the condition imposed on the nonlinear terms in Bellman's theorem.

E. M. Esipovič has proved a theorem on the instability of the trivial solution of equation (26) under somewhat more stringent conditions on the function Ψ, when at least one root of the characteristic quasipolynomial has a positive real part.

4. *Methods for finding domains of asymptotic stability of differential-difference equations that are stationary to the first order.* As we have already shown, the question of stability of differential-difference equations that are stationary to the first order reduces to the determination of the sign of the real parts of the roots of the characteristic quasipolynomial when the method of first order approximations is applicable. But the determination of the roots of the quasipolynomial is an extremely difficult task in the simplest of cases, so that it is most important to develop tests for negativity of the real parts of all roots of the quasipolynomial.

There are many tests for negativity of the real parts of all the roots of a polynomial. Of these, we can carry over to the case of the quasipolynomial those tests that are based on the generalized Sturm series (cf. Meĭman and Čebotarev [**178**]) and those tests that are based on the principle of the argument (the amplitude-phase method, the method of Mihaĭlov, etc.).

When investigating quasipolynomials with given numerical coefficients the best course normally is to use the generalized Sturm series; but if it is not necessary to define the domain of stability in the parameter space (the parameters enter into the coefficients of the quasipolynomial) it is as a rule more rewarding to use one of the methods based on the argument.

Methods based on the generalized Sturm series are explained in detail in the monograph of Meĭman and Čebotarev [**178**]; methods based on the principle of the argument are explained in papers by Ja. Z. Cypkin [**304**; **305**; **307**; **308**], N. N. Vološin [**66**], I. V. Svirskiĭ [**253**], Ju. I. Neĭmark [**201**] and others.

The amplitude-phase method, as applied to the characteristic quasipolynomial, first appeared in the papers of Ja. Z. Cypkin cited above. Let us apply this method to the characteristic quasipolynomial $f(z) = \Phi(z) + \Psi(z)e^{-\tau z}$, where $\Phi(z)$ and $\Psi(z)$ are polynomials. If the characteristic quasipolynomial corresponds to a differential equation with retarded arguments, the degree of $\Phi(z)$ is higher than that of $\Psi(z)$.

We consider, in the w-plane, the image Γ_w of the contour γ_z consisting of the interval $-R \leqq y \leqq R$ on the imaginary axis and the semicircle $|z| = R$, $\mathrm{Re}\, z > 0$, under the mapping $w = f(z)$ (Figure 50).

By the principle of the argument, if a single passage around the contour γ_z in the positive sense induces an $N - P$-fold positive passage around the point $w = 0$ by the image Γ_w (more exactly, if the linking coefficient of the contour Γ_w with the point $w = 0$ is equal to $N - P$) where P is the sum of the multiplicities of the poles of $f(z)$ inside the contour γ_z (in our case $P = 0$), then the sum of the multiplicities of the zeros of the function $f(z)$ lying inside γ_z is equal to N.

It is assumed in this connection that there are neither zeros nor poles of $f(z)$ on the contour γ_z itself.

If there are no zeros of the function $f(z)$ on the imaginary axis, and if N remains zero as $R \to \infty$, then all zeros of the function $f(z)$ lie in the left half-plane $\mathrm{Re}\, z < 0$, and therefore the stationary point $x \equiv 0$ of the original homogeneous differential equation is asymptotically stable.

It is more useful, however, to replace the function $f(z)$ by a mapping defined by $f(z)/\Phi(z) = 1 + (\Psi(z)/\Phi(z))e^{-rz}$, having the same zeros as $f(z)$ and poles coinciding with those of $\Phi(z)$ (we postulate that $f(z)$ and $\Phi(z)$ have no zeros in common), or even to replace it by the mapping $w_r(z) = -(\Psi(z)/\Phi(z))e^{-rz}$.

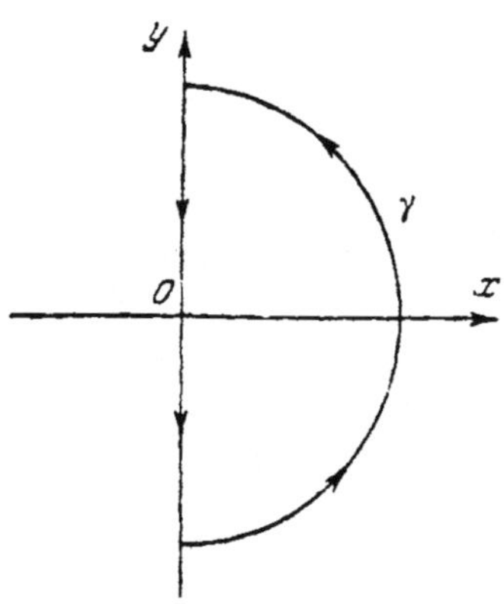

Figure 50

In the latter case we must count the number of times the image of γ_z under the mapping $w_r(z)$ encircles the point $w = 1$.

Since as $R \to \infty$ the image of the semicircle $|z| = R$, $\mathrm{Re}\, z > 0$ shrinks down to the point $w = 0$, there remains only the task of constructing the image of the imaginary axis under the mapping $w_\tau(z)$, the traversal being taken in the direction of increasing y. This image of the imaginary axis is known as the *amplitude-phase characteristic*. To determine it, one normally constructs the so-called limiting characteristic, the image of the imaginary axis under the mapping $w_0(z) = -\Psi(z)/\Phi(z)$, and then takes into account the influence of the factor $e^{-i\tau y}$, which has absolute value unity.

$$w_\tau(iy) = w_0(iy)\, e^{-i\tau y}. \tag{29}$$

For example, if we apply this method to the very simple equation $\dot{x}(t) + ax(t) + bx(t-\tau) = 0$ we obtain

$$f(z) = z + a + be^{-\tau z}, \quad \frac{f(z)}{\Phi(z)} = 1 + \frac{b}{z+a} e^{-\tau z},$$

$$w_\tau(z) = -\frac{b}{z+a} e^{-\tau z}, \quad w_0(z) = -\frac{b}{z+a}.$$

The limiting characteristic is a circle with center at the point $w = -b/2a$ and radius $|b/2a|$. It is clear that for $a > 0$ and $|b| < a$ there is no value of τ for which the amplitude-phase characteristic encircles the point $w = 1$; under these conditions, therefore, the solution $x \equiv 0$ is asymptotically stable. For $a < 0$, and also for $a > 0$ and $|b| > a$, it is necessary to take into account the influence of the factor $e^{-i\tau y}$. It is not difficult to verify that we then obtain as the region of stability the shaded area shown in Figure 51. It is bounded by the straight line $a + b = 0$ and the curve $a = -t \operatorname{ctg} \tau t$, $b = t/\sin \tau t$, $0 \leqq t \leqq \pi/\tau$.

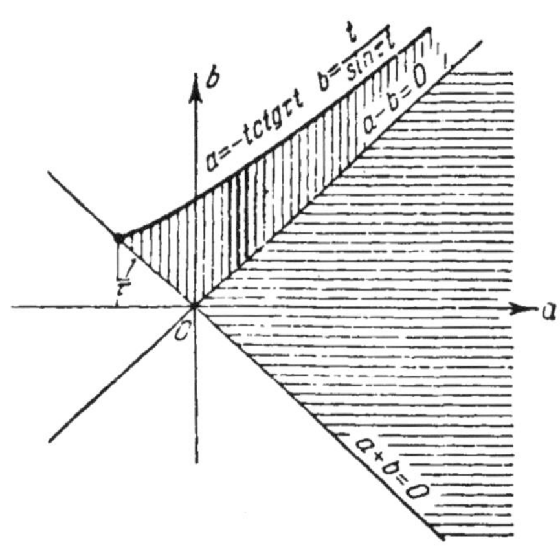

Figure 51

REMARK. It is a very frequent practice to expand the solution in powers of τ when studying the question of stability for small τ; this method, applied to the equation

$$\dot{x}(t)=f(t,\ x(t),\ x(t-\tau)), \tag{30}$$

amounts to this, that the equation becomes one without deviating arguments

$$\dot{x}(t)=f\left(t,\ x(t),\ x(t)-\tau\dot{x}(t)+\ldots+\frac{(-1)^n\tau^n}{n!}x^{(n)}(t)\right), \tag{31}$$

and we investigate the stability of the latter equation.

It is easy to see that this "method" has absolutely no foundation of any kind, since even if equations (30) and (31) had neighboring[7] solutions on the interval $t_0 \leq t \leq T$ we could say nothing about the asymptotic behavior of either, and in particular the domains of stability of equations (30) and (31) may be totally different, and in practice often are.

For example, the domain of stability of the equation

$$\dot{x}(t)+ax(t)+bx(t-\tau)=0 \tag{32}$$

is that represented in Figure 51 (page 190) whereas the domain of stability of the equation

$$\dot{x}(t)+ax(t)+b[x(t)-\tau\dot{x}(t)]=0, \tag{33}$$

which is obtained from (32) by the method of expansion in powers of τ for $n = 1$, is defined by $(a + b)/(1 - b\tau) > 0$ and is represented by the shaded area in Figure 52.

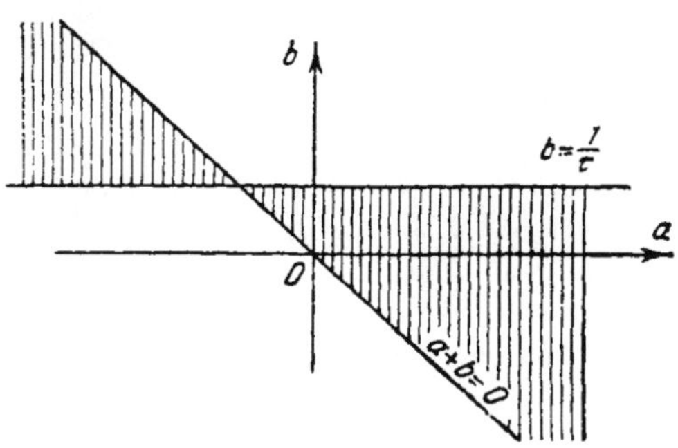

Figure 52

[7] As to the applicability of this method to approximate solutions of equations, see §2 of this chapter.

Thus even for small τ the "method" under discussion leads in this case to grossly erroneous results.

If we take $n > 1$, then even for finite intervals the solutions of (32) and (33) are in general not near to one another for small τ.

5. *Ljapunov's second method applied to differential equations with deviating arguments.* Ljapunov's second method can be applied, without significant change, to differential equations with deviating arguments. Its effectiveness, however, for the study of the stability of such equations is essentially nil. Ljapunov's second method is in fact a very general method for the study of stability of differential equations without deviating arguments, since Ljapunov's fundamental theorems, though perhaps in weakened form, admit a converse; when we come to differential-difference equations, however, this method is applicable only under very special circumstances, as we shall shortly see.

FIRST THEOREM ON LJAPUNOV'S SECOND METHOD. *If for the system of differential equations*

$$\dot{x}_i(t) = f_i(t, \ x_j(t - \tau_{js}))$$
$$(i, \ j = 1, \ 2, \ \ldots, \ n; \quad s = 1, \ 2, \ \ldots, \ m) \tag{34}$$

representing a perturbed motion it is possible to find a function $V(t, x_1, x_2, \cdots, x_n)$ of fixed sign such that its time derivative

$$\frac{dV}{dt} = \frac{\partial V}{\partial t} + \sum_{j=1}^{n} \frac{\partial V}{\partial x_j} f_j = \Phi(t, \ x_i, \ x_{js}),$$

where $x_{js} = x_j(t - \tau_{js})$, is of constant sign along the integral curves of the equation, and either the sign is opposite to that of V or the derivative is identically zero, then the unperturbed motion is stable.

The function $\Phi(t, x_i, y_{js})$ will be said to be of fixed sign, and, to fix the ideas let us say of negative sign, if $\Phi(t, x_i, y_{js}) \leqq 0$ for $t \geqq t_0$, $|x_i| < h$, $|y_{js}| < h$, where $h > 0$, and if in this connection $\tau_{js} = 0$ implies $x_j = y_{js}$ for this j and s.

PROOF. Let us agree for concreteness that $V \geqq 0$, so that for $t \geqq t_0$ and $|x_j| \leqq h$ $(j = 1, 2, \cdots, n)$ we have the inequality

$$V(t, \ x_1, \ x_2, \ \ldots, \ x_n) \geqslant W(x_1, \ x_2, \ \ldots, \ x_n),$$

where W is a positive definite function.

In this range of variation of the variables t and x_i, we have for $|x_j| < h$, $dV/dt \leqq 0$. For given $\epsilon > 0$ (we suppose that $\epsilon < h$) we choose a $\delta > 0$ so small that $V(t, \varphi_1(t), \cdots, \varphi_n(t)) < l$ on E_{t_0}, where the $\varphi_i(t)$ are the initial functions given on E_{t_0} and satisfy the inequalities $|\varphi_i| < \delta$, and l is the greatest lower bound of the values of the function $W(x_1, x_2, \cdots, x_n)$ under the conditions $\{|x_1|, |x_2|, \cdots, |x_n|\} = \epsilon$, and of course $l > 0$.

Let us also assume that $\delta < \epsilon$: then the inequalities $|x_s| < \epsilon$, which hold for the coordinates of points on the integral curves for values of t ranging over E_{t_0} will be satisfied also for all $t \geqq t_0$. This is true, since if one of these inequalities failed first at some moment $t = T$, $T > t_0$, that is if

$$\max\{|x_1(T)|, |x_2(T)|, \ldots, |x_n(T)|\} = \varepsilon,$$

then by the definition of l we would have $V(T, x_1(T), x_2(T), \cdots, x_n(T)) \geqq l$, which is impossible since $dV/dt \leqq 0$, and therefore

$$V(T, x_1(T), \ldots, x_n(T)) \leqslant V(t_0, x_1(t_0), \ldots, x_n(t_0)) < l.$$

SECOND THEOREM ON THE SECOND METHOD OF LJAPUNOV. *If, under the conditions of the preceding theorem, the derivative dV/dt is a function of fixed sign* (that is $\Phi(t, x_i, y_{js}) \leqq -W_1(x_i, y_{js}) \leqq 0$ for $t > t_0$ and arbitrary $|x_i| < h$ and $|y_{js}| < h$, where $h > 0$, and where $\tau_{js} \equiv 0$ implies $x_j = y_{js}$; the function W_1 vanishes in the region under consideration only for $x_i = 0$ $(i = 1, 2, \cdots, n)$), *and the function V admits an infinitely small upper bound, then the unperturbed motion is asymptotically stable.*

PROOF. Since the conditions of the preceding theorem are satisfied, the unperturbed motion is stable, and therefore the quantity δ considered in the preceding theorem can be so chosen that for $t > t_0$ and on E_{t_0} the solution defined by the initial functions $\varphi_1, \varphi_2, \cdots, \varphi_n$ does not leave the domain in which the conditions of the second theorem are satisfied; since dV/dt is nonpositive along the integral curves, $\lim V(x_1(t):\varphi_1, \cdots, x_n(t):\varphi_n) = \alpha$ exists, and for all $t > t_0$ the function $V(x_1(t):\varphi_1, x_2(t):\varphi_2, \cdots, x_n(t):\varphi_n) \geqq \alpha$. If $\alpha = 0$, then $\lim_{t\to\infty} x_i(t):\varphi_i = 0$ and the theorem is proved.

Suppose $\alpha > 0$; since V admits an infinitely small upper bound and $dV/dt \leqq -W_1(x_i, y_{js}) \leqq 0$, we have for all $t > t_0$ the relation $dV/dt \leqq -l_1$, where $l > 0$ is a constant. From this we obtain by integration:

$$V(t, \underset{\varphi_1}{x_1}(t), \underset{\varphi_2}{x_2}(t), \ldots, \underset{\varphi_n}{x_n}(t))$$
$$\leqslant V(t_0, \underset{\varphi_1}{x_1}(t_0), \underset{\varphi_2}{x_2}(t_0), \ldots, \underset{\varphi_n}{x_n}(t_0)) - l_1(t - t_0),$$

which for sufficiently large values of t contradicts the inequality

$$V(t, \underset{\varphi_1}{x_1}(t), \underset{\varphi_2}{x_2}(t), \ldots, \underset{\varphi_n}{x_n}(t)) \geqslant \alpha > 0.$$

THEOREM ON STABILITY. *If for the system (34) of differential equations representing a perturbed motion, there exists a differentiable function $V(t, x_1, x_2, \cdots, x_n)$ satisfying in some h-neighborhood of the origin the conditions:*

(1) $V(t, 0, 0, \cdots, 0) = 0$;

(2) *in an arbitrarily small neighborhood of the origin of coordinates there exists a domain, independent of t, which for $t \geqq t_0$ is the domain $(V > 0)$;*[8]

(3) *the function V is bounded in $(V > 0)$;*

(4) *in $(V > 0)$ the derivative satisfies*

$$\frac{dV}{dt} = \frac{\partial f}{\partial t} + \sum_{i=1}^{n} \frac{\partial V}{\partial x_i} f_i\left(t,\ x_i(t),\ x_j(t - \tau_s(t))\right)$$

$$= \Phi(t,\ x_i,\ x_{js}) > 0$$

(more exactly $\Phi > 0$ if $t > t_0$, and the points with coordinates $(x_1, x_2, \cdots, x_n)$ and $(x_{1s}, x_{2s}, \cdots, x_{ns})$ $(s = 1, 2, \cdots, m)$ lie in the domain $(V > 0)$), *and in the domain $(V \geqq \alpha)$, $\alpha > 0$ the derivative satisfies $dV/dt = \Phi(t, x_i, x_{js}) \geqq l > 0$, where l is a constant, then the unperturbed motion is unstable.*

PROOF. Let us consider the solution $x_i(t){:}\varphi_i$ $(i = 1, 2, \cdots, n)$ defined by the initial functions $\varphi_i(t) = x_i$, where x_i^0 are the coordinates of some point in the domain $V > 0$. By condition (2) the point $(x_1^0, x_2^0, \cdots, x_n^0)$ can be chosen in an arbitrarily small neighborhood of the origin of coordinates.

Since in the domain $(V > 0)$ the derivative satisfies $dV/dt > 0$, the function V must increase along the integral curves and therefore the points $x_i(t){:}\varphi_i$ $(i = 1, 2, \cdots, n)$ remain for $t \geqq t_0$ in the domain $V \geqq V_0 > 0$, so long as the solution remains within the h-neighborhood of the origin where the conditions of the theorem are satisfied.

We shall show that the trajectory $x_i(t){:}\varphi_i$ $(i = 1, 2, \cdots, n)$ must leave the h-neighborhood of the origin of coordinates. In fact, if we assume the contrary, we are led to a contradiction, since for $V \geqq V_0$ we have $dV/dt \geqq l > 0$, whence, integrating along the integral curve $x_i(t){:}\varphi_i$ $(i = 1, 2, \cdots, n)$ we obtain $V(t, x_1(t), x_2(t), \cdots, x_n(t)) \geqq V_0 + l(t - t_0)$, which is impossible, since in the h-neighborhood of the origin the function V is bounded in the domain $V > 0$.

We could also prove in a similar fashion the analogues of other theorems on instability, but, as may be easily seen, the theorem on the second

[8] We use the notation of Chapter I. The set $(V > 0)$ consists of all points of the space $x_1, x_2, \cdots, x_n$ for which $V > 0$.

method of Ljapunov is of very little import for differential equations with retarded argument.

These theorems may be applied only under very special circumstances, and as to their converses the situation is absolutely hopeless.

In fact, for the system (34) the derivative along the integral curves

$$\frac{dV}{dt}=\frac{\partial V}{\partial t}+\sum_{i=1}^{n}\frac{\partial V}{\partial x_i}f_i\left(t,\ x_j(t-\tau_{js}(t))\right)=\Phi\left(t,\ x_i,\ x_j(t-\tau_{js}(t))\right)$$

contains the arguments $x_j(t-\tau_{js}(t))$ and therefore it is only under special circumstances that we can establish the constancy or determinacy of sign of the derivative dV/dt.

If the theorems in question, the theorems on Ljapunov's second method, did in fact admit converses, the domains of stability of the solution $x\equiv 0$ in the space of the parameters μ_s would coincide for the two equations

$$\dot{x}(t)=f(t,\ x(t),\ x(t-\tau),\ \mu_s)$$

and

$$\dot{x}(t)=kf(t,\ x(t),\ \dot{x}(t)(t),\ \mu_s)$$

(where k is a positive constant), since if the first equation admits a Ljapunov function $V(t,x_1,x_2,\cdots,x_n)$ the second admits the Ljapunov function $V(kt,x_1,x_2,\cdots,x_n)$. The simplest possible examples, however, show that these two equations may have different domains of stability, e.g., the domain of stability of the two equations

$$\dot{x}(t)=ax(t)+bx(t-\tau)$$

and

$$\dot{x}(t)=k\left[ax(t)+bx(t-\tau)\right]$$

are different, as in Figure 51 on page 190.

EXAMPLES. (1) The trivial solution of the equation

$$\dot{x}(t)=-x(t)x^2(t-\tau(t))$$

is asymptotically stable, since the Ljapunov function may be taken as $V=x^2$.

(2) The trivial solution of the system

$$\dot{x}(t)=y(t)-x(t)y^2\left(t-\tau_1(t)\right),$$
$$\dot{y}(t)=-x(t)-y(t)x^2\left(t-\tau_2(t)\right),$$
$$\tau_i(t)\geqslant 0\quad (i=1,\ 2,\ \ldots)$$

is stable, since the Ljapunov function may be taken as $V = x^2 + y^2$.

The definition of stability on page 182 deals with the case of stability against momentary perturbation. The question of stability against continued perturbation for equations with retarded arguments is posed and solved just as it is for unretarded equations; in effect, in conjunction with the system

$$\dot{x}_i(t) = f_i(t, x_j(t - \tau_{js}(t)))$$
$$(i, j = 1, 2, \ldots, n;\ s = 1, 2, \ldots, m), \tag{34}$$

for which we may at once assert that $x_j(t) \equiv 0$ $(j = 1, 2, \cdots, n)$ is the desired solution; then we consider the system

$$\dot{x}_i(t) = f_i(t, x_j(t - \tau_{js}(t))) + R_i(t, x_j(t - \tau_{js}(t))), \tag{35}$$

where the indices i, j, s, assume their earlier values, and the functions R_i are sufficiently small in absolute value.

The trivial solution of the system (34) is said to be stable with respect to continued perturbations if for every $\epsilon > 0$ there exist $\delta_1(\epsilon)$ and $\delta_2(\epsilon)$ such that every solution $x_i(t):\varphi_i$ $(i = 1, 2, \cdots, n)$, for which the initial functions φ_i satisfy the inequality $\sup\{|\varphi_1(t)|, \cdots, |\varphi_n(t)|\} < \delta_1$ on the initial set E_{t_0} for arbitrary R_i subject to $|R_i| < \delta_2(\epsilon)$ for $t \geqq t_0$ and $i = 1, 2, \cdots, n$, satisfies the inequality $|x_i(t):\varphi_i| < \epsilon$ $(i = 1, 2, \cdots, n)$ for all $t \geqq t_0$.

Theorem. *If for the system* (34) *of differential equations there exists a positive definite function* $V(t, x_1, x_2, \cdots, x_n)$ *for which the total time derivative as defined by this system is a negative definite function, and if in the interval* $t \geqq t_0$, $\max\{|x_1|, |x_2|, \cdots, |x_n|\} \leqq h$ *the partial derivatives* $\partial V / \partial x_i$ $(i = 1, 2, \cdots, n)$ *are bounded, then the perturbed motion is stable with respect to continued perturbation.*

The proof is almost identical to the proof of the analogous theorem of I. G. Malkin for unretarded equations (page 134).

A large number of theorems related to Ljapunov's second method can also be proved for differential equations of neutral type, but they are essentially worthless.

If we look back over this section and sum up the results, we may say that many of the basic theorems of stability theory can be carried over without essential alteration to the case of differential equations with deviating arguments, but the fact is that up to now stability theory for equations with deviating arguments is essentially useless except for equations that are stationary to the first approximation, and even then in the noncritical case.

We can find estimates of the domain of stability for equations that are not stationary to a first approximation, but only under highly restricted conditions.

For example, the trivial solution of the equation

$$\dot{x}(t) = f\big(t, x(t), x(t-\tau_1(t)), x(t-\tau_2(t)), \ldots, x(t-\tau_n(t))\big),$$
$$\tau_i(t) \geqslant 0 \quad (i = 1, 2, \ldots, n),$$

is stable if for all $t > t_0$

$$f(t, \delta, k_1, k_2, \ldots, k_n) < 0$$

and

$$f(t, -\delta, k_1, k_2, \ldots, k_n) > 0$$

for sufficiently small $\delta > 0$ and arbitrary k_i subject only to the inequalities $|k_i| < \delta$.

In fact, for this case the trajectory of the solution cannot leave the δ-neighborhood of the trivial solution $x \equiv 0$ for $t > t_0$ since if the trajectory of the initial function is so situated in this neighborhood, then under the given conditions the integral curve cannot intersect the straight lines $x = \pm\delta$.

Remark. After this book had gone to press, N. N. Krasovskiĭ obtained results that constituted a significant advance in the theory of stability of differential equations with retarded argument.

Krasovskiĭ considered, in place of Ljapunov's function, certain functionals and showed that all the basic theorems of Ljapunov's second method can be proved, and with certain restrictions their converses.

Krasovskiĭ's method allows us in many cases to investigate the stability of the stationary points of differential equations with retarded argument that are not stationary to a first approximation.

6. *Stability with respect to perturbation of a deviating argument.* In almost all processes with aftereffects that are describable by differential equations with deviating arguments, the deviations themselves cannot be prescribed exactly; that is, at an arbitrary instant of time $t \geqq t_0$ the deviation of the argument may be influenced by any one of a number of causative factors and may undergo a small disturbance.

This phenomenon leads us naturally into the question of the stability of equations with respect to small disturbances of the deviating arguments.

If the perturbation of a deviating argument has a "momentary" character, the stability question reduces to the question of stability vis-a-vis perturbations of the initial functions, because of the validity (for a very wide class of differential equations with deviating arguments) of the theorems on the continuous dependence of the solutions on the deviations.

But if the perturbation of the deviating argument has a continuing character—if, for example, in some process or other with after effects the retardation is not precisely defined, there arises a new and wholly unsolved problem.

Together with the system of first-order differential equations with deviating arguments

$$\dot{x}_i(t) = f_i\big(t,\ x_j(t - \tau_{jk}(t))\big)$$
$$(i,\ j = 1,\ 2,\ \ldots,\ n;\ k = 1,\ 2,\ \ldots,\ m), \tag{34}$$

let us consider a second system

$$\dot{x}_i(t) = f_i\big(t,\ x_j(t - \bar{\tau}_{jk}(t))\big)$$
$$(i,\ j = 1,\ 2,\ \ldots,\ n;\ k = 1,\ 2,\ \ldots,\ m). \tag{36}$$

The solution $\dot{x}_i(t){:}\varphi_i$ $(i = 1, 2, \cdots, n)$ of the system (34) defined by the initial functions φ_i, which we shall suppose have been continuously extended by some arbitrary process from the set E_{t_0} to the set $t \leqq t_0$, is said to be stable with respect to perturbations of the deviating argument if for every $\epsilon > 0$ we can choose quantities $\delta_1(\epsilon)$ and $\delta_2(\epsilon)$ such that if (a) the inequalities

$$|\underset{\varphi_i}{x_i}(t) - \psi_i(t)| < \delta_1(\varepsilon) \qquad (i = 1,\ 2,\ \ldots,\ n)$$

hold on the initial set $E_{\bar{t}_0}$ of the system (37) for arbitrary $\bar{t}_0 \geqq t_0$, and (b) $|\bar{\tau}_{jk}(t) - \tau_{jk}(t)| < \delta_2(\epsilon)$ for

$$t \geqslant t_0\ (j = 1,\ 2,\ \ldots,\ n;\ k = 1,\ 2,\ \ldots,\ m)$$

then

$$|\underset{\psi_i}{\bar{x}_i}(t) - \underset{\varphi_i}{x_i}(t)| < \varepsilon\ (i = 1,\ 2,\ \ldots,\ n)\ \text{ for } t \geqslant t_0,$$

where $\bar{x}_i(t){:}\psi_i$ $(i = 1, 2, \cdots, n)$ is the solution of (37) determined by the initial functions $\psi_i(t)$ defined on $E_{\bar{t}_0}$.

Even when the τ_{jk} are constant, the $\bar{\tau}_{jk}(t)$ will in general be functions of t; in applied work, however, one constantly meets cases in which not only the τ_{jk} but also the $\bar{\tau}_{jk}$—the perturbed deviations—are constant. This is the case when, for example, in a process with after effects the delay τ_{jk} is constant but cannot be precisely measured.

In this case the inequality $|\bar{\tau}_{jk}(t) - \tau_{jk}(t)| < \delta_2(\epsilon)$ in the definition of stability must be replaced by the inequality $|\bar{\tau}_{jk} - \tau_{jk}| < \delta_2(\epsilon)$, where $\bar{\tau}_{jk}$ and τ_{jk} are constants.

Stability in this sense we shall call stability with respect to a continually perturbed deviating argument.

If the system (34), assumed stationary to the first approximation, admits an investigation of the stability of the first-order trivial solution, all the roots k_s of the characteristic quasipolynomial $f(z)$ corresponding to the first order approximate equation satisfy the inequality $\operatorname{Re} k_s \leqq -\alpha < 0$, all the τ_{jk} are constant and undergo the continual perturbation $\bar{\tau}_{jk} \neq 0$,[9] then the stationary point $x_i \equiv 0$ is stable with respect to continual perturbation of the deviating argument.

This assertion can be verified by applying the principle of the argument to the mapping $w = f(z)$ of the semicircle of radius R, as described on page 189, when $R \to \infty$, since for a sufficiently small change in the τ_{jk} the linking coefficient of the image of the contour γ_z with respect to the point $w = 0$ does not change.

On the boundary of the domain of stability of the system (34) small changes in the τ_{jk} may, however, affect the stability.

We note that in some cases the perturbations $\bar{\tau}_{jk} - \tau_{jk}$ of a deviating argument may all have the same sign. It is natural to refer to this case as that of stability with respect to one-sided perturbations of a deviating argument.

The case arises, for example, when a retardation is perturbed, imprecisely but by what is known to be a deficiency (or an excess). It arises also in the frequently encountered practical instance of neglect of a small retardation; we shall investigate this instance in some detail. In the simplest version, we have one first order equation with one small positive retardation, either constant or variable,

$$\dot{x}(t) = f(t,\ x(t),\ x(t - \tau(t)));\ \ x(t) = \varphi(t) \text{ on } E_{t_0} \tag{37}$$

and we neglect the retardation, writing in place of (20) the following equation:

$$\dot{x}(t) = f(t,\ x(t),\ x(t));\ \ x(t_0) = \varphi(t_0). \tag{38}$$

In order to show that we may do this for $t \geqq t_0$ we must prove that the solution of (37) is stable with respect to one-sided perturbations (since $\tau(t) \geqq 0$) of the deviation in the independent variable entering the third

[9] If some one of the τ_{jk} is not perturbed, we must take $\bar{\tau}_{jk} = \tau_{jk}$ for that one.

argument of the function $f(t, x(t), x(t))$, or, sometimes, with respect to one-sided constant perturbations of the same deviation.

Very simple examples demonstrate that without such a verification the use of (38) in place of (37) may lead to grossly erroneous results for large values of t.

For example, the equation

$$\ddot{x}(t)+a\dot{x}(t)+bx(t)=0, \tag{39}$$

which is derived from the equation

$$\ddot{x}(t)+a\dot{x}(t)+bx(t-\tau)=0 \tag{40}$$

for $\tau = 0$, has a stable trivial solution for $a = 0$ and $b > 0$, whereas the trivial solution of (40) under the same assumptions on the parameters a and b is unstable for any constant $\tau > 0$; thus, the trivial solution of equation (39) is unstable with respect to small constant positive deviations of the independent variable in the term $bx(t)$ and the replacement of (40) for $a = 0$ and $b > 0$ by (39) when we are studying stability is not permissible no matter how small τ may be.

11. **Quasilinear equations with retarded argument.** In §6 of Chapter IV we briefly developed the basic questions of the theory of quasilinear oscillations of differential equations without deviating arguments. In this section we transfer the theory, with some minor changes, to the case of quasilinear equations with retarded argument.[10]

1. *Linear homogeneous equations*. Let us consider a linear homogeneous second-order equation with real constant coefficients a_1, b_1, a_2, b_2 and constant retardations $\tau_1 > 0$, $\tau_2 > 0$:

$$\ddot{x}(t)+a_1\dot{x}(t)+b_1\dot{x}(t-\tau_1)+a_2x(t)+b_2x(t-\tau_2)=0. \tag{41}$$

An analytic periodic solution of this equation is a linear combination of the real and imaginary parts of solutions of the form $e^{i\alpha t}$, where α is a real number.

To determine the eigenfrequencies α we have the equation

$$-\alpha^2+ia_1\alpha+ib_1\alpha e^{-i\alpha\tau_1}+a_2+b_2e^{-i\alpha\tau_2}=0$$

or, separating the real and imaginary parts,

---

[10] Interesting results have been obtained in this area by the following authors: L. Gordon, N. Minorsky [181; 182], G. S. Gorelik [74], K. F. Teodorčik [263], Ja. Z. Cypkin [306], Ja. P. Terleckiĭ [261; 262], and others.

$$\alpha b_1 \sin \alpha\tau_1 + b_2 \cos \alpha\tau_2 = \alpha^2 - a_2,$$
$$b_2 \sin \alpha\tau_2 - \alpha b_1 \cos \alpha\tau_1 = a_1\alpha. \qquad (42)$$

For arbitrary given $a_1, b_1, a_2, b_2, \tau_1, \tau_2$ these equations, generally speaking are incompatible and therefore an analytic periodic solution of (41) can exist only for specifically selected values of the parameters $a_1, b_1, a_2, b_2, \tau_1, \tau_2$ (we should note that even for the unretarded equation $\ddot{x}(t) + p\dot{x}(t) + qx(t) = 0$ the solution is periodic only if $p = 0$ and not for all values of p and q).

For certain values of the parameters $a_1, b_1, a_2, b_2, \tau_1, \tau_2$ the system may have a number of real roots α_i with different absolute values.

If some p of these roots $\alpha_1, \alpha_2, \cdots, \alpha_p$ are commensurable, there corresponds to them a $2p$-parameter family of periodic solutions which are linear combinations with arbitrary constant coefficients of the real and imaginary parts of the solutions $e^{i\alpha_i t}$.

The possibility that eigenfrequencies α_i may exist, and the appearance of periodic solutions in the presence of "friction" ($a_1 \neq 0$ or $b_1 \neq 0$) are characteristic properties of equations of the type of (41).

2. *Linear inhomogeneous equations.* A linear inhomogeneous second-order equation with constant coefficients and constant retardations

$$\ddot{x}(t) + a_1\dot{x}(t) + b_1\dot{x}(t-\tau_1) + a_2x(t) + b_2x(t-\tau_2) = f(t) \qquad (43)$$

may have periodic solutions only if the term on the right is periodic.

Without loss of generality we may assume that the periodic function $f(t)$ has period 2π.

We assume further that $f(t)$ is continuous and can be expanded in a Fourier series

$$f(t) = \frac{A_0}{2} + \sum_{n=1}^{\infty} (A_n \cos nt + B_n \sin nt).$$

We obtain the periodic solutions of (43), of period 2π, in the form of a series

$$x(t) = \frac{C_0}{2} + \sum_{n=1}^{\infty} (C_n \cos nt + D_n \sin nt). \qquad (44)$$

Substituting this in (43) and equating coefficients of $\cos nt$, $\sin nt$ and of the free term on the left and right hand sides of the resulting identity, we obtain

$$\left.\begin{aligned} C_0 &= \frac{A_0}{a_2 + b_2}; \\ C_n &= \frac{P_n A_n - Q_n B_n}{P_n^2 + Q_n^2}; \\ D_n &= \frac{P_n B_n + Q_n A_n}{P_n^2 + Q_n^2}; \end{aligned}\right\} \tag{45}$$

where

$$P_n = a_2 + b_1 n \sin n\tau_1 + b_2 \cos n\tau_2 - n^2;$$
$$Q_n = a_1 n + b_1 n \cos n\tau_1 - b_2 \sin n\tau_2$$
$$(n = 1, 2, \ldots).$$

If the corresponding homogeneous equation has no integral eigenfrequencies, we have $P_n^2 + Q_n^2 \neq 0$ $(n = 1, 2, \cdots)$ (cf. (42)) and, as is easily seen, the sum of the series (44) is the desired periodic solution of equation (43).

But if there is an integral value $\alpha = n_1$ among the eigenfrequencies, then $P_{n_1} = Q_{n_1} = 0$ and there arises the so-called resonant case, for which periodic solutions can exist only if the right hand side

$$f(t) = \frac{A_0}{2} + \sum_{n=1}^{\infty} (A_n \cos nt + B_n \sin nt)$$

contains no resonant terms, that is only if $A_{n_1} = B_{n_1} = 0$, or

$$\int_0^{2\pi} f(t) \cos n_1 t \, dt = 0 \quad \text{and} \quad \int_0^{2\pi} f(t) \sin n_1 t \, dt = 0.$$

If the homogeneous equation has certain integer eigenfrequencies α_i $(i = 1, 2, \cdots, m)$ we may have a periodic solution only if the right hand side contains no resonant terms corresponding to these frequencies:

$$\int_0^{2\pi} f(t) \cos n_i t \, dt = 0 \quad \text{and} \quad \int_0^{2\pi} f(t) \sin n_i t \, dt = 0$$
$$(i = 1, 2, \ldots, m)$$

If these conditions are satisfied, there is a periodic solution, of period 2π, and it is defined by (44) and (45) for $n \neq n_i$, and the coefficients C_{n_i} and D_{n_i} are arbitrary.

3. *Nonautonomous quasilinear equations in the nonresonant case.* Let us consider the nonautonomous quasilinear system with retarded arguments

$$\begin{aligned} \ddot{x}(t) + a_1 \dot{x}(t) + b_1 \dot{x}(t - \tau_1) + a_2 x(t) + b_2 x(t - \tau_2) \\ = f(t) + \mu F(t, x(t), x(t - \tau_2), \dot{x}(t), \dot{x}(t - \tau_1), \mu), \end{aligned} \tag{46}$$

which we write in abbreviated form as

$$L(x) = f + \mu F, \tag{46}$$

where μ is a small parameter, the function F is continuous in t, has period 2π, possesses a Fourier expansion, and is analytic in the remaining arguments in the following domains: as to the second, third, fourth, and fifth, it is analytic in the regions in which these variables are allowed to vary, and as to μ it is analytic for sufficiently small values of $|\mu|$. The remaining quantities have the meanings previously defined.

For $\mu = 0$ we have the equation $L(x) = f(t)$, which is called the *generating equation* of (46).

We assume that the equation $L(x) = 0$ has no integral eigenfrequencies.

If we also assume that equation (46) has a unique periodic solution for every sufficiently small μ, then, by the theorem on the analytic dependence of the solution on a parameter,[11] we may say that this solution $x(t,\mu)$ of equation (46) can be represented in the form

$$x(t, \mu) = x_0(t) + \mu x_1(t) + \ldots + \mu^n x_n(t) + \ldots, \tag{47}$$

where the $x_k(t)$ are periodic functions of period 2π and are solutions of the linear equations

$$\left.\begin{aligned} &L(x_0(t)) = f(t),\\ &L(x_1(t)) = F(t,\ x_0(t),\ x_0(t-\tau_1),\ \dot{x}_0(t),\ \dot{x}_0(t-\tau_2)),\\ &L(x_2(t)) = \left(\frac{\partial F}{\partial x(t)}\right)_0 x_1(t) + \left(\frac{\partial F}{\partial x(t-\tau_1)}\right)_0 x_1(t-\tau_1)\\ &\qquad + \left(\frac{\partial F}{\partial \dot{x}(t)}\right)_0 \dot{x}_1(t) + \left(\frac{\partial F}{\partial \dot{x}(t-\tau_2)}\right)_0 \dot{x}_1(t-\tau_2),\\ &\cdots\cdots\cdots\cdots\cdots\cdots\cdots\cdots \end{aligned}\right\} \tag{48}$$

where the symbol $(\ )_0$ means that the function $x(t)$ enclosed in parentheses is replaced by the solution $x_0(t)$ of the generating equation and that the parameter $\mu = 0$. We obtain the equations (48) in the same way as we obtained equations (59) (page 146), by equating coefficients of like powers of μ on both sides of equation (46).

The equations (48) are linear and inhomogeneous; their periodic solutions are to be found by the methods of subsection 2 of the present section.

---

[11] The theorem on analytic dependence of the solution on a parameter for equations with retarded argument is most simply proved by the method of A. N. Tihonov [216]; aside from the analyticity of the right hand sides of the system $dx_i(t)/dt = f_i(t, x_j(t-\tau_j), \mu)$, we must also require for all $x_j(t-\tau_j)$ and μ that the initial functions be analytic in μ.

The formally determined periodic solution

$$x(t, \mu) = x_0(t) + \mu x_1(t) + \dots + \mu^n x_n(t) + \dots$$

is in fact the unique periodic solution, which must exist for sufficiently small μ, by hypothesis, and reduce to the solution of the generating equation for $\mu = 0$. It would be desirable, of course, to waive this hypothesis and prove under certain restrictions the existence of a unique periodic solution for all sufficiently small μ that would reduce to $x_0(t)$ for $\mu = 0$. An existence theorem, however, for a periodic solution has been carried through only by topological methods and under severe restrictions on the right hand side of equation (46).

4. *The resonant case in the theory of quasilinear oscillations.* Let us again consider the equation $L(x) = f + \mu F$, preserving the notation and assumptions of subsection 3, but assuming this time that the equation $L(x) = 0$ has integer-valued eigenfrequencies.[12] Just as in the nonresonant case, we find the solution in the form of a series

$$x(t, \mu) = x_0(t) + \mu x_1(t) + \dots + \mu^n x_n(t) + \dots.$$

The functions $x_n(t)$ are as before periodic, with period 2π, and defined by equations (48). In the present case, however, a periodic solution of the first of the equations (48) can exist only if the resonant terms are missing from the right hand side of this equation, i.e., under the conditions

$$\int_0^{2\pi} f(t) \cos n_i t \, dt = 0 \quad \text{and} \quad \int_0^{2\pi} f(t) \sin n_i t \, dt = 0$$

$$(i = 1, 2, \dots, m),$$

where the n_i $(i = 1, 2, \cdots, m)$ are integer-valued eigenfrequencies of the equation $L(x) = 0$.

If these conditions are satisfied, there exists a family of periodic solutions of the equation $L(x_0(t)) = f(t)$, depending on $2m$ parameters (see subsection 2)

$$x_0(t) = \sum_{i=1}^{m} (C_{0i} \cos n_i t + D_{0i} \sin n_i t) + \varphi_0(t).$$

The parameters C_{0i} and D_{0i} are in general determined from the condition

[12] Just as for quasilinear equations without deviating arguments, the case of "small disturbances," i.e., the case when some eigenfrequency has a value close to an integer, is easily reduced to the case of precise resonance.

that the right hand side of the second equation of the system (48) should contain no resonant terms, that is from the conditions

$$\left.\begin{aligned}\int_0^{2\pi} F(t, x_0(t), x_0(t-\tau_1), \dot{x}_0(t), \dot{x}_0(t-\tau_2), 0)\cos n_i t\, dt = 0,\\ \int_0^{2\pi} F(t, x_0(t), x_0(t-\tau_1), \dot{x}_0(t), \dot{x}_0(t-\tau_2), 0)\sin n_i t\, dt = 0.\end{aligned}\right\} \qquad (49)$$

If these conditions for the absence of resonant terms are satisfied, there exists a $2m$-parameter family of periodic solutions of the second equation of the system (48)

$$x_1(t) = \sum_{i=1}^{m} (C_{1i}\cos n_i t + D_{1i}\sin n_i t) + \varphi_1(t).$$

The arbitrary constants C_{1i} and D_{1i} are in general determined from the condition that there be no resonant terms in the third equation of the system (48), and so on.

We remark that if the Jacobian of the left hand sides of the equations (49) taken with respect to the variables C_{0i} and D_{0i} $(i = 1, 2, \cdots, m)$ is different from zero, then the corresponding Jacobian of the left hand sides of the analogous equations defining the constants C_{ni} and D_{ni} $(n = 1, 2, \cdots;$ $i = 1, 2, \cdots, m)$, will also be different from zero, as may be seen after a simple calculation.

5. *Resonance of nth order in quasilinear equations with retardations.* In quasilinear equations with retardations

$$L(x) = f + \mu F, \qquad (46)$$

where we preserve the previous notation, there may exist periodic solutions, of periodic $2\pi n$, i.e., with a period which is a multiple of the period 2π of the right hand side of equation (46).

Let certain frequencies of the proper oscillations of equation (41), $L(x) = 0$, have the form k_i/n $(i = 1, 2, \cdots, p)$, where the k_i are positive whole numbers, $k_i < n$. Using the method of subsection 2 we obtain a periodic solution $\varphi_0(t)$ of the equation $L(x_0) = f(t)$, with period 2π. Then the periodic solution with period $2\pi n$ of the same equation will have the form

$$x_0(t) = \sum_{i=1}^{p}\left(A_{0i}\cos\frac{k_i t}{n} + B_{0i}\sin\frac{k_i t}{n}\right) + \varphi_0(t), \qquad (50)$$

where A_{0i} and B_{0i} are arbitrary constants. The periodic solution of period 2π belonging to the equation $L(x) = f + \mu F$, and reducing for $\mu = 0$ to $x_0(t)$, is found in the same way as in subsection 4.

We obtain the solution in the form

$$x(t, \mu) = x_0(t) + \mu x_1(t) + \dots + \mu^n x_n(t) + \dots, \tag{47}$$

where the $x_i(t)$ are the desired functions of period $2\pi n$.

Substituting (47) into (46) and equating coefficients of like powers of μ on the left and right hand sides of the resulting identity, we obtain the system (48) (page 203). The first of these equations has the solution (50), and the arbitrary constants A_{0i} and B_{0i} are determined by the condition that there be no resonant terms in the right hand sides of the second equation of the system (48):

$$\int_0^{2\pi n} (F)_0 \cos \frac{k_i t}{n}\, dt = 0$$

and

$$\int_0^{2\pi n} (F)_0 \sin \frac{k_i t}{n}\, dt = 0 \qquad (i = 1, 2, \dots, p).$$

The arbitrary constants appearing in the solution of the second equation of the system (48) are determined by the condition that there be no resonant terms in the right hand side of the third equation of (48) and so on.

6. *Autonomous quasilinear equations*. Let us consider the equation

$$L(x(t)) = \mu F(x(t), x(t-\tau_1), \dot{x}(t), \dot{x}(t-\tau_2), \mu), \tag{51}$$

where we preserve the earlier notation and assumptions. In order to simplify the calculations, we assume that the equation $L(x(t)) = 0$ has only one proper frequency α. We suppose that equation (51) has a periodic solution for sufficiently small μ and that its initial function is analytic in μ.

Since F does not depend explicitly on t, the period $T(\mu)$ of the desired periodic solution will in general be a function of μ. It may be shown that under the above restrictions the period $T(\mu)$ is an analytic function of μ. Although the period of the desired solution depends on μ it would be pointless to look for a solution $x(t, \mu)$ of equation (51) as a power series in μ

$$x(t, \mu) = x_0(t) + \mu x_1(t) + \dots + \mu^n x_n(t) + \dots \tag{47}$$

since the coefficients $x_i(t)$ will not in general be periodic with period $T(\mu)$, and therefore cannot be determined by the methods given above.

Therefore we make a preliminary transformation of the independent variable in such a way that the transformed equation will have period 2π independently of the value of μ, and then we may look for a solution in the form (47).

Since the period of the desired solution $T(\mu)$ is analytic in μ, and reduces for $\mu = 0$ to $2\pi/\alpha$, we have

$$T(\mu) = \frac{2\pi}{\alpha}(1 + \mu x_1 + h_2\mu^2 + \dots + h_n\mu^n + \dots),$$

and if we make the substitution $t_1 = \alpha t/h$, where $h = 1 + h_1\mu + h_2\mu^2 + \cdots + h_n\mu^n + \cdots$, we transform equation (51) into

$$\bar{L}(x(t_1)) = \mu\bar{F}(x(t_1), \dot{x}(t_1 - \bar{\tau}_1), \dot{x}(t_1), \dot{x}(t_1 - \bar{\tau}_2), \mu), \tag{52}$$

where the operator $\bar{L}$ differs from the operator L in its coefficients and in the values of the constant retardations $\bar{\tau}_1$ and $\bar{\tau}_2$, and the function $\bar{F}$ has the properties of the function F indicated earlier.

We look for a period solution of (52), which is bound to have the period 2π, in the form of a power series (47) and, to determine the functions $x_k(t_1)$, we obtain an equation of the form of (48), but with different coefficients, depending on h, and with right hand sides in which the function F is replaced by the function $\bar{F}$, and with $f(t) \equiv 0$. The generating equation $\bar{L}(x_0(t)) = 0$ has a family of periodic solutions $x_0(t_1) = A\cos(t_1 - t_0)$.

Periodic solutions of the equation defining $x_1(t_1)$

$$\bar{L}(x_1(t_1)) = \Phi(h, x_0(t_1), x_0(t_1 - \bar{\tau}_1), \dot{x}_0(t_1), \dot{x}_0(t_1 - \bar{\tau}_2)) \tag{53}$$

exist only if there are no resonant terms in the right hand side.

From the two conditions for the absence of resonant terms in the right hand side of equation (53) we can in general define the constant A, the radius of the circle in the phase plane x, $\dot{x}$ within which periodic trajectories of (52) appear for small μ, and the constant h_1, which determines an approximate period of the desired solution

$$T(\mu) \approx \frac{2\pi}{\alpha}(1 + h_1\mu).$$

We must re-emphasize the fact that the formally obtained series

$$x(t, \mu) = x_0(t) + \mu x_1(t) + \dots + \mu^n x_n(t) + \dots$$

as derived in this section represents the desired solution only under the assumption that there exists a unique periodic solution, tending as $\mu \to 0$

to a periodic solution of the generating equation, and that the proof of the existence of such a solution has been carried through only under very specialized circumstances.

We note that for quasilinear equations of first order

$$\dot{x}(t)+ax(t)+bx(t+\tau)=f(t)+\mu F(t,\ x(t),\ x(t-\tau))$$

the same methods allow us to make a more complete investigation, extending even to a proof of the existence for sufficiently small $|\mu|$ of periodic solutions tending as $\mu \to 0$ to solutions of the generating equation, and including results on the stability of periodic solutions.

12. **Equations of neutral type.** Differential-difference equations of neutral type are of many different forms.

For many of them the boundary value problem must be posed in a way quite different from the case of equations with retarded argument.

In this section we shall consider several properties of these classes of differential equations of neutral type which admit a theory close to the theory of differential equations with retarded argument. Like equations with retarded arguments, these classes of differential equations of neutral type can describe processes with aftereffects, and in this connection are often met with in practice.

The simplest equation of this type has the form

$$\dot{x}(t)=f(t,\ x(t),\ x(t-\tau(t)),\ \dot{x}(t-\tau(t))). \tag{54}$$

We seek to determine a continuous solution $x(t)$ of equation (54) in the interval $t_0 \leqq t \leqq B$ under the condition that $x(t) = \varphi_0(t)$ on E_{t_0}. Assuming that the set $F_{\bar{t}}$, $t_0 \leqq \bar{t} \leqq B$, does not reduce to a single point, and applying the method of succesive integrations (the method of steps), that is, sequentially integrating the equations

$$\dot{x}(t)=f(t,\ x(t),\ \varphi_{i-1}(t-\tau(t)),\ \dot{\varphi}_{i-1}(t-\tau(t)));$$
$$\gamma_{i-1}(t_0)\leqslant t\leqslant\gamma_i(t_0)\quad(i=1,\ 2,\ \ldots), \tag{55}$$

where we have preserved the notation of the preceding sections, we arrive at the following conclusions:

(1) If the function f is continuous in the appropriate domains, and the initial function φ_0 has a continuous derivative on E_{t_0}, the solution of the given boundary value problem exists, since equation (55) meets all the conditions for the existence of a solution satisfying the initial conditions $x(\gamma_{i-1}(t_0)) = \varphi_{i-1}(\gamma_{i-1}(t_0))$.

(2) If the functions f and φ_0 are such that equation (55) satisfies the

conditions of the existence and uniqueness theorem (for example, if its right hand side is continuous and satisfies a Lipschitz condition on the second argument) the solution of the given boundary value problem for equation (54) is unique.

(3) If f is k times differentiable, and if φ_0 is $k+1$ times differentiable on E_{t_0} and $\tau'(t) < 1$ for $t \geqq t_0$, then at the inner points of the intervals $\gamma_{i-1}(t_0) \leqq t \leqq \gamma_i(t_0)$ the solution $x(t)$ of equation (54) is also $k+1$ times differentiable, but for $t = \gamma_i(t_0)$ $(i = 0, 1, 2, \cdots)$ the solution $x(t)$ will in general have an angular point.

For in fact, the solution $x(t)$ has in general an angular point $t = t_0$, since a solution of the equation $\dot{x}(t) = f(t, x(t), \varphi_0(t - \tau(t)), \dot{\varphi}_0(t - \tau(t)))$ that satisfies $x(t_0) = \varphi_0(t_0)$ can only under exceptional circumstances also satisfy the condition

$$\dot{x}(t_0 + 0) = \dot{\varphi}_0(t_0 - 0).$$

Writing in succession $t = \gamma_0(t_0)$, $t = \gamma_1(t_0), \cdots, t = \gamma_n(t_0), \cdots$, we note that the derivative $\dot{x}(t)$ at these points will in general be discontinuous, since $\dot{x}(t) = f(t, x(t), x(t - \tau(t)), \dot{x}(t - \tau(t)))$ and for these values of t, generally speaking, the last argument, $\dot{x}(t - \tau(t))$, of the function f is discontinuous.

If the initial function $\varphi_0(t)$ is so chosen that at the point t_0 the derivatives of the solution $x(t)$ are continuous up to order k and the functions f and φ_0 are differentiable up to the number of times indicated above, then the solution will be differentiable k times, as an intermediate consequence of equation (55).

Thus the solutions of the neutral equation (54) do not become more smooth with increasing t, but the degree of smoothness does not decrease.

For the second order equation

$$\ddot{x}(t) = f(t, x(t), \dot{x}(t), x(t - \tau(t)), \dot{x}(t - \tau(t)), \ddot{x}(t - \tau(t))) \tag{56}$$

under similar assumptions regarding the functions f and φ_0 and regarding the initial set E_{t_0}, we also find that the solution $x(t)$ does not smooth out with increasing t, but as opposed to the solutions of (54), the solutions of (56) have continuous first derivatives at the points $\gamma_i(t_0)$ and in general have discontinuous second derivatives. This follows immediately from the fact that the solution of the equation

$$\ddot{x}(t) = f(t, x(t), \dot{x}(t), \varphi_0(t - \tau(t)), \dot{\varphi}_0(t - \tau(t)), \ddot{\varphi}_0(t - \tau(t))),$$

defined by the initial conditions $x(t_0) = \varphi_0(t), \dot{x}(t_0 + 0) = \dot{\varphi}_0(t_0 - 0)$, does not in general satisfy the condition $\ddot{x}(t_0 + 0) = \ddot{\varphi}_0(t_0 - 0)$.

In the special case, when F_{t_0} reduces to a single point, the solution of equation (54) may fail to exist, even under severe constraints on the function f, and if it does exist it will not in general be unique.

For in fact, for $t = t_0$ we obtain from (54) in this case:

$$\dot{x}_0 = f(t_0, x_0, x_0, \dot{x}_0) \tag{57}$$

where x_0 is given.

The following cases are possible:

(1) Equation (57) has no real root $\dot{x}_0$: then (54) has no continuous solution satisfying the condition $x(t_0) = x_0$.

(2) Equation (57) has k real roots $\dot{x}_{0j}$; then under certain rather stringent conditions governing the function f, there exist k continuous solutions of equation (54).[13]

(3) Equation (57) reduces to an identity; then under similar conditions there exists a one-parameter family of continuous solutions of equation (54).

Among all the approximate methods for integrating equation (54), the presence of angular points of the solution forces us to recommend one method only, namely Euler's method, with the obligatory inclusion of the points $\gamma_k(t_0)$ $(k = 1, 2, \cdots)$ in the calculation grid; this in general leads to computations with a variable step. For the same reason, we can recommend for equation (56) only those methods that are based on approximation by second-order parabolas, and the grid must include the points $\gamma_k(t_0)$ $(k = 1, 2, \cdots)$.

The method of successive approximations, and also the method of expansion of the solution in powers of τ, are not in general applicable.

The theory of the dependence of the solution of the neutral equation

$$\mu\dot{x}(t) = f\big(t, x(t), x(t-\tau(t)), \dot{x}(t-\tau(t))\big), \tag{58}$$

$$x = \varphi(t) \text{ on } E_{t_0}$$

and its generalizations, on the small parameter μ in the nonspecial case (F_{t_0} does not reduce to a single point) is no different from the theory for similar equations with retarded arguments, since the application of the method of successive integrations (method of steps) to this equation leads also to differential equations without deviating arguments and with a small coefficient of the principal derivative.

In the special case the solution of (58) will not converge to the solution of the generating equation as $\mu \to 0$ except under highly restricted conditions.

---

[13] The conditions on f that make assertions (2) and (3) valid have been studied by G. A. Kamenskiĭ.

As we have already shown in §10, Chapter V, the stability of the solutions of differential-difference equations of neutral type can be successfully attacked by studying the first-order stability if the system of first order approximations is stationary, whereas Ljapunov's second method is applicable only under restricted circumstances.

13. **Equations with advanced arguments.** Differential equations with advanced arguments have been very little studied, and we can therefore offer only a few observations regarding them.

Let us consider differential equations with advanced arguments of the following very simple types:

$$x(t) = f\big(t,\ x(t-\tau(t)),\ \dot{x}(t-\tau(t))\big) \tag{59}$$

and

$$\dot{x}(t) = f\big(t,\ x(t),\ x(t-\tau(t)),\ \dot{x}(t-\tau(t)),\ \ddot{x}(t-\tau(t))\big) \tag{60}$$

with the usual initial conditions $x(t) = \varphi_0(t)$ on E_{t_0}, $\tau(t) \geqq 0$, the functions f and τ being continuous and the function φ twice continuously differentiable; we are to find a solution for (59) or (60) for $t_0 \leqq t \leqq B$ (if we were to try to find a solution for $t \leqq t_0$, the transformation of variables $t = -\bar{t}$ would lead to the fundamental boundary value problem for equations with retarded argument).

We assume that the set $F_{\bar{t}_0}$ does not reduce to a single point for any value of $\bar{t}_0$ in the interval $t_0 \leqq \bar{t}_0 \leqq B$. In this case, applying the method of steps, we obtain for (59) a solution in the form

$$x(t) = f\big(t,\ \varphi_k(t-\tau(t)),\ \dot{\varphi}_k(t-\tau(t))\big) \tag{61}$$

for $\gamma_k(t_0) \leqq t \leqq \gamma_{k+1}(t_0)$; $k = 0, 1, 2, \cdots$; $\gamma_0(t_0) = t_0$, and for equation (60) a solution determined by the first order differential equation without retardations

$$\dot{x}(t) = f\big(t,\ x(t),\ \varphi_k(t-\tau(t)),\ \dot{\varphi}_k(t-\tau(t)),\ \ddot{\varphi}_k(t-\tau(t))\big)$$

$$\text{for}\quad \gamma_k(t_0) \leqslant t \leqslant \gamma_{k+1}(t_0),\quad k = 0,\ 1,\ 2,\ \ldots, \tag{62}$$

where $\varphi_k(t)$ is a solution of the same equation in the interval $\gamma_{k-1}(t_0) \leqq t \leqq \gamma_k(t_0)$. We remark that the conditions laid on f and φ_0 guarantee the existence of a solution only for $k = 0$, since $\dot{\varphi}_1(t)$ for equation (61) and $\ddot{\varphi}_1(t)$ for equation (62) need not necessarily exist. It is obvious that at the points $\gamma_p(t_0)$ $(p = 0, 1, 2, \cdots)$ the solution of equation (59) is in general discontinuous; even if we so prescribe the initial functions that the solution and all its derivatives up to and including order k are continuous at the point

t_0 and also assume that f and φ_0 are sufficiently smooth, we nevertheless find that with increasing t the solution continually loses its smoothness and at the points $\gamma_s(t_0)$, beginning with $s = k + 1$, the solution will in general have discontinuities.

A similar phenomenon is experienced in connection with the solutions of equation (60) with one difference only, that there exists a continuous solution of (60) which has angular points at the points $\gamma_p(t_0)$ $(p = 0, 1, 2, \cdots)$. The presence of angular points of the solution of (60) leads to the necessity of applying to the integration of (62) the usual methods of numerical integration in every interval $\gamma_k(t_0) \leqq t \leqq \gamma_{k+1}(t_0)$.

The method of expansion in powers of τ and the method of successive approximations are not applicable to the solution of the fundamental boundary value problem for equations with advanced arguments (unless we impose some highly special conditions). The dependence of the solution of the equation

$$\mu\dot{x}(t) = f\big(t,\ x(t),\ x(t-\tau(t)),\ \dot{x}(t-\tau(t)),\ \ddot{x}(t-\tau(t))\big),$$
$$x(t) = \varphi_0(t) \text{ on } E_{t_0},$$

on the small positive parameter μ, with the assumption that F_{t_1} does not consist of the single point $t_1 \geqq t_0$, reduces to the analogous problem for the unretarded equation

$$\mu\dot{x}(t) = f\big(t,\ x(t),\ \varphi_k(t-\tau(t)),\ \dot{\varphi}_k(t-\tau(t)),\ \ddot{\varphi}_k(t-\tau(t))\big),$$
$$x\big(\gamma_k(t_0)\big) = \varphi_k\big(\gamma_k(t_0)\big),\ \gamma_k(t_0) \leqslant t \leqslant \gamma_{k+1}(t_0).$$

Many theorems of the general theory of linear equations with advanced argument $L(x(t)) = f(t)$ differ very little from the corresponding theorems for linear equations of neutral type, or for equations with retarded argument, since they are based on the fundamental properties of all linear operators: $L(cx(t)) = cL(x(t))$, where c is a constant and $L(x_1(t) + x_2(t)) = L(x_1(t)) + L(x_2(t))$. Some linear equations that have constant coefficients and constant retardations with advanced argument can be studied by operational methods, and Euler's method can be applied to the corresponding homogeneous equations for the determination of their particular solutions. We remark that the characteristic equation, for linear homogeneous equations of arbitrary order with advanced argument, constant coefficients, and constant deviations, will always have roots with positive real parts, as already mentioned on page 185, and therefore the trivial solution of such differential equations is never stable.

This conclusion follows immediately from results due to L. S. Pontrjagin

[**230**], generalized in the text of N. G. Čebotarev and N. N. Meĭman [**178**], on the roots of quasipolynomials, according to which a quasipolynomial $\sum_{i,s=0}^{m,n} a_{is} z^s e^{\lambda_i z}$, where a_{is} and λ_i are real numbers, which lacks a leading term always has a root with positive real part.

14. **Differential-difference equations involving partial derivatives.** The theory of differential-difference equations involving partial derivatives has received almost no attention. The most interesting of the existence theorems are those of I. M. Gul′,[14] which prove the existence and uniqueness of a solution to Cauchy's problem for a system of differential-difference equations in partial derivatives, in the so-called nonspecial case, i.e., when the differential-difference equation in partial derivatives can be reduced to a partial difference equation without deviating arguments. After such a reduction, one applies the usual constraints guaranteeing the existence and uniqueness of the solution to Cauchy's problem. The question of whether the solution can be continued has also been studied. Also, Gul′ has shown that the method of characteristics is applicable to a certain class of first order equations.

In solving a number of boundary value problems with retarded argument in mathematical physics it has been possible to resort to separation of variables and to expansion of the solution in a series of eigenfunctions. In problems of mathematical physics the retardation generally appears only as a time-lag. This case arises, for example, when we are studying the free oscillations of a homogeneous string, on which we suppose that some delayed elastic restoring force is at work:

$$\frac{\partial^2 U(t, x)}{\partial t^2} = a^2 \frac{\partial^2 U(t, x)}{\partial x^2} + b^2 \frac{\partial^2 U(t-\tau, x)}{\partial x^2}.$$

A similar instance of delay can be found in the study of diffusion phenomena; in the simplest case we have the equation

$$\frac{\partial U(t, x)}{\partial t} = a^2 \frac{\partial^2 U(t, x)}{\partial x^2} + b^2 \frac{\partial^2 U(t-\tau, x)}{\partial x^2}.$$

In this and in similar linear homogeneous problems with linear homogeneous boundary conditions, with constant or variable retardation $\tau = \tau(t) \geqq 0$, with the initial condition

$$U(t, x) = \varphi(t, x),$$

[14] Gul′ I. M., *Cauchy's problems for some partial differential equations with functional arguments*, Uspehi Mat. Nauk **10**, no 2(64), 153-156 (1955). (Russian)

where t varies over the initial[15] set E_{t_0} consisting of the point $t = t_0$ and of the values of $t - \tau(t)$ which are less than t_0 for $t \geqq t_0$, we may apply the method of separation of variables and of decomposition of the solution by eigenfunctions. In fact, putting $U(t,x) = X(x)T(t)$ and separating variables (assuming that the separation is possible) we obtain for the determination of $X(x)$ a standard Sturm-Liouville problem without retardation for the determination of $T(t)$ a linear homogeneous first or second order equation with retarded argument

$$L[T(t),\ T(t-\lambda),\ \lambda] = 0.$$

Representing the solution in the form of a series in the eigenfunctions $X_n(x)$ corresponding to the eigenvalues λ_n of the Sturm-Liouville equation we have

$$U(t,\ x) = \sum_{n=1}^{\infty} B_n(t)\, X_n(x),$$

and requiring that the initial condition

$$\varphi(t,\ x) = \sum_{n=1}^{\infty} B_n(t)\, X_n(x)$$

be satisfied on E_{t_0} we determine the Fourier coefficients $B_n(t)$ of the function $\varphi(t,x)$ on the same set E_{t_0}. Knowing these, we may now find the solutions $T_n(t)$ for $t \geqq t_0$ of the equation $L[T(t), T(t-\tau), \lambda] = 0$, corresponding to the eigenvalues $\lambda = \lambda_n$ and the initial functions

$$T_n(t) = B_n(t) \text{ on } E_{t_0}.$$

After this, if the usual conditions for twofold term-by-term differentiability are satisfied, the solution can be represented in the form

$$U(t,\ x) = \sum_{n=1}^{\infty} T_n(t)\, X_n(x).$$

Obviously, this method can also be applied in the case of a larger number of independent variables, provided the retardations apply only to the time.

---

[15] We suppose that the point t_0 is not isolated in E_{t_0}; in the contrary case we must prescribe the values of the derivatives with respect to t up to and including the order $(n-1)$ at the point t_0, where the highest order of differentiation with respect to t is n.

Chapter VI

Variational Problems with Retarded Argument

1. **Statement of the elementary problem.** Variational problems arising in control theory may turn out to be complicated by the effects of retardation. Retardations appear sometimes through the time delay involved in signal transmission, but more often arise because of simplifying assumptions to the effect that the action of intermediate transmitting and amplifying links in the control system can be represented as a retarded signal transmission.

The theory of variational problems involving retarded arguments is almost totally undeveloped. We must confine ourselves to presenting the fundamental outlines of the theory; qualitative methods play an insignificant role in this chapter, but even though we do not limit ourselves to purely qualitative investigations, the chapter does deal with a number of problems of a qualitative nature.

The simplest variational problem can be stated as follows: determine the extremum of the functional

$$\int_{t_0}^{t_1} F\big(t,\ x(t),\ x(t-\tau),\ \dot{x}(t),\ \dot{x}(t-\tau)\big)\,dt$$

under the conditions $x(t) = \varphi(t)$ for $t_0 - \tau \leqq t \leqq t_0$ and $x(t_1) = x_1$. Here τ is a positive constant retardation, the function F will be supposed three times differentiable, and the function φ differentiable. As an example of a variational problem of this type, we consider the problem of the shortest time required to displace the point $M(x,y)$ from the position $A(x_0, y_0)$ to the position $B(x_1, y_1)$ if the velocity $V(x, y(x), y(x-\tau))$ depends not only on the coordinates x, y of the point M, but also on $y(x-\tau)$. This can happen, for instance, if the velocity is controlled as a function of the position of the point, and the control signal is transmitted with a certain delay. The problem reduces to the study of the extremum of the functional

$$t = \int_{x_0}^{x_1} \frac{\sqrt{1+y'(x)}\,dx}{V(x,\ y(x),\ y(x-\tau))},$$

and in this connection it is natural to assume that for $x_0 - \tau \leqq x \leqq x_0$ the value $y(x)$ is given by $y(x) = \varphi(x)$.

2. **The fundamental lemmas.** Before we can establish the fundamental necessary conditions for an extremum, we must prepare the ground by proving two lemmas analogous to the lemmas of Lagrange and du Bois-Reymond.

LEMMA I. *Let the function* $\eta(t)$ *satisfy the following conditions:*
(1) $\eta(t_0 - \tau_i(t_0)) = 0$ *and* $\eta(t_1 - \tau_i(t_1)) = 0$ $(i = 1, 2, \cdots, n)$;
(2) $\eta(t)$ *has continuous derivatives up to order* p *in the interval*

$$t_0 - \tau^0 \leqq t \leqq t_1,$$

where

$$\tau^0 = \max[\tau_1(t_0), \tau_2(t_0), \ldots, \tau_n(t_0)];$$

(3) $|\eta^{(k)}(t)| < \epsilon$ $(k = 0, 1, 2, \cdots, s)$, $s \leqq p$;
then if for arbitrary functions η *the integral*

$$\int_{t_0}^{t_1} \sum_{i=1}^{n} \Phi_i(t)\eta\big(t - \tau_i(t)\big)\,dt = 0,$$

where $\Phi_i(t)$ *is continuous in the interval* t_0, t_1 *and* $\tau_i(t)$ *is a non-negative differentiable function and the* $\tau_i(t)$ *are non-negative continuous differentiable functions, with first derivatives that satisfy the inequalities* $\tau_i'(t) < d < 1$ $(i = 1, 2, \cdots, n)$, *then*

$$\sum_{i=1}^{n} \frac{\Phi_i(\gamma_i(t))}{1 - \tau_i'(\gamma_i(t))} = 0$$

for $t_0 - \tau^0 \leqq t \leqq t_1$, *where* $\gamma_i(t)$ *is inverse to the function* $s = t - \tau_i(t)$. *Outside the interval* $t_0 \leqq t \leqq t_1$ *all the functions* $\Phi_i(t)$ *are assumed to vanish.*

PROOF. Let us suppose that at some point $t = \bar{t}$

$$\sum_{i=1}^{n} \frac{\Phi_i(\gamma_i(\bar{t}))}{1 - \tau_i'(\gamma_i(\bar{t}))} \neq 0,$$

where $t_0 - \tau^0 < \bar{t} < t_1$. Without restricting the generality of our proof, we may assume that

$$\bar{t} \neq t_0 - \tau_i(t_0) \quad \text{and} \quad \bar{t} \neq t_1 - \tau_i(t_1) \qquad (i = 1, 2, \ldots, n).$$

Then in view of the continuity of the function

$$\sum_{i=1}^{n} \frac{\Phi(\gamma_i(t))}{1 - \tau_i'(\gamma_i(t))} \quad \text{we have} \quad \sum_{i=1}^{n} \frac{\Phi_i(\gamma_i(\bar{t}_i))}{1 - \tau_i'(\gamma_i(\bar{t}_i))} \neq 0,$$

where the $\bar{t}_i$ are arbitrary values of t in the interval $(\bar{t}-\epsilon, \bar{t}+\epsilon)$ for sufficiently small ϵ. Let us choose the function $\eta(t)$ to be identically zero outside the ϵ-neighborhood $(\bar{t}-\epsilon, \bar{t}+\epsilon)$ of the point $\bar{t}$ and to be of constant sign in that neighborhood; for example, we define $\eta(t)$ by

$$\eta(t) = k(t-\bar{t}+\varepsilon)^{2m}(t-\bar{t}-\varepsilon)^{2m},$$

in the ϵ-neighborhood, where k is a constant so chosen that conditions (2) and (3) of the lemma are satisfied. With such a choice of $\eta(t)$ we arrive at a contradiction to the fundamental condition of the lemma. In fact,

$$\int_{t_0}^{t_1}\sum_{i=1}^{n}\Phi_i(t)\eta(t-\tau_i(t))\,dt = \sum_{i=1}^{n}\int_{\gamma_i(\bar{t}-\varepsilon)}^{\gamma_i(\bar{t}+\varepsilon)}\Phi_i(t)\eta(t-\tau_i(t))\,dt$$

$$= \sum_{i=1}^{n}\int_{\bar{t}-\varepsilon}^{\bar{t}+\varepsilon}\frac{\Phi_i(\gamma_i(s))\eta(s)\,ds}{1-\tau_i'(\gamma_i(s))} = \sum_{i=1}^{n}\frac{\Phi_i(\gamma_i(\bar{t}_i))}{1-\tau_i'(\gamma_i(\bar{t}_i))}\int_{\bar{t}-\varepsilon}^{\bar{t}+\varepsilon}\eta(s)\,ds \neq 0,$$

where the $\bar{t}_i$ are values of t intermediate between $\bar{t}-\epsilon$ and $\bar{t}+\epsilon$.

Lemma II. *If the integral* $\int_{t_0}^{t_1}\sum_{i=1}^{n}\Phi_i(t)\eta'(t-\tau_i(t))dt = 0$, *where the functions* τ_i *and* η *satisfy the conditions of Lemma* I, *then*

$$\sum_{i=1}^{n}\frac{\Phi_i(\gamma_i(t))}{1-\tau_i'(\gamma_i(t))} = C,$$

where C *is a constant and* $t_0-\tau^0 \leqq t \leqq t_1$, *while all the functions* $\Phi_i(t) = 0$ *outside the interval.*

Proof. We suppose that

$$\sum_{i=1}^{n}\frac{\Phi_i(\gamma_i(\bar{t}))}{1-\tau_i'(\gamma_i(\bar{t}))} \neq \sum_{i=1}^{n}\frac{\Phi_i(\gamma_i(\bar{\bar{t}}))}{1-\tau_i'(\gamma_i(\bar{\bar{t}}))},$$

where $\bar{t}$ and $\bar{\bar{t}}$ are not equal to $t_0-\tau_i(t_0)$ and $t_1-\tau_i(t_1)$ $(i = 1, 2, \cdots, n)$. In view of the continuity of the functions $\sum_{i=1}^{n}\Phi_i(t)/(1-\tau_i'(t))$ and $\gamma_i(t)$ we have

$$\sum_{i=1}^{n}\frac{\Phi_i(\gamma_i(\bar{t}_i))}{1-\tau_i'(\gamma_i(\bar{t}_i))} \neq \sum_{i=1}^{n}\frac{\Phi_i(\gamma_i(\bar{\bar{t}}_i))}{1-\tau_i'(\gamma_i(\bar{\bar{t}}_i))},$$

where $\bar{t}_i$ and $\bar{\bar{t}}_i$ are arbitrary values of t in the sufficiently small intervals $(\bar{t}-\epsilon, \bar{t}+\epsilon)$ and $(\bar{\bar{t}}-\epsilon, \bar{\bar{t}}+\epsilon)$ respectively.

Let us choose the function $\eta(t)$ so that our initial assumption leads to a contradiction: $\eta'(t) = 0$ outside the intervals $(\bar{t} - \epsilon, \bar{t} + \epsilon)$ and $(\bar{\bar{t}} - \epsilon, \bar{\bar{t}} + \epsilon)$, and within these intervals $\eta(t)$ is so chosen that

$$\int_{\bar{t}-\epsilon}^{\bar{t}+\epsilon} \eta'(t)\,dt = -\int_{\bar{\bar{t}}-\epsilon}^{\bar{\bar{t}}+\epsilon} \eta'(t)\,dt \neq 0.$$

It is known that the function $\eta(t)$ can be chosen so as to satisfy these conditions as well as the conditions of the lemma.[1] Then

$$\int_{t_0}^{t_1} \sum_{i=1}^{n} \Phi_i(t)\,\eta'(t - \tau_i(t))\,dt$$

$$= \sum_{i=1}^{n} \int_{\gamma_i(\bar{t}-\epsilon)}^{\gamma_i(\bar{t}+\epsilon)} \Phi_i(t)\,\eta'(t - \tau_i(t))\,dt + \sum_{i=1}^{n} \int_{\gamma_i(\bar{\bar{t}}-\epsilon)}^{\gamma_i(\bar{\bar{t}}+\epsilon)} \Phi_i(t)\,\eta'(t - \tau_i(t))\,dt$$

$$= \sum_{i=1}^{n} \int_{\bar{t}-\epsilon}^{\bar{t}+\epsilon} \Phi_i(\gamma_i(s))\,\eta'(s) \frac{ds}{1 - \tau_i'(\gamma_i(s))} + \sum_{i=1}^{n} \int_{\bar{\bar{t}}-\epsilon}^{\bar{\bar{t}}+\epsilon} \Phi_i(\gamma_i(s))\,\eta'(s) \frac{ds}{1 - \tau_i'(\gamma_i(s))}$$

$$= \sum_{i=1}^{n} \frac{\Phi_i(\gamma_i(\bar{t}_i))}{1 - \tau_i'(\gamma_i(\bar{t}_i))} \int_{\bar{t}-\epsilon}^{\bar{t}+\epsilon} \eta'(s)\,ds + \sum_{i=1}^{n} \frac{\Phi_i(\gamma_i(\bar{\bar{t}}_i))}{1 - \tau_i'(\gamma_i(\bar{\bar{t}}_i))} \int_{\bar{\bar{t}}-\epsilon}^{\bar{\bar{t}}+\epsilon} \eta'(s)\,ds$$

$$= \left[\sum_{i=1}^{n} \frac{\Phi_i(\gamma_i(\bar{t}_i))}{1 - \tau_i'(\gamma_i(\bar{t}_i))} - \sum_{i=1}^{n} \frac{\Phi_i(\gamma_i(\bar{\bar{t}}_i))}{1 - \tau_i'(\gamma_i(\bar{\bar{t}}_i))}\right] \int_{\bar{t}-\epsilon}^{\bar{t}+\epsilon} \eta'(s)\,ds \neq 0.$$

REMARK I. If under the conditions of Lemmas I or II we also require that $\eta(t) = 0$ for $t_0 - \tau^0 \leqq t \leqq \tilde{t}_0$, where $\tilde{t}_0 < t_1$, the lemma obviously remains valid for $\tilde{t}_0 \leqq t \leqq t_1$.

REMARK II. If under the conditions of Lemma I or Lemma II we also

---

[1] See for example L. A. Ljusternik and M. A. Lavrent'ev, *Course in the calculus of variations*, 2d. ed., GITTL. Moscow, 1950. (Russian)

require that $\eta(t) = 0$ for $\tilde{t}_1 \leqq t \leqq t_1$, the functions $\Phi_i(t)$ in (1) and (2) must be taken to vanish outside the interval $(t_0, \tilde{t}_1)$.

3. **Fundamental necessity conditions for an extremum.** It is clear that for functionals with retarded arguments as well, the fundamental necessity condition for an extremum, in the case where two-sided variation is possible, is that the variation of the functional should be zero. As applied to the functional

$$V[x(t)] = \int_{t_0}^{t_1} F\big(t,\ x(t-\tau_1(t)), \ldots$$

$$\ldots,\ x(t-\tau_n(t)),\ \dot{x}(t-\tau_1(t)),\ \ldots,\ \dot{x}(t-\tau_n(t))\big)\, dt,$$

$$x\big(t_0-\tau_i(t_0)\big) = x_{i0},\ x\big(t_1-\tau_i(t_1)\big) = x_{i1} \quad (i = 1,\ 2,\ \ldots,\ n)$$

this condition has the form

$$\delta V = \int_{t_0}^{t_1} \sum_{i=1}^{n} \big[F'_{x(t-\tau_i(t))}\, \delta x\big(t-\tau_i(t)\big)$$

$$+ F'_{\dot{x}(t-\tau_i(t))}\, \delta\dot{x}\big(t-\tau_i(t)\big)\big]\, dt = 0. \qquad \text{(A)}$$

Integrating the second term under the summation sign by parts and taking the boundary condition into account, we obtain:

$$\int_{t_0}^{t_1} \sum_{i=1}^{n} \left(F'_{x(t-\tau_i(t))} - \frac{d}{dt} \frac{F'_{\dot{x}(t-\tau_i(t))}}{1-\tau'_i(t)} \right) \delta x\big(t-\tau_i(t)\big)\, dt = 0.$$

Applying Lemma I, we obtain:

$$\sum_{i=1}^{n} \left. \frac{F'_{x(s-\tau_i(s))} - \dfrac{d}{ds} \dfrac{F'_{\dot{x}(s-\tau_i(s))}}{1-\tau'_i(s)}}{1-\tau'_i(s)} \right|_{s=\gamma_i(t)} = 0. \qquad (1)$$

The foregoing argument assumes the existence of $\ddot{x}(t)$ on a curve realizing the extremum. In order to dispense with this requirement, we may integrate the first term under the summation sign in (A) by parts, obtaining:

$$\int_{t_0}^{t_1} \sum_{i=1}^{n} \left(\int_{t_0}^{t} F'_{x(t-\tau_i(t))}\, dt - F'_{\dot{x}(t-\tau_i(t))} \right) \delta\dot{x}\big(t-\tau_i(t)\big) = 0$$

and, applying Lemma II, we obtain:

$$\sum_{i=1}^{n} \frac{\left(1-\tau_i'(s)\right)\int\limits_{t_0}^{s} F'_{x(t-\tau_i(t))}\,dt - F'_{\dot{x}(s-\tau_i(s))}}{1-\tau_i'(s)}\Bigg|_{s=\gamma_i(t)} = C,$$

or

$$\sum_{i=1}^{n}\left[\int\limits_{t_0}^{s} F'_{x(t-\tau_i(t))}\,dt - \frac{F'_{\dot{x}(s-\tau_i(s))}}{1-\tau_i(s)}\right]_{s=\gamma_i(t)} = C. \qquad (2)$$

As applied to the very simple functional

$$V[x(t)] = \int\limits_{t_0}^{t_1} F\left(t,\ x(t),\ x(t-\tau),\ \dot{x}(t),\ \dot{x}(t-\tau)\right) dt,$$

where τ is a constant, the fundamental necessity condition (1) for an extremum will have the form

$$\left(F'_{x(s)} - \frac{d}{ds} F'_{\dot{x}(s)}\right)_{s=t} + \left(F'_{x(s-\tau)} - \frac{d}{dt} F'_{\dot{x}(s-\tau)}\right)_{s=t+\tau} = 0. \qquad (3)$$

Let us consider several particular examples, in which this equation becomes simple:

(1) Suppose that

$$F = \Phi\left(t,\ x(t),\ \dot{x}(t)\right) + \Psi\left(t,\ x(t-\tau),\ \dot{x}(t-\tau)\right).$$

In this case we obtain an unretarded equation

$$\Phi'_{x(t)} - \frac{d}{dt}\Phi'_{\dot{x}(t)} + \left(\Psi'_{x(s-\tau)} - \frac{d}{ds}\Psi'_{\dot{x}(s-\tau)}\right)_{s=t+\tau} = 0,$$

which is understandable, since the integral

$$\int\limits_{t_0}^{t_1} F\,dt = \int\limits_{t_0}^{t_1} \Phi\,dt + \int\limits_{t_0}^{t_1} \Psi\left(t,\ x(t-\tau),\ \dot{x}(t-\tau)\right) dt$$

$$= \int\limits_{t_0}^{t_1} \Phi\,dt + \int\limits_{t_0-\tau}^{t_1-\tau} \Psi\left(t+\tau,\ x(t),\ \dot{x}(t)\right) dt,$$

that is, the problem has been reduced to one of studying a functional without retardations. For example:

$$V[x(t)] = \int\limits_{0}^{a} [\dot{x}^2(t) + x^2(t-\tau)]\,dt; \quad x(0) = 0,\ x(a) = 0.$$

The extremal equations have the form

$$\ddot{x}(t)+x(t)=0, \quad 0\leqslant t\leqslant a-\tau, \quad x(t)=C_1\sin t;$$
$$\ddot{x}(t)=0, \quad a-\tau\leqslant t\leqslant a, \quad x(t)=C_2(t-a);$$

C_1 and C_2 may be determined from the conditions of continuity and smoothness of the function $x(t)$ at the point $a-\tau$.

(2) If $F=\Phi(t,x(t-\tau),\ \dot{x}(t-\tau))$, then the interval $t_0\leqq t\leqq t_1-\tau$ we obtain an equation without retardation, and on the interval $[t_1-\tau,t_1]$ we obtain an identity; on this interval the function $x(t)$ may be chosen arbitrarily, which is understandable since the choice of the function $x(t)$ in this interval has no influence whatever on the value of the functional $\int_{t_0}^{t_1}\Phi\,dt$.

(3) $F=x(t)\Phi(t,x(t-\tau),\ \dot{x}(t-\tau))+x(t)\Psi(t,x(t-\tau),\ \dot{x}(t-\tau))$. In the interval $(t_1-\tau,t_1)$ the equation of the extremal takes the form $\Phi(t,x(t-\tau),$ $\dot{x}(t-\tau))-(d/dt)\Psi(t,x(t-\tau),x(t-\tau))=0$ and this is an equation without retardation.

(4) If $F=\Phi(t,x(t),x(t-\tau))$, equation (3) has the form

$$\Phi'_{x(t)}(t,\ x(t),\ x(t-\tau))+\Phi'_{x(t-\tau)}(t+\tau,\ x(t+\tau),\ x(t))=0$$

which is a finite difference equation. The continuous solutions of this equation, if they exist, will not in general satisfy the boundary conditions. Generally speaking, continuous solutions of this problem do not exist.

(5) $F=\Phi(t,x(t))\dot{x}(t)+\Psi(t,x(t-\tau))\cdot\dot{x}(t-\tau)+\chi(t,x(t),x(t-\tau))$. We obtain a finite equation or an identity:

$$\chi'_{x(t)}-\Phi'_t+[\chi'_{x(t-\tau)}-\Psi'_t]_{t+\tau}=0.$$

In the first case a continuous solution, generally speaking, does not exist; in the second case the value of the functional does not depend on the path of integration.

(6) $F=F(t,\dot{x}(t),\dot{x}(t-\tau))$; then equation (3) admits a first integral

$$F_{\dot{x}(t)}+(F_{\dot{x}(s-\tau)})_{s=t+\tau}=C.$$

4. **Further necessity conditions.** For the functionals considered in the preceding section we may obtain further conditions necessary for the existence of an extremal, similar to the Legendre conditions, arising from the following theorem.

THEOREM. *If on the extremal $x=x(t)$ the functional*

$$V[x(t)] = \int_{t_0}^{t_1} F\big(t,\ x(t-\tau_1(t)), \ldots$$

$$\ldots, x(t-\tau_n(t)),\ \dot{x}(t-\tau_1(t)),\ \ldots,\ \dot{x}(t-\tau_n(t))\big)\,dt$$

attains a minimum (or maximum), then on that extremal

$$\begin{vmatrix} F''_{\dot{x}_1\dot{x}_1} & \ldots & F''_{\dot{x}_1\dot{x}_k} \\ \cdot & \cdot & \cdot \\ \cdot & \cdot & \cdot \\ F''_{\dot{x}_n\dot{x}_1} & \ldots & F''_{\dot{x}_n\dot{x}_n} \end{vmatrix} \geqslant 0 \qquad (k = 1, 2, \ldots, n)$$

(in the case of the maximum, the sense of the inequality must be reversed) where

$$t_0 \leqslant t \leqslant t_1 - \tau^1, \quad x_j = x(t-\tau_j(t)), \quad \dot{x}_j = \dot{x}_j(t-\tau_j(t)),$$

$$\tau^1 = \max \tau_j(t_1) \qquad (j = 1, 2, \ldots, n).$$

We subject the functions F, φ, and τ_j to the same conditions as in Lemma I, and, moreover, require that $\gamma_i(t) \neq \gamma_j(t)$ for $i \neq j$.

PROOF.

$$\delta^2 V = \int_{t_0}^{t_1} \sum_{i,j=1}^{n} \big(F''_{x_i x_j}\delta x_i \delta x_j + 2F''_{\dot{x}_i x_j}\delta\dot{x}_i\delta x_j + F''_{\dot{x}_i\dot{x}_j}\delta\dot{x}_i\delta\dot{x}_j\big)\,dt.$$

In order that the extremal be a minimum, it is necessary that $\delta^2 V \geqq 0$, and this in turn implies that

$$\sum_{i,j=1}^{n} F''_{\dot{x}_i\dot{x}_j}\delta\dot{x}_i\delta\dot{x}_j \geqslant 0.$$

The assertion of the theorem is also a necessary and sufficient condition for non-negativity of the quadratic form $\sum_{i,j=1}^{n} F''_{\dot{x}_i\dot{x}_j}\delta\dot{x}_i\delta\dot{x}_j$ under arbitrary variations $\delta\dot{x}_i$ and $\delta\dot{x}_j$. In the present case, even though $\delta\dot{x}_i$ and $\delta\dot{x}_j$ are not independent, the necessary and sufficient conditions for non-negativity of the quadratic form remain the same, since at some point

$$t = \bar{t}, \quad t_0 < \bar{t} < t_1 - \tau^1$$

we may give δx_i and δx_j the arbitrary values α and β respectively.

In fact, we choose the variation $\delta x(t)$ so that at the points $t = \gamma_i(\bar{t})$ and

$t = \gamma_j(\bar{t})$, $\delta\dot{x}(t)$ shall be equal respectively to α and β; then for $t = \bar{t}$, $\delta\dot{x}_i = \alpha$ and $\delta\dot{x}_j = \beta$.

5. **Generalizaticn to functionals of more complicated type.** The fundamental necessity conditions for an extremal can easily be obtained for functionals of more complicated type. For example, for functionals of the form

$$V = \int_{t_0}^{t_1} F\big(t,\ x_j(t-\tau_i(t)),\ \dot{x}_j(t-\tau_i(t))\big)\,dt,$$

$$(j = 1,\ 2,\ \ldots,\ m;\quad i = 1,\ 2,\ \ldots,\ n),$$

where F is a function of the argument t and of the variables $x_j(t - \tau_i(t))$ ($j = 1, 2, \cdots, m$; $i = 1, 2, \cdots, n$), and of their derivatives, the values of all the variables at the points $t_0 - \tau_i(t_0)$ and $t_1 - \tau_i(t_1)$ being prescribed; varying only one of the $x_j(t)$ we obtain:

$$\sum_{i=1}^{n}\left[\frac{F'_{x_j(s-\tau_i(s))} - \dfrac{d}{ds}\dfrac{F'_{\dot{x}_j(s-\tau_i(s))}}{1-\tau_i'(s)}}{1-\tau_i'(s)}\right]_{s=\gamma_i} = 0$$

for every $j = 1, 2, \cdots, m$.

For the functional

$$V = \int_{t_1}^{t_1} F\big(t,\ x(t-\tau_i(t)),\ \dot{x}(t-\tau_i(t)),\ \ldots,\ x^{(m)}(t-\tau_i(t))\big)\,dt$$

$$(i = 1,\ 2,\ \ldots,\ n),$$

where we retain the abbreviated notation for the variables and assume that at the points

$$t_0 - \tau_i(t_0) \quad \text{and} \quad t_1 - \tau_1(t_1) \qquad (i = 1,\ 2,\ \ldots,\ n)$$

the values of the function $x(t)$ and its derivatives up to order $m - 1$ inclusive are prescribed, we obtain:

$$\delta V = \int_{t_0}^{t_1} \sum_{j=1}^{n} \sum_{k=0}^{m} F'_{x^{(k)}(t-\tau_j(t))}\, \delta x^{(k)}\big(t-\tau_j(t)\big)\,dt = 0.$$

If we postulate that the function $x(t)$ is differentiable $2m$ times, and integrate the general term under the integral sign by parts k-times and apply the first lemma,[2] we obtain:

$$\sum_{i=1}^{n}\left[\frac{F'_{x(s-\tau_i(s))}-\frac{d}{ds}\frac{F'_{\dot{x}(s-\tau_i(s))}}{1-\tau_i'(s)}+\ldots+(-1)^m\frac{d^m}{ds^m}\frac{F'_{x^{(m)}(s-\tau_i(s))}}{[1-\tau_i'(s)]^m}}{1-\tau_i'(s)}\right]_{s=\gamma_i(t)}=0.$$

For functionals depending on several functions $x_j(t-\tau_i(t))$ we obtain analogous equations.

In studying extremals of multiple integrals with retardations in one of the variables we can use the same methods to obtain equations analogous to **Ostrogradskii's** equations.

We remark that methods of solution, or even of qualitative investigation, of boundary value problems for generalized Euler equations, as considered in §§3 and 5, are almost altogether undeveloped.

6. **Variational problems with moving boundaries.** Let us consider the problem of the extremal of the functional

$$V=\int_{t_0}^{t_1}F\big(t,\ x\left(t-\tau_i(t)\right),\ \dot{x}\left(t-\tau_i(t)\right)\big)\,dt$$

$$(i=1,\ 2,\ \ldots,\ n)$$

under the assumption that some or all of the points with coordinates

$$t_0-\tau_i(t_0)\quad\text{and}\quad t_1-\tau_i(t_1)\qquad(i=1,\ 2,\ \ldots,\ n)$$

are movable.

Since in this case the class of admissible functions is wider than in the case of nonmovable boundaries, the function $x(t)$ that realizes the extremum in the present case must satisfy the fundamental necessity condition for the case of nonmovable boundary points, i.e., it must satisfy the Euler equation

$$\sum_{i=1}^{n}\left[F'_{x(s-\tau_i(s))}-\frac{d}{ds}\frac{F'_{\dot{x}(s-\tau_i(s))}}{1-\tau_i'(s)}\right]_{s=\gamma_i(t)}=0.$$

[2] We can of course eliminate the requirement that the derivative $x^{(2m)}(t)$ exist if we make use of the generalization of the second lemma and obtain an extremal equation in integral form.

Assuming that the value of the functional is calculated on the extremals only, we shall have:

$$\Delta V = \int_{t_0+\delta t_0}^{t_1+\delta t_1} F(t, x_i+\delta x_i, \dot{x}_i+\delta\dot{x}_i)\,dt$$

$$-\int_{t_0}^{t_1} F(t, x_i, \dot{x}_i)\,dt = \int_{t_1}^{t_1+\delta t_1} F\,dt - \int_{t_0}^{t_0+\delta t_0} F\,dt$$

$$+\int_{t_0}^{t_1} [F(t, x_i+\delta x_i, \dot{x}_i+\delta\dot{x}_i) - F(t, x_i, \dot{x}_i)]\,dt,$$

$$\delta V = (F)_1\,\delta t_1 - (F)_0\,\delta t_0 + \left(\sum_{i=1}^{n} \frac{F'_{\dot{x}(t-\tau_i(t))}}{1-\tau_i'(t)}\,\delta x\,(t-\tau_i(t))\right)_0^1 .$$

Since

$$(\delta x\,(t-\tau_i(t)))_1 = \delta x\,(t_1-\tau_i(t_1)) - \dot{x}\,(t_1-\tau_i(t_1))\,\delta t_1$$

and

$$(\delta x\,(t-\tau_i(t)))_0 = \delta x\,(t_0-\tau_i(t_0)) - \dot{x}\,(t_0-\tau_i(t_0))\,\delta t_0,$$

as may be easily shown by the usual method, we find

$$\delta V = \left[F - \sum_{i=1}^{n} \dot{x}\,(t-\tau_i(t))\,\frac{F'_{\dot{x}(t-\tau_i(t))}}{1-\tau_i'(t)}\right]_1 \delta t_1$$

$$+\left[\sum_{i=1}^{n} \frac{F'_{\dot{x}(t-\tau_i(t))}}{1-\tau_i'(t)}\,\delta x\,(t_1-\tau_i(t_1))\right]_1$$

$$-\left[F - \sum_{i=1}^{n} \dot{x}\,(t-\tau_i(t))\,\frac{F'_{\dot{x}(t-\tau_i(t))}}{1-\tau_i'(t)}\right]_0 \delta t_0$$

$$-\left[\sum_{i=1}^{n} \frac{F'_{\dot{x}(t-\tau_i(t))}}{1-\tau_i'(t)}\,\delta x\,(t_0-\tau_i(t_0))\right]_0 .$$

If all the variations δx_i and δt are independent, their coefficients along the extremal curves must be zero. But if the boundary values are constrained by certain equations, then the part of the variation due to δx_i and δt may be expressed in terms of the other variables, and the coefficients of the remaining, now independent, variables must reduce to zero.

We remark that when we calculated the variation we did not take into account variations of the initial, or more precisely the boundary functions, since we did not put the boundary value problem in concrete form and tacitly assumed that the boundary functions were invariable; but in varying the boundary points one must generally plan to vary the boundary functions, and this means setting up a variation δV which is different from the form considered above.

In this section, as in the preceding sections, we have avoided a concrete formulation of the boundary conditions, since for the solution of equations of the form of (1) and (3), and of their generalizations, there can be many formulations of the boundary value problem which are quite distinct from the mathematical point of view, while up to now there is no way of knowing which formulation we want for applications.

If we consider the problem of the extremum of the functional

$$V=\int_{t_0}^{t_1} F\bigl(t,\ x(t-\tau_i(t)),\ \dot{x}(t-\tau_i(t))\bigr)\,dt$$

in the class of piecewise smooth admissible functions, then, applying the above forms for the variation δV, we easily obtain expressions like the Weierstrass-Erdmann conditions.

7. **Conditional extrema.** The problem of conditional extrema with finite or differential constraints and retarded arguments has scarcely been looked at up to the present time; we therefore pause only briefly to take up the isoperimetric problem.

The simplest isoperimetric problem with retarded argument consists in the study of the extremal of the functional

$$V=\int_{t_0}^{t_1} F\bigl(t,\ x(t),\ x(t-\tau),\ \dot{x}(t),\ \dot{x}(t-\tau)\bigr)\,dt$$

with fixed boundaries, constant retardation τ, and with the auxiliary condition

$$\int_{t_0}^{t_1} G\bigl(t,\ x(t),\ x(t-\tau),\ \dot{x}(t),\ \dot{x}(t-\tau)\bigr)\,dt=l, \tag{4}$$

where l is a constant.

Varying the functional V at the two points $\bar{t}$ and $\bar{\bar{t}}$ we obtain:

$$\delta V = \left(F_{x(t)} - \frac{d}{dt} F_{\dot{x}(t)}\right)_{\bar{t}} \sigma_1 + \left(F_{x(t-\tau)} - \frac{d}{dt} F_{\dot{x}(t-\tau)}\right)_{\bar{t}+\tau} \sigma_1$$
$$+ \left(F_{x(t)} - \frac{d}{dt} F_{\dot{x}(t)}\right)_{\bar{\bar{t}}} \sigma_2 + \left(F_{x(t-\tau)} - \frac{d}{dt} F_{\dot{x}(t-\tau)}\right)_{\bar{\bar{t}}+\tau} \sigma_2,$$

where

$$\sigma_1 = \int_{\bar{t}-\eta}^{\bar{t}+\eta} \delta x(t)\, dt, \quad \sigma_2 = \int_{\bar{\bar{t}}-\eta}^{\bar{\bar{t}}+\eta} \delta x(t)\, dt.$$

To determine the relationship between σ_1 and σ_2 we vary (4):

$$\left(G_{x(t)} - \frac{d}{dt} G_{\dot{x}(t)}\right)_{\bar{t}} \sigma_1 + \left(G_{x(t-\tau)} - \frac{d}{dt} G_{\dot{x}(t-\tau)}\right)_{\bar{t}+\tau} \sigma_1$$
$$+ \left(G_{x(t)} - \frac{d}{dt} G_{\dot{x}(t)}\right)_{\bar{\bar{t}}} \sigma_2 + \left(G_{x(t-\tau)} - \frac{d}{dt} G_{\dot{x}(t-\tau)}\right)_{\bar{\bar{t}}+\tau} \sigma_2 = 0, \qquad (5)$$

whence under the assumption that the desired extremum is not simultaneously an extremal of the functional $\int_{t_0}^{t_1} G\,dt$ we obtain:

$$\sigma_2 = -\frac{\left(G_{x(t)} - \frac{d}{dt} G_{\dot{x}(t)}\right)_{\bar{t}} + \left(G_{x(t-\tau)} - \frac{d}{dt} G_{\dot{x}(t-\tau)}\right)_{\bar{t}+\tau}}{\left(G_{x(t)} - \frac{d}{dt} G_{\dot{x}(t)}\right)_{\bar{\bar{t}}} + \left(G_{x(t-\tau)} - \frac{d}{dt} G_{\dot{x}(t-\tau)}\right)_{\bar{\bar{t}}+\tau}} \sigma_1,$$

where t is so chosen that the denominator is different from zero. Replacing σ_2 in (5) by σ_1 and using the fundamental necessary condition for the extremal, $\delta V = 0$, we find:

$$\Bigg\{\left(F_{x(t)} - \frac{d}{dt} F_{\dot{x}(t)}\right)_{\bar{t}} + \left(F_{x(t-\tau)} - \frac{d}{dt} F_{\dot{x}(t-\tau)}\right)_{\bar{t}+\tau}$$
$$+ \frac{\left(F_{x(t)} - \frac{d}{dt} F_{\dot{x}(t)}\right)_{\bar{\bar{t}}} + \left(F_{x(t-\tau)} - \frac{d}{dt} F_{\dot{x}(t-\tau)}\right)_{\bar{\bar{t}}+\tau}}{\left(G_{x(t)} - \frac{d}{dt} G_{\dot{x}(t)}\right)_{\bar{\bar{t}}} + \left(G_{x(t-\tau)} - \frac{d}{dt} G_{\dot{x}(t-\tau)}\right)_{\bar{\bar{t}}+\tau}}$$
$$\times \left[\left(G_{x(t)} - \frac{d}{dt} G_{\dot{x}(t)}\right)_{\bar{t}} + \left(G_{x(t-\tau)} - \frac{d}{dt} G_{\dot{x}(t-\tau)}\right)_{\bar{t}+\tau}\right]\Bigg\} \sigma_1 = 0.$$

Taking $\bar{\bar{t}}$ as fixed and $\bar{t}$ as arbitrary in the interval (t_0, t_1) we define

$$\lambda = -\frac{\left(F_{x(t)} - \frac{d}{dt} F_{\dot{x}(t)}\right)_{\bar{t}} + \left(F_{x(t-\tau)} - \frac{d}{dt} F_{\dot{x}(t-\tau)}\right)_{\bar{t}+\tau}}{\left(G_{x(t)} - \frac{d}{dt} G_{\dot{x}(t)}\right)_{\bar{t}} + \left(G_{x(t-\tau)} - \frac{d}{dt} G_{\dot{x}(t-\tau)}\right)_{\bar{t}+\tau}}.$$

Since the variation of σ_1 is arbitrary, we have finally

$$\left(F_{x(t)} - \frac{d}{dt} F_{\dot{x}(t)}\right)_{\bar{t}} + \left(F_{x(t-\tau)} - \frac{d}{dt} F_{\dot{x}(t-\tau)}\right)_{\bar{t}+\tau}$$
$$+ \lambda\left[\left(G_{x(t)} - \frac{d}{dt} G_{\dot{x}(t)}\right)_{\bar{t}} + \left(G_{x(t-\tau)} - \frac{d}{dt} G_{\dot{x}(t-\tau)}\right)_{\bar{t}+\tau}\right] = 0.$$

Thus, the Lagrange multiplier rule turns out to be valid for the isoperimetric problem even with retarded argument: the extremals of the functional $V = \int_{t_0}^{t_1} F\,dt$ under the isoperimetric condition $\int_{t_0}^{t_1} G\,dt = l$ coincide with the extremals of the functional $V^* = \int_{t_0}^{t_1} (F + \lambda G)\,dt$, where λ is a constant.

This result is easily generalized to functionals of more complicated form, with one or several isoperimetric conditions.

8. **Direct methods.** In recent years the problem of the applicability of fundamental direct methods to variational questions without deviating arguments has been significantly developed by L. V. Kantorovič [**109**], M. A. Krasnosel'skiĭ, and others.

Using the apparatus of functional analysis and topology, they have succeeded in proving the convergence of sequences obtained by Ritz's method (or other similar methods) in solving variational problems, and have even succeeded in estimating the rate of convergence in a broad class of cases.

These results have a general character that can be applied without essential change to problems of variation with retarded argument; yet no systematic investigation of this problem has been undertaken. Work in this connection has been confined to various studies of the applicability of Ritz's method to quadratic functionals of a special type.

Direct methods for the solution of variational problems with retarded argument are especially important in view of the fact that up to now we have no well-developed methods for solving boundary value problems for differential equations of the forms (1), (2), and (3) (see pages 219-220), to which the solution of the most elementary boundary value problems can be reduced. For such problems we can apply the method of steps, which for small values of τ leads to a very tedious calculation, or we can apply the method of B. G. Galerkin.

9. **Estimates of the number of solutions of a variational problem.** The methods of estimating the number of solutions of a variational problem which we described in §8, Chapter I are wholly applicable to the boundary value problem with retarded argument; they are complicated only by the topological structure of the function space in which the functionals are defined. This complication arises from the presence of complicated boundary conditions.

For example, in the most elementary spatial variation problems with constant retardations τ_i $(i = 1, 2, \cdots, n)$ and movable boundary points we must consider not just the motion of two boundary points A and B on the manifolds M_0 and M_1, but $2n + 2$ points (cf. page 223) which vary on given manifolds M_{0i} and M_{1i} $(i = 0, 1, \cdots, n)$, and, moreover, the space of the initial functions may have a complicated topological structure.

In the simplest cases it is worthwhile to calculate or estimate the topological invariants such as the Betti classes or numbers of the space

$$M_{00} \times M_{01} \times \cdots \times M_{0n} \times M_{10} \times M_{11} \times \cdots \times M_{1n} \times \Omega \times \omega,$$

where Ω is the space of admissible curves in the problem with fixed boundary points and fixed initial functions, and ω is the space of initial functions.

Bibliography

[1] P. S. Aleksandrov, *Combinatorial topology*, OGIZ, Moscow, 1947; English transl., I, II, Graylock Press, Rochester, N. Y., 1956, 1957. MR **10**, 55; MR **17**, 882; MR **19**, 759.

[2] ———, *On homological situation properties of complexes and closed sets*, Izv. Akad. Nauk SSSR Ser. Mat. **6** (1942), 227-282. (Russian. English summary) MR **4**, 249.

[3] ———, *General duality theorems for nonclosed sets in n-dimensional space*, Mat. Sb. (N.S.) **21**(**63**) (1947), 161-232. (Russian) MR **9**, 456.

[4] S. I. Al'ber, *Homologies of a space of planes and their applications to the calculus of variations*, Dokl. Akad. Nauk SSSR **91** (1953), 1237-1240. (Russian) MR **15**, 457.

[5] L. Amerio, *Studio asintotico del moto di un punto su una linea chiusa, per azione di forze independenti dal tempo*, Ann. Scuola Norm. Sup. Pisa (3) **3** (1949), 19-57 (1950). MR **12**, 180.

[6] A. A. Andronov, *Les cycles limites de Poincaré et la théorie des oscillations auto-entretenues*, C. R. Acad. Sci. Paris **189** (1929), 559-562.

[7] A. A. Andronov and L. S. Pontrjagin, *Coarse systems*, Dokl. Akad. Nauk SSSR **14** (1937), 247-250. (Russian)

[8] A. A. Andronov and S. E. Chaikin, *Theory of oscillations*. I, Moscow, 1937; English transl., Princeton Univ. Press, Princeton, N. J., 1949. MR **10** 535.

[9] A. A. Andronov and A. G. Maier, *The simplest linear systems with retardation*, Avtomat. i Telemeh. **7** (1946), 95-106. (Russian) MR **8**, 517.

[10] ———, *Vyšnegradskii's problem in the theory of direct regulation*. I. *The theory of the regulator of direct action in the presence of Coulomb and viscous friction*, ibid. **8** (1947), 314-334. (Russian) MR **12**, 413.

[11] A. A. Andronov and A. Vitt, *The mathematical theory of capture*, Z. Prikl. Fiz. **7** (1930), 1. (Russian)

[12] ———, *Ljapunov stability*, Ž. Èksper. Teoret. Fiz. **5** (1933). (Russian)

[13] ———, *The mathematical theory of self-oscillatory systems with two degrees of freedom*, Ž. Tehn. Fiz. **4** (1934), 122-143. (Russian)

[14] N. A. Artem'ev, *Realizable motions*, Izv. Akad. Nauk SSSR Ser. Mat. **1939**, 351-370 or 429-448. (Russian)

[15] ———, *Über realisierbare Trajektorien*, ibid. **1939**, 429-448. (Russian. German summary) MR **1**, 281.

[16] ———, *Die Bestimmung der Realisierbarkeit der periodischen Bewegungen*, ibid. **5** (1941), 127-158. (Russian. German summary) MR **2**, 326.

[17] G. Ascoli, *Osservazioni sopra alcune questioni di stabilità*. I, Atti Accad. Naz. Lincei Rend. Cl. Sci. Fis. Mat. Nat. (8) **9** (1950), 129-134. MR **12**, 705.

[18] E. A. Barbašin and N. N. Krasovskiĭ, *On the existence of Ljapunov functions in the case of asymptotic stability in the large*, Prikl. Mat. Meh. **18** (1954), 345-350. (Russian) MR **15**, 957.

[19] E. A. Barbašin, *On σ-coverings of spaces*, Mat. Sb. (N.S.) **18 (60)** (1946), 423-428. (Russian. English summary) MR **8**, 47.

[20] E. A. Barbašin and N. N. Krasovskiĭ, *On stability of motion in the large*, Dokl. Akad. Nauk SSSR **86** (1952), 453-456. (Russian) MR **14**, 646.

[21] E. A. Barbašin, *The method of sections in the theory of dynamical systems*, Mat. Sb. (N.S.) **29 (71)** (1951), 233-280. (Russian) MR **13**, 756.

[22] ———, *On the behavior of points under homeomorphic transformations of a space. Generalizations of theorems of Birkhoff*, Ural. Politehn. Inst. Trudy **51** (1954), 4-11. (Russian) MR **17**, 1230.

[23] M. I. Bat', *Forced vibrations in a system of hysteresis*, Prikl. Mat. Meh. **4** (1940), 13-30. (Russian)

[24] N. N. Bautin, *The behavior of dynamical systems near the boundaries of the domain of stability*, Gostehizdat, Moscow, 1950. (Russian)

[25] ———, *On the number of limit cycles appearing with variation of the coefficients from an equilibrium state of the type of a focus or a center*, Mat. Sb. (N.S.) **30 (72)** (1952), 181-196. (Russian) MR **13**, 652.

[26] M. Bebutov, *On dynamical systems which are stable in the sense of Ljapunov*, Dokl. Akad. Nauk SSSR **18** (1938). (Russian)

[27] R. Bellman, *Stability theory of differential equations*, McGraw-Hill, New York, 1953; Russian transl., IL, Moscow, 1954. MR **15**, 794; MR **17**, 734.

[28] ———, *On the boundedness of solutions of nonlinear differential and difference equations*, Trans. Amer. Math. Soc. **62** (1947), 357-386. MR **9**, 436.

[29] ———, *On the existence and boundedness of solutions of non-linear differential-difference equations*, Ann. of Math. (2) **50** (1949), 347-355. MR **10**, 715.

[30] ———, *A survey of the theory of the boundedness, stability and asymptotic behavior of solutions of linear and nonlinear differential and difference equations*, Office of Naval Research, Washington, D. C., 1949. MR **11**, 31.

[31] I. Bendixon, *On curves defined by means of differential equations*, Uspehi Mat. Nauk **9** (1941). (Russian)

[32] G. D. Birkhoff, *Surface transformations and their dynamical applications*, Acta Math. **43** (1922), 1-119.

[33] ———, *Dynamical systems*, Amer. Math. Soc., Providence, R. I., 1927, 1960.

[34] G. D. Birkhoff and O. D. Kellog, *Invariant points in function space*, Trans. Amer. Math. Soc. **23** (1922), 96-115.

[35] M. F. Bokštein, *Über die Homologiegruppen der Vereinigung zweier Komplexe*, Mat. Sb. (N.S.) **9** (**51**) (1941), 365-376. MR **3**, 60.

[36] K. Borsuk, *Über den Ljusternik-Šnirel'mannschen Begriff der Kategorie*, Fund. Math. **26** (1936), 123-136.

[37] S. Bochner and W. T. Martin, *Several complex variables*, Princeton Mathematical Series Vol. 10, Princeton Univ. Press, Princeton, N. J., 1948. MR **10**, 366.

[38] B. V. Bulgakov, *Self-oscillations of regularized systems*, Dokl. Akad. Nauk SSSR **38** (1942), 283-285. (Russian)

[39] ———, *Oscillations*, GITTL, Moscow, 1954. (Russian) MR **16**, 1024.

[40] B. van der Pol, *The non-linear theory of electric oscillations*, Proc. Inst. Radio Engr. **22** (1934), 1051-1086.

[41] A. B. Vasil'eva, *On differentiation of solutions of systems of differential equations containing a small parameter*, Dokl. Akad. Nauk SSSR **75** (1950), 483-486. (Russian) MR **12**, 412.

[42] ———, *On differentiation with respect to a small parameter of solutions of systems of differential equations*, ibid. **78** (1951), 845-848. (Russian) MR **13**, 236.

[43] ———, *On the differentiability of the solutions of differential equations containing a small parameter*, Mat. Sb. (N.S.) **28** (**70**) (1951), 131-146. (Russian) MR **13**, 37.

[44] ———, *On the differentiation of solutions of systems of differential equations with respect to the largest of small parameters*, Dokl. Akad. Nauk SSSR **77** (1951), 781-784. (Russian) MR **13**, 236.

[45] ———, *On differential equations containing small parameters*, Mat. Sb. (N.S.) **31** (**73**) (1952), 587-644. (Russian) MR **14**, 1086.

[46] W. R. Wasow, *On the construction of periodic solutions of singular perturbation problems*, pp. 313-350, Annals of Mathematics Studies No. 20, Princeton Univ. Press, Princeton, N. J., 1950. MR **12**, 29.

[47] W. R. Wasow and K. O. Friedrichs, *Singular perturbations of nonlinear oscillations*, Duke Math. J. **13** (1946), 367-381. MR **8**, 272.

[48] M. M. Vaĭnberg, *On the characteristic elements of a class of nonlinear operators*, Dokl. Akad. Nauk SSSR **75** (1950), 609-612. (Russian) MR **12**, 713.

[49] ———, *On the variational theory of characteristic values of nonlinear integral equations*, ibid. **80** (1951), 309-312. (Russian) MR **13**, 353.

[50] ———, *On the differential and gradient of functionals*, Uspehi Mat. Nauk **7** (1952), no. 3 (49), 139-143. (Russian) MR **14**, 55.

[51] ———, *Some questions of the differential calculus in linear spaces*, ibid. **7** (1952), no. 4 (50), 55-102. (Russian) MR **14**, 384.

[52] J. Weier, *Fixpunkttheorie in topologischen Mannigfaltigkeiten*, Math. Z. **59** (1953), 171-190. MR **15**, 337.

[53] R. È. Vinograd, *On an assertion of K. P. Persidskiĭ*, Uspehi Mat. Nauk **9** (1954), no. 2 (60), 125-128. (Russian) MR **16**, 360.

[54] ———, *Instability of characteristic exponents of regular systems*, Dokl. Akad. Nauk SSSR **91** (1953), 999-1002. (Russian) MR **16**, 132.

[55] ———, *On the limit behavior of an unbounded integral curve*, Moskov. Gos. Univ. Učen. Zap. **155** (1952), Mat. 5, 94-136. (Russian) MR **18**, 482.

[56] ———, *On a criterion for instability in the sense of Ljapunov of the solutions of a linear system of ordinary differential equations*, Dokl. Akad. Nauk SSSR **84** (1952), 201-204. (Russian) MR **14**, 376.

[57] ———, *Some criteria of boundedness of the solutions of a system of two linear differential equations*, ibid. **85** (1952), 265-268. (Russian) MR **14**, 276.

[58] I. M. Volk, *A generalization of the small parameter method in the theory of non-linear oscillations of non-autonomous systems*, ibid. **51** (1946), 437-440. (Russian) MR **8**, 69.

[59] ———, *On periodic solutions of non-autonomic systems depending upon the small parameter*, Prikl. Mat. Meh. **10** (1946), 559-574. (Russian. English summary) MR **8**, 330.

[60] ———, *Generalizations of the small parameter method in the theory of periodic motions of non-autonomous systems*, ibid. **11** (1947), 433-444. (Russian. English summary) MR **9**, 185.

[61] ———, *On periodic solutions of autonomic systems*, ibid. **12** (1948), 29-38. (Russian) MR **9**, 588.

[62] V. M. Volosov, *On differential equations with a small parameter in the highest derivative*, Dokl. Akad. Nauk SSSR **73** (1950), 873-876. (Russian) MR **12**, 101.

[63] ———, *Nonlinear differential equations of the second order with a small parameter with the highest derivative*, Mat. Sb. (N.S.) **30** (**72**) (1952), 245-270. (Russian) MR **14**, 276.

[64] ———, *On the theory of nonlinear differential equations of higher orders with a small parameter in the highest derivative*, ibid. **31** (**73**) (1952), 645-674. (Russian) MR **14**, 1086.

[65] ———, *On solutions of some differential equations of the second order depending upon a parameter*, ibid. **31** (**73**) (1952), 675-686. (Russian) MR **14**, 1086.

[66] N. N. Vološin, *An account of the phenomenon of retardation*, Avtomat. i Telemeh. **9** (1948), 285-292. (Russian)

[67] V. Volterra, *Leçons sur les équations intégrales et les équations intégro-différentielles*, Gauthier-Villars, Paris, 1913.

[68] ———, *Teoria de las functionales y de las ecuationes integrales e integro-differenciales*, Madrid, 1927.

[69] ———, *Theory of functionals and of integral and integro-differential equations*, London, 1930.

[70] Hartogs, *Über die aus singulären Stellen einer analytischer Funktion mehrerer Veränderlichen bestehenden Gebilde*, Acta Math. **32** (1909), 57-79.

[71] S. G. Gerasimov, *Theoretical foundations of automatic regulation of heat processes*, Gostehizdat, Moscow, 1949. (Russian)

[72] H. Hopf, *Vektorfelder in n-dimensionalen Mannigfaltigkeiten*, Math. Ann. **96** (1926), 225-250.

[73] I. I. Gordon, *On the minimal number of critical points of a real function defined on a manifold*, Mat. Sb. (N.S.) **4** (**46**) (1938), 105-113. (Russian)

[74] G. S. Gorélik, *On the theory of a retarded inverse connection*, Ž. Tehn. Fiz. **9** (1939), 450. (Russian)

[75] I. S. Gradšteĭn, *On the behavior of the solutions of systems of linear differential equations with constant coefficients, degenerating in the limit*, Izv. Akad. Nauk SSSR Ser Mat. **13** (1949), 253-280. (Russian) MR **10**, 709.

[76] ———, *Differential equations with small coefficients for the derivatives and Ljapunov's theory of stability*, Dokl. Akad. Nauk SSSR **65** (1949), 789-792. (Russian) MR **10**, 708.

[77] ———, *Linear equations with variable coefficients and small parameters in the highest derivatives*, Mat. Sb. (N.S.) **27** (**69**) (1950), 47-68; English transl., Amer. Math. Soc. Transl. no. 82, 1953; reprint, Amer. Math. Soc. Transl. (1) **4** (1962), 300-330. MR **12**, 260; MR **14**, 645.

[78] ———, *An application of A. M. Ljapunov's stability theory to the theory of differential equations with small coefficients of the derivatives*, Dokl. Akad. Nauk SSSR **81** (1951), 985-986. (Russian) MR **13**, 460.

[79] ———, *Differential equations in which various powers of a small parameter appear as coefficients of the derivatives*, ibid. **82** (1952), 5-8. (Russian) MR **13**, 557.

[80] ———, *On solutions on a time half-line of differential equations with small multipliers of the derivatives*, Mat. Sb. (N.S.) **32** (**74**) (1953), 533-544. (Russian) MR **14**, 1086.

[81] D. P. Grossman, *An estimation of the category of Ljusternik-Šnirel'man*, Dokl. Akad. Nauk SSSR **54** (1946), 109-112. (Russian) MR **8**, 334.

[82] D. M. Grobman, *Systems of differential equations analogous to linear ones*, ibid. **86** (1952), 19-22. (Russian) MR **14**, 274.

[83] ———, *Characteristic exponents of systems next to linear ones*, Mat. Sb. (N.S.) **30** (**72**) (1952), 121-166. (Russian) MR **13**, 652.

[84] W. Hurewicz, *Beiträge zur Topologie der Deformationen*, Akad. Wetensch. Amsterdam Proc. **38** (1935), 521-528 and **39** (1936), 117-126.

[85] H. F. DeBaggis, *Dynamical systems with stable structures*, Contributions to the Theory of Non-linear Oscillations Vol. 2, pp. 37-59, Princeton Univ. Press, Princeton, N. J., 1952. MR **14**, 557.

[86] R. Deheuvels, *Points critiques d'une fonctionnelle*, C. R. Acad. Sci. Paris **236** (1953), 1847-1849. MR **14**, 1109.

[87] A. DeCastro, *Soluzioni periodiche di una equazione differenziale del secondo ordine*, Boll. Un. Mat. Ital. (3) 8 (1953), 26-29. MR **14**, 874.

[88] ———, *Sopre l'equazione differenziale delle oscillazioni nonlineari*, Riv. Mat. Univ. Parma **4** (1953), 133-143. MR **15**, 127.

[89] B. P. Demidovič, *On certain sufficient conditions for the existence of an integral invariant*, Mat. Sb. (N.S.) **3** **(45)** (1938), 291-310. (Russian)

[90] ———, *On stability in the sense of Ljapunov of a linear system of ordinary differential equations*, ibid. **28** **(70)** (1951), 659-684. (Russian) MR **13**, 460.

[91] ———, *On some averaging theorems for ordinary differential equations*, ibid. **35** **(77)** (1954), 73-92. (Russian) MR **16**, 361.

[92] G. N. Dubošin, *On the stability of solutions of canonical systems*, Dokl. Akad. Nauk SSSR **1** (1935), 273-279. (Russian)

[93] ———, *On stability of motion with respect to constant perturbations*, Trudy Gos. Astronom. Inst. Šternberg **14** (1940), no. 1. (Russian)

[94] ———, *Foundations of the theory of stability of motion*, Izdat. Moskov. Gos. Univ., Moscow, 1952. (Russian) MR **14**, 471.

[95] M. I. El'šin, *The method of comparison in the qualitative theory of an incomplete differential equation of second order*, Mat. Sb. (N.S.) **34** **(76)** (1954), 323-330. (Russian) MR **15**, 957.

[96] ———, *A qualitative system of two linear homogeneous equations of second order*, Dokl. Akad. Nauk SSSR **94** (1954), 5-8. (Russian)

[97] N. P. Erugin, *Reducible systems*, Trudy Mat. Inst. Steklov. **13** (1946). (Russian. English summary) MR **9**, 509.

[98] ———, *On asymptotically stable solutions of certain systems of differential equations*, Prikl. Mat. Meh. **12** (1948), 157-164. (Russian) MR **9**, 589.

[99] ———, *Generalizations of a theorem of Ljapunov*, ibid. **12** (1948), 632-638. (Russian) MR **10**, 456.

[100] ———, *On certain questions of stability of motion and the qualitative theory of differential equations in the large*, ibid. **14** (1950), 459-512. (Russian) MR **12**, 412.

[101] ———, *Some general questions of the theory of stability of motion*, ibid. **15** (1951), 227-236. (Russian) MR **12**, 705.

[102] E. M. Esipovič, *On stability of solutions of a class of differential equations with retarded argument*, ibid. **15** (1951), 601-608. (Russian) MR **13**, 466.

[103] H. Seifert and Wm. Threlfall, *Calculus of variations in the large (the Morse theory)*, IL, Moscow, 1947.

[104] ———, *Topologie*, Teubner, Leipzig, 1934.

[105] I. A. Zeideman, *On the fundamental group of the sum of two connected polyhedra with nonconnected intersection*, Moskov. Gos. Univ. Učen. Zap. **163** (1952), Mat. 6. (Russian)

[106] F. John, *Über die Vollständigkeit der Relation von Morse*, Math. Ann. **109** (1934), 381-394.

[107] G. V. Kamenkov, *On stability of motion*, Kazan. Trudy Astronom. Inst. **6** (1937). (Russian)

[108] G. A. Kamenskiĭ, *On the asymptotic behavior of solutions of linear differential equations of the second order with retarded argument*, Moskov. Gos. Univ. Učen. Zap. **165** (1954), Mat. 7, 195-204. (Russian) MR **16**, 829.

[109] L. V. Kantorovič, *Functional analysis and applied mathematics*, Uspehi Mat. Nauk **3** (1948), no. 6 (28), 89-185. (Russian) MR **10**, 380.

[110] A. N. Kolmogorov, *On dynamical systems with an integral invariant on the torus*, Dokl. Akad. Nauk SSSR **93** (1953), 763-766. (Russian) MR **16**, 36.

[111] ———, *On conservation of conditionally periodic motions for a small change in Hamilton's function*, ibid. **98** (1954), 527-530. (Russian) MR **16**, 924.

[112] M. A. Krasnosel'skiĭ, *On the estimation of the number of critical points of functionals*, Uspehi Mat. Nauk **7** (1952), no. 2 (48), 157-164. (Russian) MR **14**, 55.

[113] ———, *On an elementary topological theorem*, ibid. **6** (1951), no. 2 (42), 160-164. (Russian) MR **13**, 150.

[114] M. A. Krasnosel'skiĭ and S. G. Kreĭn, *On proof of the theorem on category of a projective space*, Ukrain. Mat. Ž. **1** (1949), no. 2, 99-102. (Russian) MR **14**, 72.

[115] M. A. Krasnosel'skiĭ, *Some problems of nonlinear analysis*, Uspehi Mat. Nauk **9** (1954), no. 3 (61), 57-114. (Russian) MR **17**, 769.

[116] ———, *On a fixed point principle for completely continuous operators in functional spaces*, Dokl. Akad. Nauk SSSR **73** (1950), 13-15. (Russian) MR **12**, 111.

[117] ———, *Convergence of Galerkin's method for nonlinear equations*, ibid. **73** (1950), 1121-1124. (Russian) MR **12**, 187.

[118] N. N. Krasovskiĭ, *On the inversion of the theorems of A. M. Ljapunov and N. G. Četaev on instability for stationary systems of differential equations*, Prikl. Mat. Meh. **18** (1954), 513-532. (Russian) MR **16**, 473.

[119] ———, *On stability of solutions of a system of second order in critical cases*, Dokl. Akad. Nauk SSSR **93** (1953), 965-967. (Russian) MR **15**, 795.

[120] M. G. Kreĭn, *A generalization of some investigations of A. M. Ljapunov on linear differential equations with periodic coefficients*, ibid. **73** (1950), 445-448. (Russian) MR **12**, 100.

[121] ———, *On certain questions related to the ideas of Ljapunov in the theory of stability*, Uspehi Mat. Nauk **3** (1948), 166-169. (Russian) MR **10**, 128.

[122] I. M. Krylov and N. N. Bogoljubov, *Introduction to nonlinear mechanics*, Kiev, 1939. (Russian)

[123] ———, *New methods in nonlinear mechanics*, Gostehizdat, Moscow, 1934. (Russian)

[124] K. L. Cooke, *The asymptotic behavior of the solutions of linear and non-linear differential-difference equations*, Trans. Amer. Math. Soc. **75** (1953), 80-105. MR 15, 629.

[125] I. S. Kukles, *On two fundamental groups of singular points*, Dokl. Akad. Nauk SSSR **42** (1944), 262-264. (Russian)

[126] N. Levinson, *Perturbations of discontinuous solutions of non-linear systems of differential equations*, Acta Math. **82** (1950), 71-106. MR **11**, 722.

[127] ———, *An ordinary differential equation with an interval of stability, a separation point, and an interval of instability*, J. Mathematical Phys. **28** (1950), 215-222. MR **11**, 722.

[128] ———, *The asymptotic nature of solutions of linear systems of differential equations*, Duke Math. J. **15** (1948), 111-126. MR **9**, 509.

[129] ———, *On stability of non-linear systems of differential equations*, Colloq. Math. **2** (1949), 40-45. MR **12**, 412.

[130] E. A. Leontovič, *On the generation of limiting cycles from separatrices*, Dokl. Akad. Nauk SSSR **78** (1951), 641-644. (Russian) MR **13**, 132.

[131] A. F. Leont'ev, *Differential-difference equations*, Mat. Sb. (N.S.) **24** (**66**) (1949), 347-374. (Russian) MR **11**, 113.

[132] ———, *On functions represented by series of Dirichlet polynomials*, Izv. Akad. Nauk SSSR Ser. Mat. **13** (1949), 221-230. (Russian) MR **10**, 695.

[133] J. Leray and J. Schauder, *Topology and functional equations*, Uspehi Mat. Nauk **1** (1946), no. 3-4 (13-14), 71-95. (Russian) MR **9**, 606.

[134] V. L. Lossievskiĭ, *Automatic control*, Izdat. Akad. Nauk SSSR, Moscow, 1946. (Russian)

[135] ———, *Foundations of automatic control of technological processes*, Moscow, 1949. (Russian)

[136] L. A. Ljusternik, *Application of topology to extremal problems*, Proc. 2nd All-Union Math. Conf. Vol. 1, Moscow, 1936. (Russian)

[137] ———, *A class of nonlinear operators in Hilbert space*, Izv. Akad. Nauk SSSR Ser. Mat. (1939), 257-264. (Russian)

[138] ———, *Intersections dans les espaces localement linéaires*, Dokl. Akad. Nauk SSSR **27** (1940), 771-774. MR **2**, 179.

[139] ———, *Structure topologique d'un espace fonctionnel*, ibid. **27** (1940), 775-777. MR **2**, 179.

[140] ———, *Ring of intersections in a functional space*, ibid. **38** (1943), 59-61. (Russian) MR **5**, 273.

[141] ———, *Families of arcs with common end points on the sphere*, ibid. **39** (1943), 88-90. (Russian) MR **5**, 273.

[142] ———, *On dimensionality of critical sets*, ibid. **39** (1943), 371-372. (Russian)

[143] ———, *On categories of some arc families*, ibid. **40** (1943), 147-148. (Russian)

[144] ———, *On the number of solutions of a variational problem*, ibid. **40** (1943), 243-245. (Russian)

[145] ———, *A new proof of the theorem about the three geodesics*, ibid. **41** (1943), 3-4, (Russian) MR **6**, 105.

[146] ———, *Topology and the calculus of variations*, Uspehi Mat. Nauk **1** (1946), no. 1 (11), 30-56. (Russian)

[147] ———, *Topology of functional spaces and calculus of variations in the large*, Trudy Mat. Inst. Steklov. **19** (1947). (Russian. English summary) MR **9**, 596.

[148] ———, *Sur quelques méthodes topologiques dans la géométrie différentielle*, Atti Congr. Mat. Bologna 4 (1928), 291-296.

[149] ———, *Topologische Grundlagen der allgemeinen Eigenwerttheorie*, Monatsh. Math. **37** (1930), 125-130.

[150] ———, *Über die topologischen Eigenschaften der Kurvenfamilien auf Flächen*, Mat. Sb. (N.S.) **38** (1931), 59-65.

[151] L. A. Ljusternik and L. G. Šnirel'man, *Topological methods for problems in the calculus of variations*, Izdat. Moskov. Gos. Univ., Moscow, 1930. (Russian)

[152] A. M. Ljapunov, *General problem in the stability of motion* (first published Kharkov, 1892), GITTL, Moscow, 1950; French transl., Annals of Mathematics Studies No. 17, Princeton Univ. Press, Princeton, N. J., 1947. MR **12**, 612; MR **9**, 34.

[153] A. G. Maier, *Trajectories on the closed orientable surfaces*, Mat. Sb. (N.S.) **12** (**54**) (1943), 71-84. (Russian. English summary) MR **5**, 156.
[154] ———, *Proof of the existence of limit cycles for the equations of Rayleigh and van der Pol*, Učen. Zap. Gor'kovsk. 2 (1953), 19-25. (Russian)
[155] ———, *Sur un problème de Birkhoff*, Dokl. Akad. Nauk SSSR **55** (1947), 447-480. MR **8**, 590.
[156] ———, *On trajectories in three dimensional space*, ibid. **55** (1947), 583-586. (Russian)
[157] A. D. Maizel', *On stability of solutions of systems of differential equations*, Ural. Politehn. Inst. Trudy **51** (1954), 20-50. (Russian) MR **17**, 738.
[158] I. G. Malkin, *The problem of the existence of Ljapunov functions*, Izv. Kazan. Fiz.-Mat. **5** (1931), 63-84. (Russian)
[159] ———, *On stability to the first approximation*, Sb. Trudov. Kazan. Astronom. **3** (1935). (Russian)
[160] ———, *On stability of motion to the first approximation*, Dokl. Akad. Nauk SSSR **18** (1938), 159-161. (Russian)
[161] ———, *On stability of motion in the sense of Ljapunov*, Mat. Sb. (N.S.) **3** (**45**) (1938), 47-101. (Russian)
[162] ———, *Stability of periodic motions of dynamic systems*, Prikl. Mat. Meh. **8** (1944), 327-331. (Russian. English summary) MR **6**, 226.
[163] ———, *Stability in the case of constantly acting disturbances*, ibid. **8** (1944), 241-245. MR **7**, 298.
[164] ———, *The methods of Ljapunov and Poincaré in the theory of nonlinear oscillations*, OGIZ, Moscow, 1949. (Russian) MR **12**, 28.
[165] ———, *A theorem on stability to the first approximations*, Dokl. Akad. Nauk SSSR **76** (1951), 783-784. (Russian) MR **12**, 827.
[166] ———, *On the solution of a stability problem in the case of two purely imaginary roots*, Prikl. Mat. Meh. **15** (1951), 255-257. (Russian) MR **13**, 38.
[167] ———, *On a method of solution of the problem of stability in the critical case of a pair of purely imaginary roots*, ibid. **15** (1951), 473-484. (Russian) MR **13**, 346.
[168] ———, *Solution of some critical cases of the problem of stability of motion*, ibid. **15** (1951), 575-590. (Russian) MR **13**, 346.
[169] ———, *On characteristic values of linear differential equations*, ibid. **16** (1952), 3-14. (Russian) MR **13**, 651.
[170] ———, *On a theorem concerning stability of motion*, Dokl. Akad. Nauk SSSR **84** (1952), 877-878. (Russian) MR **14**, 275.
[171] ———, *On the reversibility of Ljapunov's theorem on asymptotic stability*, Prikl. Mat. Meh. **18** (1954), 129-138. (Russian) MR **15**, 873.

[172] L. I. Mandel′štam and N. D. Papaleksi, *On resonance phenomena of the nth kind*. Ž. Tehn. Fiz. **2** (1932), no. 7-8. (Russian)

[173] A. A. Markov, *Stabilität im Liapounoffschen Sinne und Fastperiodizität*, Math. Z. **36** (1933), 708-738.

[174] ———, *On the existence of an integral invariant*, Dokl. Akad. Nauk SSSR **17** (1937), 455-458. (Russian)

[175] ———, *On a general property of minimal Birkhoff sets*, Bjull. Astronom. Inst. **32** (1934), 147-152. (Russian)

[176] I. L. Massera, *Remarks on the periodic solutions of differential equations*, Bol. Fac. Ingen. Agrimens. Montevideo. **4** (Año 14), 37-45 = Fac. Ingen. Montevideo. Publ. Inst. Mat. Estadíst. **2** (1950), 45-53. (Spanish) MR **13**, 944.

[177] ———, *On Ljapunov's condition of stability*, Ann. of Math. (2) **50** (1949), 705-721. MR **11**, 721.

[178] N. N. Meĭman and N. G. Čevotarev, *The Routh-Hurwitz problem for polynomials and entire functions*, Trudy Mat. Inst. Steklov. **26** (1949). (Russian) MR **11**, 509.

[179] J. Milnor, *The characteristics of a vector field on the two-sphere*, Ann. of Math. (2) **58** (1953), 253-257. MR **15**, 336.

[180] N. Minorsky, *Introduction to non-linear mechanics. Topological methods. Analytical methods. Non-linear resonance. Relaxation oscillations*, J. W. Edwards, Ann Arbor, Mich., 1947. MR **8**, 583.

[181] ———, *Control problems*, J. Franklin Inst. **232** (1941), 519-551. MR **3**, 93.

[182] ———, *Self-excited oscillations in dynamical systems*, J. Appl. Mech. **9** (1942), 65-71.

[183] A. V. Mihaĭlov, *The method of harmonic analysis in control theory*, Avtomat. i Telemeh. **3** (1948). (Russian)

[184] N. D. Monseev, *On the probability of stability in the sense of Ljapunov*, Dokl. Akad. Nauk SSSR **1** (1936), 211-213. (Russian)

[185] ———, *On the construction of domains of continuous stability and instability in the sense of Ljapunov*, ibid. **20** (1938), 419-423. (Russian)

[186] ———, *On phase domains of continuous stability and instability*, ibid. **20** (1938), 423-425. (Russian)

[187] ———, *On certain methods in the theory of technical stability*, Trudy Voenno-vozd. Akad. **135** (1945). (Russian)

[188] M. Morse, *Topological methods in the theory of functions of a complex variable*, Annals of Mathematics Studies No. 15, Princeton Univ. Press, Princeton, N. J., 1947. MR **9**, 20.

[189] ———, *Relations between the critical points of a real function of n variables*, Trans. Amer. Math. Soc. **27** (1925), 345-396.
[190] ———, *The critical points of a function of n variables*, ibid. **33** (1931), 72-91.
[191] ———, *The calculus of variations in the large*, Amer. Math. Soc. Colloq. Publ. No. 18, Amer. Math. Soc., Providence, R. I., 1934.
[192] A. D. Myškis, *Hystero-differential equations*, Uspehi Mat. Nauk **4** (1949), no. 1 (29), 190-193. (Russian) MR **11**, 365.
[193]* ———, *General theory of differential equations with retarded arguments*, ibid. **4** (1949), no. 5 (33), 99-141. (Russian) MR **11**, 365.
[194] ———, *On the solutions of linear homogeneous differential equations of the first order of unstable type with a retarded argument*, Dokl. Akad. Nauk SSSR **70** (1950), 953-956. (Russian) MR **11**, 522.
[195] ———, *Investigation of a class of differential equations with retarded arguments by means of a generalized Fibonacci series*, ibid. **71** (1950), 13-16. (Russian) MR **11**, 726.
[196] ———, *On solutions of linear homogeneous differential equations of the second order of periodic type with a retarded argument*, Mat. Sb. (N.S.) **28** (**70**) (1951), 15-54. (Russian) MR **13**, 43.
[197] ———, *On solutions of linear homogeneous differential equations of the first order of stable type with a retarded argument*, ibid. **28** (**70**) (1951), 641-658. (Russian) MR **13**, 246.
[198] ———, *Linear differential equations with retarded argument*, GITTL, Moscow, 1951. (Russian) MR **14**, 52.
[199] ———, *Generalizations of the theorem on a fixed point of a dynamical system inside of a closed trajectory*, Mat. Sb. (N.S.) **34** (**76**) (1954), 525-540. (Russian) MR **15**, 978.
[200] Ju. I. Neĭmark, *On the determination of the values of the parameters for which a system of automatic regulation is stable*, Avtomat. i Telemeh. **9** (1948), 190-203. (Russian) MR **12**, 498.
[201] ———, *The structure of the D-decomposition of the space of quasi-polynomials and the diagrams of Vyšnegradskiĭ and Nyquist*, Dokl. Akad. Nauk SSSR **60** (1948), 1503-1506. (Russian) MR **10**, 37.
[202] M. Nagumo, *Über das Verhalten der Integrale von* $\lambda y'' + f(x, y, y', \lambda) = 0$ *für* $\lambda \to 0$, Proc. Phys.-Math. Soc. Japan **21** (1939), 529-534. MR **1**, 177.
[203] V. V. Nemyckiĭ, *The method of fixed points in analysis*, Uspehi Mat. Nauk **1** (1936), 141-175. (Russian)

* This article contains an extensive bibliography which is continued in: *Supplementary bibliographical material to the paper "General theory of differential equations with retarded arguments,"* Uspehi Mat. Nauk 5 (1950), no. 2 (36), 148-154. (Russian) MR 12, 416.

[204] ——, *Sur les systèmes de courbes remplissant un espace métrique (Généralisation des théorèmes de Birkhoff)*, Mat. Sb. (N.S.) **6** (48) (1939), 283-292. MR **1**, 242.

[205] ——, *Systèmes dynamiques sur une multiplicité intégrale limite*, Dokl. Akad. Nauk SSSR **47** (1945), 535-538. MR **7**, 255.

[206] ——, *Intégration qualitative du système d'équations différentielles $dx/dt = Q(x,y)$; $dy/dt = P(x,y)$*, Mat. Sb. (N.S.) **16**(58) (1945), 307-344. (Russian. French summary) MR **7**, 298.

[207] ——, *Les systèmes dynamiques généraux*, Dokl. Akad. Nauk SSSR **53** (1946), 491-494. MR **8**, 280.

[208] ——, *The structure of one-dimensional limiting integral manifolds in the plane and three-dimensional space*, Vestnik Moskov. Univ. **1948**, no. 10, 49-61. (Russian) MR **12**, 344.

[209] ——, *Über vollständig unstabile dynamische Systeme*, Ann. di Math. **14** (1936), 275-286.

[210] V. V. Nemyckiĭ and V. V. Stepanov, *Qualitative theory of differential equations*, OGIZ, Moscow, 1947. (Russian) MR **10**, 612.

[211] V. V. Nemyckiĭ, *Problems of the qualitative theory of differential equations*, Vestnik Moskov. Univ. Ser. Fiz.-Mat. Estest. Nauk **1952**, no. 8, 19-39. (Russian) MR **14**, 753.

[212] ——, *Some problems of the qualitative theory of differential equations. (Survey of contemporary literature)*, Uspehi Mat. Nauk **9** (1954), no. 3 (61), 39-56. (Russian) MR **16**, 130.

[213] O. A. Oleĭnik, *On equations of elliptic type with a small parameter in the highest derivatives*, Mat. Sb. (N.S.) **31** (73) (1952), 104-117. (Russian) MR **14**, 560.

[214] O. A. Oleĭnik and A. I. Žižina, *On a boundary problem for the equation $\epsilon y'' = F(x,y,y')$ for small ϵ*, ibid. **31** (73) (1952), 709-717. (Russian) MR **14**, 875.

[215] N. F. Otrokov, *On the stability of periodic integrals*, Gorkiĭ Učen. Zap. **6** (1938), 125-128. (Russian)

[216] ——, *Sur le nombre de cycles limites au voisinage d'un foyer*, Dokl. Akad. Nauk SSSR **43** (1944), 98-101. (Russian) MR **6**, 154.

[217] O. Perron, *Über die Gestalt der Integralkurven einer Differentialgleichung erster Ordnung in der Umgebung eines singulären Punktes*, Math. Z. **15** (1922).

[218] ——, *Die Ordnungszahlen der Differentialgleichungssysteme*, ibid. **31** (1929).

[219] K. P. Persidskiĭ, *On stability of motion to the first approximation*, Mat. Sb. (N.S.) **40** (1933), 284-293. (Russian)

[220] ———, *On stability of motion*, ibid. **42** (1935), 37-42. (Russian)
[221] ———, *On the theorem of Ljapunov*, Dokl. Akad. Nauk SSSR **14** (1937), 541-544. (Russian)
[222] ———, *On the theory of stability of integrals of systems of differential equations*, Izv. Kazan. Fiz.-Mat. **11** (1939), 29-45. (Russian)
[223] ———, *On the theory of stability of solutions of differential equations*, Uspehi Mat. Nauk **1** (1946), no. 5-6 (15-16), 250-255. (Russian) MR **10**, 456.
[224] ———, *On eigenvalues of differential equations*, Izv. Akad. Nauk Kaz. SSR Ser. Mat. i Meh. **1** (1947), 5-47. (Russian)
[225] I. G. Petrovskiĭ, *Über das Verhalten der Integralkurven eines Systems gewöhnlicher Differentialgleichungen in der Nähe eines singulären Punktes*, Mat. Sb. (N.S.) **41** (1934), 107-156 and **42** (1935), 403.
[226] ———, *Lectures on partial differential equations*, 2nd ed., GITTL, Moscow, 1953. (Russian) MR **16**, 133.
[227] E. B. Pearson and G. I. Lingwood, *A technique for rapid determination of the harmonic response characteristics of a position control servomechanism*, Instrum. Practice **7** (1935), no. 5.
[228] L. S. Pontrjagin, *On dynamical systems which are close to Hamiltonian*, Ž. Èksper. Teoret. Fiz. **4** (1934), 883-885. (Russian)
[229] ———, *Characteristic cycles on manifolds*, Dokl. Akad. Nauk SSSR **35** (1942), 34-37. (Russian) MR **4**, 147.
[230] ———, *On zeros of some transcendental functions*, Izv. Akad. Nauk SSSR Ser. Mat. **6** (1942), 115-134. (Russian. English summary) MR **4**, 214.
[231] ———, *On some topologic invariants of Riemannian manifolds*, Dokl Akad. Nauk SSSR **43** (1944), 91-94. (Russian) MR **6**, 182.
[232] ———, *Characteristic cycles*, ibid. **47** (1945), 242-245. (Russian) MR **7**, 138.
[233] ———, *Topological duality theorems*, Uspehi Mat. Nauk 2 (1947), no. 2 (18), 21-44. (Russian) MR **10**, 56.
[234] ———, *Foundations of combinatorial topology*, OGIZ, Moscow, 1947. (Russian) MR **11**, 450.
[235] ———, *On the zeros of certain elementary transcendental functions (supplement)*, Dokl. Akad. Nauk SSSR **91** (1953), 1279-1280. (Russian) MR **16**, 23.
[236] H. Poincaré, *On curves defined by differential equations*, Gostehizdat, Moscow, 1947.
[237] ———, *Les méthodes nouvelles de la mécanique c éleste.* I, no. 19, Gauthier-Villars, Paris, 1892.

[**238**] E. M. Wright, *Linear difference-differential equations*, Proc. Cambridge Philos. Soc. **44** (1948), 179-185. MR **10**, 125.

[**239**] ——, *The linear difference-differential equation with asymptotically constant coefficients*, Amer. J. Math. **70** (1948), 221-238. MR **9**, 592.

[**240**] ——, *The linear difference-differential equations with constant coefficients*, Proc. Roy. Soc. Edinburgh Sect. A **62** (1949), 387-393. MR **11**, 182.

[**241**] ——, *The stability of solutions of non-linear difference-differential equations*, ibid. **63** (1950), 18-26. MR **12**, 106.

[**242**] I. M. Rapoport, *On linear differential equations with periodic coefficients*, Dokl. Akad. Nauk SSSR **76** (1950), 793-795. (Russian) MR **12**, 827.

[**243**] ——, *On the stability oscillations of material systems*, ibid. **77** (1951), 25-28. (Russian) MR **13**, 38.

[**244**] ——, *On some asymptotic methods in the theory of differential equations*, Izdat. Akad. Nauk Ukrain. SSR, Kiev, 1954. (Russian) MR **17**, 734.

[**245**] L. L. Rauch, *Oscillation of a third order nonlinear autonomous system*, Contributions to the theory of Nonlinear Oscillations, pp. 39-88, Annals of Mathematics Studies No. 20, Princeton Univ. Press, Princeton, N. J., 1950. MR **11**, 665.

[**246**] P. I. Romanovskii, *Successive approximations for functional equations*, Trudy Mat. Inst. Steklov. **24** (1953). (Russian)

[**247**] E. Rothe, *Zur Theorie der topologischen Ordnung und der Vektorfelde in Banachschen Räumen*, Compositio Math. **5** (1937), 177-197.

[**248**] ——, *Gradient mappings and extrema in Banach spaces*, Duke Math. J. **15** (1948), 421-431. MR **10**, 548.

[**249**] ——, *Leray-Schauder index and Morse type numbers in Hilbert space*, Ann. of Math. (2) **55** (1952), 433-467. MR **14**, 185.

[**250**] V. A. Rohlin, *Homotopy groups*, Uspehi Mat. Nauk **1** (1946), no. 5-6 (15-16), 175-223. (Russian) MR **10**, 393.

[**251**] M. Sammerfil'd, *Nonstationary heating in the ŽRD chamber*, Questions of Rocket Technique (9) **3** (1952), 52-61. (Russian)

[**252**] G. Sansone, *Le equazioni delle oscillazioni non lineari-resultati analitici*, Atti del Quarto Congresso dell'Unione Matematica Italiana (Taormina, 1951) Vol. 1, pp. 186-217, Casa Editrice Perrella, Rome, 1953. MR **15**, 32.

[**253**] I. V. Svirskii, *The determination of the number of roots lying in the right half-plane for a function of the form $F(e^z, z)$, where $F(e^z, z)$ is a rational function of the arguments e^z and z, and an application of the results to the investigation of automatic regulation of steam turbines*, Izv. Kazan. Filial. Akad. Nauk SSSR Ser. Fiz.-Mat. Tehn. Nauk **1** (1948), 51-61. (Russian) MR **14**, 150.

[254] J.-P. Serre, *Homologie singulière des espaces fibrés. Applications*, Ann. of Math. (2) **54** (1951), 425-505. MR **13**, 574.

[255] V. I. Sobolev, *Sur les éléments caractéristiques de certains opérateurs*, Dokl. Akad. Nauk SSSR **31** (1941), 735-737. MR **3**, 208.

[256] Ju. K. Solncev, *On the asymptotic behaviour of integral curves of a system of differential equations*, Izv. Akad. Nauk SSSR Ser. Mat. **9** (1945), 233-240. (Russian) MR **7**, 117.

[257] V. V. Stepanov and A. N. Tihonov, *Über die Räume der fastperiodschen Funktionen*, Mat. Sb. (N.S.) **41** (1934), 166-178.

[258] V. V. Stepanov, *On the definition of the probability of stability*, Dokl. Akad. Nauk SSSR **18** (1938), 151-154. (Russian)

[259] ———, *Sur une extension du théorème ergodique*, Compositio Math. **3** (1936), 239-253.

[260] J. J. Stoker, Jr., *Nonlinear vibrations in mechanical and electrical systems*, Interscience, New York, 1950. MR **11**, 666.

[261] Ja. P. Terleckiĭ, *Calculations of the conditions for the excitation of an electronic-radiating generator in a braking field*, Moskov. Gos. Univ. Učen. Zap. **3** (1945), Mat. 77, 142-151. (Russian)

[262] ———, *Computation of the stationary amplitude for the simplest electronic-radiating generator with braking field*, ibid. **3** (1945), Mat. 77, 151. (Russian)

[263] K. F. Teodorčik, *Self-oscillating systems*, Gostehizdat, Moskow, 1952. (Russian)

[264] A. N. Tihonov, *Ein Fixpunktsatz*, Math. Ann. **111** (1935), 767-776.

[265] ———, *Functional equations of Volterra type and their applications to certain problems of mathematical physics*, Bjull. Moskov. Gos. Univ. A1 (1938), 1-25. (Russian)

[266] ———, *On the stability of inverse problems*, Dokl. Akad. Nauk SSSR **39** (1943), 176-179. (Russian) MR **5**, 184.

[267] ———, *On the dependence of the solutions of differential equations on a small parameter*, Mat. Sb. (N.S.) **22** (**64**) (1948), 193-204. (Russian) MR **9**, 588.

[268] ———, *On systems of differential equations containing parameters*, ibid. **27** (**69**) (1950), 147-156. (Russian) MR **12**, 181.

[269] ———, *Systems of differential equations containing small parameters in the derivatives*, ibid. **31** (**73**) (1952), 575-586. (Russian) MR **14**, 1085.

[270] A. I. Fet, *Integral homology of the space of closed curves on a sphere*, Dokl. Akad. Nauk SSSR **66** (1949), 569-570. (Russian) MR **11**, 47.

[271] ———, *The homology ring of the space of closed rectifiable curves on the sphere*, ibid. **66** (1949), 347-350. (Russian) MR **11**, 47.

[272] ———, *Variational problems on closed manifolds*, Mat. Sb. (N.S.) **30 (72)** (1952), 271-316; English transl., Amer. Math. Soc. Transl. no. 90, 1953; reprint, Amer. Math. Soc. Transl. (1) **6** (1962), 147-206. MR **13**, 955; MR **15**, 41.

[273] ———, *On the algebraic number of closed extremals on a manifold*, Dokl. Akad. Nauk SSSR **88** (1953), 619-621. (Russian) MR **14**, 992.

[274] ———, *A connection between the topological properties and the number of extremals on a manifold*, ibid. **88** (1953), 415-417. (Russian) MR **14**, 992.

[275] A. I. Fet and L. A. Ljusternik, *Variational problems on closed manifolds*, ibid. **81** (1951), 17-18. (Russian) MR **13**, 474.

[276] A. F. Filippov, *A sufficient condition for the existence of a stable limit cycle for an equation of the second order*, Mat. Sb. (N.S.) **30 (72)** (1952), 171-180. (Russian) MR **13**, 944.

[277] C. Floquet, *Sur les équations différentielles linéaires à coefficients périodiques*, Ann. Sci. École Norm. Sup. **13** (1883).

[278] S. V. Fomin, *Finite invariant measures in the flows*, Mat. Sb. (N.S.) **12 (54)** (1943), 99-108. (Russian. English summary) MR **5**, 101.

[279] K. O. Friedrichs, *On nonlinear vibrations of third order*, Studies in Nonlinear Vibration Theory, pp. 65-103, Inst. Math. Mech., New York Univ., New York, 1946. MR **8**, 329.

[280] S. V. Frolov and L. È. Èl′sgol′c, *Limite inférieure pour le nombre des valeurs critiques d'une fonction, donnée sur une variété*, Mat. Sb. (N.S.) **42** (1935), 637-643.

[281] M. Frommer, *Integral curves of an ordinary differential equation of first order in the neighborhood of a singular point of rational character*, Uspehi Mat. Nauk **9** (1941).

[282] B. A. Fuks, *Theory of analytic functions of several complex variables*, OGIZ, Moscow, 1948; English transl., Transl. Math. Monographs Vol. 8, Amer. Math. Soc., Providence, R. I., 1963. MR **12**, 328.

[283] F. B. Fuller, *Note on trajectories in a solid torus*, Ann. of Math. (2) **56** (1952), 438-439. MR **14**, 556.

[284] F. Haas, *Poincaré-Bendixon type theorems for two-dimensional manifolds different from the torus*, ibid. (2) **59** (1954), 292-299. MR **15**, 793.

[285] ———, *The global behavior of differential equations on n-dimensional manifolds*, Proc. Nat. Acad. Sci. U.S.A. **39** (1953), 1258-1260. MR **15**, 793.

[286] E. Hilb, *Zur Theorie der linearen funktionalen Differentialgleichungen*, Math. Ann. **78** (1917), 137-170.

[287] G. F. Hil′mi, *On a property of minimal sets*, Dokl. Akad. Nauk SSSR **14** (1937), 261-262. (Russian)

[288] ———, *Metrically non-decomposable sets of motions*, ibid. **15** (1937), 421-424. (Russian)

[289] ———, *On the ergodic theorem*, ibid. **24** (1939), 213-216. (Russian)
[290] A. Ja. Hinčin, *Eine Verschärfung der Poincarié'schen "Wiederkehrsatzes,"* Compositio. Math. **1** (1934), 177-179.
[291] ———, *Zur Birkhoffs Lösung des Ergodenproblems*, Math. Ann. **107** (1932), 485-488.
[292] ———, *The method of spectral reduction in classical dynamics*, Proc. Math. Acad. Washington **19** (1933), 567-578.
[293] ———, *Sur le problème ergodique de la mécanique quantique*, Izv. Akad Nauk SSSR Ser. Mat. **7** (1943), 167-184. (Russian. French summary) MR **5**, 280.
[294] ———, *Mathematical foundations of statistical mechanics*, GITTL, Moscow, 1933; English transl., Dover, New York, 1949. MR **10**, 666.
[295] V. N. Capyrin, *On the problem of Hurwitz for transcendental equations*, Prikl. Mat. Meh. **12** (1948), 301-328. (Russian) MR **10**, 241.
[296] ———, *The Routh-Hurwitz problem for a quasipolynomial for $s = 1$, $r = 5$*, Inžen. Sb. **15** (1953), 201-206. (Russian) MR **16**, 807.
[297] A. I. Cvetkova, *On a theorem of L. S. Pontrjagin on the displacement of cycles*, Dokl. Akad. Nauk SSSR **57** (1947), 331-334. (Russian) MR **9**, 152.
[298] È. S. Citlanadze, *On certain problems concerning eigenvalues for nonlinear operators in the Hilbert space*, ibid. **53** (1946), 307-309. (Russian) MR **8**, 386.
[299] ———, *Certains problèmes de l'extrême relatif et de la théorie des valeurs caractéristiques*, ibid. **56** (1947), 15-18. MR **9**, 95.
[300] ———, *On a question about characteristic values of nonlinear completely continuous operators in Hilbert space*, ibid. **57** (1947), 879-881. (Russian) MR **9**, 447.
[301] ———, *Some problems of nonlinear operators and the calculus of variations in spaces of Banach type*, Uspehi Mat. Nauk **5** (1950), no. 4 (38), 141-142. (Russian) MR **12**, 110.
[302] ———, *Existence theorems for minimax points in Banach spaces and their applications*, Trudy Moskov. Mat. Obšč. **2** (1953), 235-274. (Russian) MR **14**, 1094.
[303] Ja. Z. Cypkin and P. V. Bromberg, *On the degree of stability of linear systems*, Izv. Akad. Nauk SSSR **1945**, 1163-1168 (1945). (Russian) MR **7**, 519.
[304] Ja. Z. Cypkin, *Stability of a system with retardation with feed-back*, Avtomat. i Telemeh. **7** (1946), 107-129. (Russian)
[305] ———, *Systems with retarded feed-back*, Trudy NISO **24** (1947). (Russian)
[306] ———, *On the theory of the klystron*, Radiotehnika **2** (1947), 49. (Russian)
[307] ———, *Stability of systems of automatic control*, Izdat. VZEI, Moscow, 1953. (Russian)

[308] ———, *Stability of a class of systems of automatic regulation with distributed parameters*, Avtomat. i Telemeh. **9** (1948), 176-189. (Russian) MR **11**, 508.

[309] N. G. Četaev, *A theorem on instability*, Dokl. Akad. Nauk SSSR **1** (1934), 529-530. (Russian)

[310] ———, *On the equations of Poincaré*, Prikl. Mat. Meh. **5** (1941), 253-262. (Russian. English summary) MR **4**, 225.

[311] ———, *Theorem concerning the non-stability of regular systems*, ibid. **8** (1944), 323-326. (Russian. English summary) MR **6**, 225.

[312] ———, *On the smallest characteristic number*, ibid. **9** (1945), 193-196. (Russian. English summary) MR **7**, 300.

[313] ———, *Stability of motion*, 2nd ed., GITTL, Moscow, 1955. (Russian) MR **17**, 1087.

[314] ———, *Concerning the stability and instability of irregular systems*, Prikl. Mat. Meh. **12** (1948), 639-642. (Russian) MR **11**, 249.

[315] G. S. Čogošvilli, *Change in the Betti numbers of a moving level surface*, Dokl. Akad. Nauk SSSR **22** (1939), 297-310. (Russian)

[316] ———, *On level surfaces and domains of smaller values of a function defined on a bounded manifold*, ibid. **24** (1939), 635-639. (Russian) MR **1**, 320.

[317] ———, *On Šnirel'man's transformations*, ibid. **30** (1941), 199-203. (Russian) MR **2**, 325.

[318] J. Schauder, *Der Fixpunktsatz in Funktionalräumen*, Studia Math. **2** (1930).

[319] A. A. Šestakov, *On the distribution of singular points of a system of differential equations*, Trudy Kazan. Astronom. **27** (1953), 41-50. (Russian)

[320] S. N. Šimanov, *On the theory of oscillations of quasilinear systems*, Prikl. Mat. Meh. **18** (1954), 155-162. (Russian) MR **16**, 131.

[321] E. Schmidt, *Über eine Klasse linearer funktionaler Differentialgleichungen*, Math. Ann. **70** (1911), 499-524.

[322] L. G. Šnirel'man, *Über eine neue kombinatorische Invariante*, Math. Monatsh. **37** (1930), 131-134.

[323] R. Ja. Šostak, *Integration of linear differential-difference equations by the method of expansion of operators into infinite series*, Trudy MVTU (1948), 89-164. (Russian)

[324] F. Schürer, *Eine gemeinsame Methode zur Behandlung gewisser Funktionalgleichungprobleme*, Leipziger Berichte (M.-Ph.) **70** (1918), 185-240.

[325] L. È. Èl'sgol'c, *Zur Theorie der Invarianten, die zur Bestimmung der unteren Grenze der Anzahl der kritischen Punkte einer stetigen Funktion, die auf einer Mannigfaltigkeit bestimmt ist, dienen können*, Mat. Sb (N.S.) **5** (47) (1939), 551-558. (Russian. German summary) MR **1**, 319.

[326] ______, *Die Länge einer Mannigfaltigkeit und ihre Eigenschaften*, ibid. **5 (47)** (1939), 565-571. (Russian. German summary) MR **1**, 317.

[327] ______, *Die Änderung der Bettischen Zahlen der Niveauflächen einer stetigen Funktion, die auf einer Mannigfaltigkeit definiert ist*, ibid. **5 (47)** (1939), 559-564. (Russian. German summary) MR **1**, 320.

[328] ______, *Zu der Frage über die Bestimmung der unteren Grenze der Anzahl der kritischen Punkte einer stetigen Funktion, die auf einem Raum, der keine Mannigfaltigkeit ist, bestimmt ist*, ibid. **8 (50)** (1940), 455-461. (Russian. German summary) MR **2**, 325.

[329] ______, *Zur Theorie der Änderung der topologischen Invarianten der Niveauflächen*, ibid. **8 (50)** (1940), 463-470. (Russian. German summary) MR **3**, 61.

[330] ______, *Sur la variation du groupe fondamental du domaine des valeurs inférieures d'une fonction définie sur une multiplicité*, ibid. **19 (61)** (1946) 237-238. (Russian. French summary) MR **8**, 525.

[331] ______, *The variation of the topological structure of level surfaces*, ibid. **23 (65)** (1948), 399-418. (Russian) MR **10**, 392.

[332] ______, *An estimate for the number of singular points of a dynamical system defined on a manifold*, ibid. **26 (68)** (1950), 215-223; English transl., Amer. Math. Soc. Transl. no. 68, 1952; reprint, Amer. Math. Soc. Transl. (1) **5** (1962), 498-510. MR **11**, 671; MR **13**, 850.

[333] ______, *Estimation of the number of critical points*, Uspehi Mat. Nauk **5** (1950), no. 6 (40), 52-87. (Russian) MR **12**, 721.

[334] ______, *Genaue Abschätzung der Anzahl der geometrisch-und analytischverschiedenen kritischen Punkte*, Moskov. Gos. Univ. Ucen. Zap. **163** (1952), Mat. 6, 61-68. (Russian. German summary) MR **17**, 397.

[335] ______, *Variational problems with retarded argument*, Vestnik Moskov. Univ. Ser. Fiz.-Mat. Estest. Nauk **1952**, no. 10, 57-62. (Russian) MR **15**, 41.

[336] ______, *Approximative methods of integration of differential-difference equations*, Uspehi Mat. Nauk **8** (1953), no. 4 (56), 81-93. (Russian) MR **15**, 629.

[337] ______, *Estimation of the number of critical points of a continuous mapping of a manifold onto a circle*, Moskov. Gos. Univ. Učen. Zap. **165** (1954), Mat. 7, 34-38. (Russian) MR **16**, 503.

[338] ______, *Remark on the estimation of the number of points of rest of dynamical systems with retarded argument*, ibid. **165** (1954), Mat. 7, 221-222. (Russian) MR **16**, 473.

[339] ______, *Stability of solutions of differential-difference equations*, Uspehi Mat. Nauk **9** (1954), no. 4 (62), 95-112. (Russian) MR **17**, 44.